T0304658

Bias and Causation

Bias and Causation

Models and Judgment for Valid Comparisons

HERBERT I. WEISBERG

Correlation Research, Inc.
Needham, MA

A JOHN WILEY & SONS, INC., PUBLICATION

Published by John Wiley & Sons, Inc., Hoboken, New Jersey
Published simultaneously in Canada

Brief explanation of the cover photo for Bias and Causation:
In the quest for evidence of causation, the reseacher must follow a confusing and circuitous path, avoiding false steps that can easily lead to bias.

For general information on our other products and services or for technical support, please contact our Customer Care Department within the United States at (800) 762-2974, outside the United States at (317) 572-3993 or fax (317) 572-4002.

Wiley publishes in a variety of print and electronic formats and by print-on-demand. Some material included with standard print versions of this book may not be included in e-books or in print-on-demand. If this book refers to media such as a CD or DVD that is not included in the version you purchased, you may download this material at http://booksupport.wiley.com. For more information about Wiley products, visit www.wiley.com.

Library of Congress Cataloging-in-Publication Data:

Weisberg, Herbert I., 1944–
   Bias and causation : models and judgment for valid comparisons / Herbert I. Weisberg.
      p. cm.—(Wiley series in probability and statistics)
   Includes bibliographical references and index.
   ISBN 978-0-470-28639-5 (cloth)
 1. Discriminant analysis. 2. Paired comparisons (Statistics) I. Title.
   QA278.65.W45 2010
   519.5′35–dc22

                                                                    2009054242

Printed in Singapore

10 9 8 7 6 5 4 3 2 1

To Nina, Alexander, and Daniel

Contents

Preface

Throughout a long career as a statistician, I have frequently found myself wrestling, in one way or another, with issues of bias and causation. As a methodologist, researcher, consultant, or expert witness, I have had to propose, justify, or criticize many varieties of causal statements. My training in mathematics and statistics prepared me well to deal with many aspects of the diverse, and occasionally bizarre, problems I have chanced to encounter. However, the statistical theory I studied in graduate school did not deal explicitly with the subject of causal inference, except within the narrow confines of randomized experimentation.

When I entered the "real world" of statistical research and consulting, the problems I regularly faced were not amenable to strict experimental control. They typically involved causal effects on human health and behavior in the presence of observational data subject to many possible sources of bias. To attack these problems, I needed analytic weapons that were not in my statistical arsenal. Little by little, I found myself being transformed into a practitioner of some dark art that involved statistics, but that drew as well on intuition, logic, and common sense.

The nature of this evolution can be best illustrated by an anecdote. The first legal case in which I provided statistical expertise was an employment discrimination lawsuit against a Boston-based Fortune 500 company. The plaintiffs were convinced that black workers were being systematically prevented from rising to higher-level positions within the manufacturing division of the company. A young associate in a large Boston law firm representing the plaintiffs had somehow been referred to me. Knowing next to nothing about the industry or the relevant employment law, I peppered this attorney with questions of all sorts. Eventually, I even got around to requesting some data from the company's human resources department. I dutifully subjected the data to various standard analyses, searching for an effect of race on promotion rates, but came up empty. Despite repeated failures, I harbored a nagging suspicion that something important had been overlooked.

I began to scrutinize listings of the data, trying to discern some hidden pattern behind the numbers. Preliminary ideas led to further questions and to discussions with some of the plaintiffs. This interactive process yielded a more refined understanding of personnel decision making at the company. Eventually, it became clear to me what was "really" going on. Up to a certain level in the hierarchy of positions, there was virtually no relationship between race and promotion. But for a particular level midway up the organizational ladder, very few workers were being promoted from within the company when openings arose. Rather, these particular jobs were being filled primarily through outside hires, and almost always by white applicants. Moreover, these external candidates were sometimes less qualified than the internally available workers. We came to call this peculiar dynamic "the bottleneck."

This subtle pattern, once recognized, was supported anecdotally in several ways. The statistical data, coupled with qualitative supporting information, was eventually presented to the defendant company's attorneys. The response to our demonstration of the bottleneck phenomenon was dramatic: a sudden interest in negotiation after many months of intransigence. Within weeks, a settlement of the case was forged.

The methods of data analysis I had employed in this instance did not conform to any textbook methods I had been taught. Indeed, I felt a bit guilty that the actual statistical techniques were quite simple and did not require any advanced mathematical knowledge. Moreover, the intensive "data dredging" in which I had engaged was highly unorthodox. But the results made perfect sense! Besides leading to a practical solution, the answer was intellectually satisfying, connecting all the dots of a previously inexplicable data pattern.

The kinds of data analyses that have proved most useful in my work have often displayed this same quality of making sense in a way that is intuitive and logically compelling. This plausibility derives not only from statistical criteria, such as levels of significance, but also from broader considerations that are harder to articulate. I have come to believe that statisticians tend to be uncomfortable with causal inference in part because the issues cannot be settled with technical skill alone. Substantive knowledge and expert judgment are also necessary, in ways that are often difficult to quantify. Thus, at least until very recently, statisticians have been content to cede most methodological questions related to bias and causation to other academic disciplines. This situation has certainly started to change for the better. However, it remains unclear how far the statistical profession is prepared to stretch to meet the real challenges of causal puzzles.

When I began writing this book, I envisioned something much less ambitious than this effort has turned out to be. Fortunately, I was foolish enough to rush in where others apparently feared to tread, not realizing that I would be drawn into the subject so deeply. Rather than a "handy reference" on the types of bias, with some causal modeling framework in the background, the tail (causation) has come to wag the dog (bias). It seems to me that to really understand bias, a clear counterfactual framework for formulating the issues is necessary. This framework provides the foundation upon which potential solutions, whether quantitative or qualitative, may rest.

This book is intended primarily for practicing researchers and methodologists, and for students with a reasonably solid grounding in basic statistics and research methods. The mathematics used involves nothing beyond elementary algebra and basic statistics and probability theory. The few more complicated derivations are relegated to appendices that will be of interest only to the more mathematically sophisticated. The main value of the book is conceptual, not technical. The purpose of the mathematical models is to provide insight, rather than methods, although some methods have been and can be built upon the conceptual foundations.

I have provided very little detail on the traditional statistical methods that address problems of random variability. When conducting actual research, these problems need to be addressed in tandem with those related to bias. In reading this book, it may be helpful to imagine that we are dealing with extremely large samples in which random variability can be ignored. Of course, in reality, finite-sampling issues are usually very important. Much of the recent research by statisticians pertaining to causal inference concentrates on estimating causal effects based on finite samples. The statistical principles and methods they apply are well established and have been expounded in numerous texts. My subject is the poor stepchild of statistics: systematic error that cannot be cured by obtaining a large-enough sample.

This book was not written specifically as a textbook. However, it may be found useful as a central or secondary resource in a graduate-level research methods course in epidemiology or the social sciences. I believe that much of the material is best learned in the context of real research. So, teachers may wish to supplement the limited set of examples in the book with articles and reports relevant to their particular interests and areas of application. Some teachers may find it useful to treat this book as a reference for selected topics, or to approach the topics covered in a different order than I have presented them. This is certainly their prerogative, but I would caution that material in later chapters depends strongly on concepts and terminology introduced earlier.

Chapter 1 provides an introduction to the problems considered, and brief summaries of six "case studies" that are referenced throughout the book. Chapter 2 discusses the counterfactual framework for causal inference, and some important concepts and terminology pertaining to bias and causation. Chapter 3 contains a brief exposition of several methodological issues that are central to the difficulties posed by bias. Chapter 4 summarizes the various types of bias as viewed in the biomedical and the social sciences. These four chapters form an extended introduction and review of the issues addressed in the remainder of the text.

Chapter 5 deals with the problem of selection bias—a term used in different ways, and a source of considerable confusion. Selection bias is approached from the perspective of traditional statistical modeling as well as from a causal modeling viewpoint. Chapters 6 and 7 both focus on the problem of confounding. Chapter 6 lays out the various ways in which this central, but enigmatic, concept has been defined. Chapter 7 offers an explanation of confounding that is based on the causal (counterfactual) framework. Chapter 8 discusses intermediate causal factors, which can engender bias even in a randomized controlled trial. Chapter 9 considers the topic of information bias, the distortion of a causal effect that can result when the exposure and/or the outcome is measured with error. Chapters 5 through 9 are the most challenging, both conceptually and mathematically.

Chapters 10 and 11 are more practical and less technical than the preceding four chapters. Chapter 10 offers a preliminary organization of bias sources. I define a source of bias as a real-world condition that affects a comparative study and can lead to bias. I define and describe 20 general sources of bias. This list includes, at a high level of generality, most of the common sources of bias that arise in practice. Finally, Chapter 11 considers the different ways in which we can attempt to cope with bias. I argue that the standard statistical paradigm is most appropriate for a narrow (albeit important) range of problems related to bias and causation. There is a vast uncharted territory of research problems for which this paradigm is impractical, and sometimes even inappropriate.

In this sense, I am a pessimist when it comes to the ability of standard approaches, and highly sophisticated mathematical extensions of them, to solve the problems of causal inference. I believe that a broader paradigm for data analysis is needed, one that focuses much more on individual variability and that meshes qualitative and quantitative sources of information more effectively. I am most definitely an optimist, however, about the enormous potential of research to improve human health and well-being.

I would like to acknowledge the contributions, direct and indirect, of many colleagues and friends with whom I have been privileged to work. Many of you have helped to shape my thinking during the course of our collaborations on various projects over the years. In this regard, I highlight especially the following (in alphabetical order): Tony Bryk, Xiu Chen, Richard Derrig, Mike Dolan, Eric Garnick, Mike Grossman, Vanessa Hayden, Peter Höfler, Jarvis Kellogg, Eric Kraus, Tom Marx, Mike Meyer, Bruce Parker, Victor Pontes, Sam Ratick, David Rogosa, Peter Rousmaniere, David Schwartz, and Terry Tivnan.

I offer my sincere thanks for important direct contributions to this book by several individuals. Richard Derrig and Jarvis Kellogg participated in several discussions that helped to sharpen my focus and improve the presentation of ideas. Vanessa Hayden and Victor Pontes provided insightful feedback as early versions of the causal model presented in this book were being hatched, and collaborated as coauthors on an article that introduced the basic conceptual framework. Jay Kadane and Dave Sackett reviewed early drafts of several chapters and made many valuable comments. Tom Marx and Terry Tivnan deserve special commendations for plowing through the entire manuscript and offering their helpful (and always tactful) editorial and substantive suggestions. My editor at Wiley, Steve Quigley, was enormously helpful at all stages of the process. His practical advice at critical junctures, delivered with wry, self-effacing humor, was just what I needed to see this project through to completion.

Finally, I wish to acknowledge the support and encouragement supplied by many family members, including my sons, Alex and Dan, my sister, Sally Goldberg, and my sisters-in-law, Barbara Irving and Abigail Natenshon. Alex also skillfully translated my rough drafts of the pie-chart graphs into electronic form. Finally, my deepest sense of gratitude goes to my wife, Nina, who has always given me the confidence to trust my instincts, and the courage to pursue my dreams.

Herbert I. Weisberg
Needham, Massachusetts
March 2010

CHAPTER 1

What Is Bias?

Two Red Sox fans were discussing the finer points of baseball strategy one day while driving to Fenway Park in Boston. Burt had read a statistical study about the effectiveness of the sacrifice bunt. In this maneuver, the batter tries to advance a base-runner from first to second base by tapping the ball a few feet in front of home plate. He is willing to be thrown out at first base in exchange for helping the runner to reach second base safely. The data in the study revealed that a runner on first base scored less frequently when the batter attempted to bunt. This implied, Burt insisted, that a batter should never attempt to sacrifice. Harry disagreed. Situations in which managers called for a sacrifice bunt, he argued, were not the same as those in which batters were allowed to swing away. Somehow, Harry knew intuitively that he was right and that some deeper principle of logic was involved, but he was never able to convince his friend.

Burt was unaware that by comparing the frequency of scoring between two different sets of at-bats, he was making a biased comparison. A lower success rate observed after attempting to bunt than when "swinging away" would not necessarily mean that bunting always, or even sometimes, *causes* a decrease in the probability of scoring the runner. Perhaps less proficient batters often bunt, whereas stronger hitters nearly always swing away. Then the success rate of the bunters would have been lower even if they had not bunted. So, was the lower observed success rate really caused by bunting?

The remainder of this book focuses on more consequential (and often controversial) analyses of causation that arise in many scientific contexts. In particular, we will concentrate on the problem of trying to reach a valid conclusion about some *factor* that might affect human health, behavior, or well-being. Sometimes we will denote this causal factor as F. For example, F might be an innovative educational program, and the outcome of interest some measure of academic achievement. Mathematically, we will treat F as

Bias and Causation: Models and Judgment for Valid Comparisons, By Herbert I. Weisberg
Copyright © 2010 John Wiley & Sons, Inc.

an indicator variable, with $F = 1$ if the causal factor is present and $F = 0$ if it is not.

This introductory chapter defines the problem of bias in a general sense. Bias is intrinsically a problem related to causality. We explain how traditional statistical methods are severely limited as a way to address causality in general, and questions related to bias in particular. Consequently, a new approach to data analysis is needed. Subsequent chapters describe a theoretical framework within which such a "new paradigm" has begun to evolve. For concreteness, this chapter includes six illustrative case studies that motivate and provide context for the ideas developed throughout the book.

1.1 APPLES AND ORANGES

Typically, the scientific community weighs the evidence provided by one or more *comparative studies* in order to decide whether a causal relationship between F and the outcome exists and to measure the strength of this effect. A comparative study examines some relevant aspect of a specified population of individuals. The aim is to ascertain whether and how a particular characteristic of individuals in the population (e.g., academic achievement) tends to respond when the factor is introduced, either deliberately (intervention) or unintentionally (risk factor). To provide statistical evidence, the study obtains data on individuals under two alternative conditions: exposure to the factor and nonexposure. Of course, the actual sets of individuals in the two groups being compared will differ. So, the critical question becomes whether the two *study groups* are sufficiently similar for the comparison to be interpreted as a *causal* effect of the factor, not as an "apples-to-oranges" comparison.

In a comparative study, a difference between groups that is not attributable to the factor under study can result from either *random* or *systematic* variability. Random variability can occur for a variety of reasons, but does not tend to favor the exposed or unexposed group. In large groups, these random variations tend to even out. If we imagine the size of the groups to increase without limit, the error in estimating the causal effect eventually becomes negligible. Moreover, in smaller groups, the amount of variability can at least be calculated and taken into account. Therefore, uncertainty related to random variability can be "managed" through statistical methods based on probability theory. These methods (significance testing, confidence intervals, regression modeling, Bayesian posterior distributions, etc.) represent the principal triumph of twentieth-century statistical theory.

Our focus in this book will be on the ways in which a comparison can be *systematically* (i.e., nonrandomly) distorted. An estimated effect that deviates systematically from the actual causal effect of interest is said to be biased. Unlike random variability, *bias* is a structural tendency that does not balance out, even with extremely large study groups. Unlike random error, bias cannot be reduced by increasing the sample size. In our baseball example, effects of random variation could be virtually eliminated if many thousands of at-bats were included in the analysis. We could therefore obtain a very precise answer to the question of whether runners tend to score less often after a bunt or not. But this information by itself would be of little practical value to a baseball manager, who wants to know when calling for a sacrifice will have a causal effect on the chances of scoring a run.

Throughout this book, the term *bias* will mean the extent to which a particular measure of a *causal effect* has been systematically distorted. Forms of bias that fall under this umbrella derive from shortcomings of research design, implementation, and analysis, and they can thus be considered *methodological biases*. To say that a particular study is biased is to assert that the research methods employed have resulted in systematic error in the estimation of a causal effect. Systematic error, or nonrandom error, is inherent in the research process itself. The magnitude and direction of bias do not depend on random variation across the particular sample of subjects included in the study.

When scientists refer to research bias, they generally mean methodological bias. However, discussions of bias are sometimes confusing because this term also has several other connotations. To a mathematical statistician, bias is a technical property of an estimate. An estimate of some parameter, such as the mean of a given population, is biased if the estimate "on average" deviates from the true value of the parameter. To a social scientist, bias may pertain to aspects of human behavior or psychology. Do certain individuals or groups tend to think or act in a predetermined manner in a specified situation? In addition, bias may suggest a negative or prejudicial attitude toward a particular group or ideology. As used throughout this book, bias is only incidentally related to any of these other interpretations.

Because they result from systematic and not random distortion, methodological biases are generally not amenable to correction by mathematical formulas. An understanding of potential biases in practice requires not only quantitative sophistication, but also a solid grounding in the relevant scientific context. The topic of bias resides in a kind of no-man's-land between the discipline of statistics and the various scientific fields in which research takes place. This orphan status may help to explain why a comprehensive theory of bias has yet to emerge.

1.2 STATISTICS VS. CAUSATION

We have defined bias as a systematic error in estimating a causal effect based on statistical data. Attempts to estimate causal effects represent one of the most common, and arguably the most important, application of statistical methods. However, statistical theory, at least until quite recently, has been almost exclusively concerned with the implications of *random* error. As a result, classical statistical methods are applicable to a very narrow range of problems related to causal inference. Indeed, it is a universal mantra that statistical association, or correlation, does not necessarily imply causation. To the layperson, it must seem odd that statistics has so little to offer for learning about causal effects. To explain this irony, we must understand the primary problem that statistical methods were originally designed to address.

Classical statistical methods were devised primarily to deal with uncertainty that arises from the limited nature of the available data. Intuitively, it was recognized long ago that a small set of observations of some quantity was generally less reliable as a guide to action than a larger sample. For instance, a farmer might wish to learn how many apples he could expect to obtain from his orchard in a typical year, or perhaps in relation to factors such as rainfall and soil quality. Having data from many farms would provide much better information than relying on only a few. But better in what sense and by how much? The genius of modern statistical theory lies largely in its conceptual framework for formalizing and answering such questions.

Central to this conceptualization was the idea that the set of units in hand (e.g., apple orchards) could be imagined to comprise a representative "sample" randomly drawn from a much larger (virtually infinite) population of units. In principle, this hypothetical infinite population would include all of the units that could *potentially* have been observed, whether or not they were actually observed in the available sample. Furthermore, this population is assumed to possess a particular *distribution* of characteristics that can be described by the values of different variables (yield per acre, soil conditions, moisture, wind, etc.). This distribution essentially describes the proportions (or probabilities) of various possible values of the variables. The aim of statistical inference then becomes to describe the parameters (e.g., mean, median, variance, correlation) pertaining to this hypothetical population's distribution. For example, the farmer might wish to know the average yield of apples per acre and how this yield relates to the amount of rainfall during the growing season.

This statistical paradigm has by now become so familiar that it is hard to appreciate that it embodies certain *assumptions* about the world. First and foremost, there is the mental construct of a hypothetical infinite population. Moreover, the distribution of variables is often assumed to have a particular mathematical form, such as the (Gaussian) "normal" distribution. Buried even deeper, however, is another critical assumption: the probability distribution is regarded as stable, reflecting a fixed set of underlying conditions. Chance and uncertainty enter through the (assumed) process of randomly sampling from the population. However, because this variability is now subject to well-established mathematical rules of probability theory, a world of statistical inference opens up. For instance, the uncertainty associated with a small-sample estimate can be expressed as a confidence interval or a Bayesian posterior distribution. As long as the statistical model of the world remains fixed, inferences based on probability theory will be valid.

In particular, the implicit supposition of a stable universe allows the possibility of making accurate *predictions*. Our farmer may measure various conditions early in the growing season and then try to predict what his yield is likely to be. If relevant circumstances remain stable, and if he has a substantial database of prior observations, he can make a reliable forecast. This could be accomplished by effectively *conditioning* on the measured values he has observed. Suppose that the farmer magically knew the full distribution of yields per acre for the hypothetical apple-orchard population. Then he could identify all orchards that have approximately the same characteristics as his own. He could, for example, compute the average yield per acre *for this subgroup of the population.* That would be a logical value to predict for his current crop. In general, using these *conditional probabilities* is the basic idea underlying many sophisticated techniques for prediction. But the stability of the population distribution is what makes reliable prediction based on conditioning possible.

Now let us consider the problem of causal inference. Causal inference is also about making predictions. However, causation is not concerned primarily with random variation under a stable set of circumstances. Rather, causation pertains to what systematic alteration would occur if the circumstances were to change in a specified manner. For example, our farmer might be deciding whether to introduce an irrigation system. He wants to know what change in yield this innovation would *cause.* In effect, he envisions two hypothetical populations: one without irrigation and one with irrigation. The (causal) parameter of interest would then become the difference between the average yields produced in these two populations.

To answer causal questions, the classical statistical machinery just described is still necessary to cope with random variability. However, the strategy of conditioning is not adequate for causal inference. Conditioning for prediction depends on a stable set of circumstances, but analysis of causation entails consideration of a real or hypothetical modification of at least one important circumstance. Consequently, conditional probabilities within a fixed population cannot tell us what would happen under such alteration. For that, we must carry out (or at least envision) a *manipulation* of the circumstances. Classical statistical methods address random variation by invoking a stable hypothetical infinite population consisting of all units that *might have been* observed, whether or not they actually were observed. Similarly, causal inference requires a way to conceptualize what *might have been* observed under different specified circumstances. This central concept will be elaborated at length in Chapter 2. The key point for now is that the stable population assumed by traditional statistical methods can only reveal how various factors are *associated*, but it does not by itself disclose how a change in one factor would produce changes in some other factor of interest.

Bias in comparative studies has traditionally been either ignored by statisticians or addressed solely within the classical statistical framework. The result has been a failure to develop data-analytic methods capable of dealing appropriately with this pervasive methodological problem. Consequently, a conceptual framework for causal thinking that *extends* classical statistical theory is necessary to obtain a deeper understanding of bias. Such a causal framework has been evolving for roughly the past 30 years and has provided many of the building blocks needed for understanding the nature and sources of bias. A goal of this book is to draw together and elaborate those strands of this causal theory that pertain to the problem of bias in comparative studies.

Unlike some applications of this new theory, our primary goal is not to "solve" the problem of bias by offering more complicated or mathematically sophisticated statistical methods. Indeed, comprehending the nature and sources of bias can help to clarify why improved technology based on ever more complex mathematical analysis can be counterproductive. More mathematically "advanced" methods can even become an impediment to insight, because they remove the data analyst further from direct contact with the data. Consequently, the analyst may be forced to accept on faith that the assumptions underlying the statistical model are consistent with the data.

We will take the position that the sort of data-analytic tools required are those that will facilitate the exercise of logic and scientific judgment to reach conclusions that are supported by the weight of available evidence. Such methods typically cannot provide the degree of certainty or quantification familiar to us in managing random variability via standard statistical techniques. The successful development and application of causal knowledge ultimately depend on cultivation of sound scientific judgment, as well as basic mathematical facility, to discover what is likely, though by no means certain, to be true and real:

> A scientist's actions are *guided*, not determined, by what has been derived from theory or established by experiment, *as is his advice to others*. The judgment with which isolated results are put together to guide action or advice in the usual situation, which is too complex for guidance to be *deduced* from available knowledge, will often be a mixture of individual and collective judgments, but judgment will play a crucial role. Scientists know that they will sometimes be wrong; they try not to err too often, but they accept some insecurity as the price of wider scope. Data analysts must do the same. (Tukey, 1962, 9)

1.3 BIAS IN THE REAL WORLD

Statistics textbooks present mathematical techniques that can be applied in a variety of scientific areas. These statistical tools are almost exclusively devoted to the management of uncertainty attributable to random variability. It is therefore quite natural to consider these techniques in the abstract, with only minimal reference to the details of their application. A thorough knowledge of the context in which a particular procedure will be applied is not essential to understanding how and why it works. Training in the *application* of statistical methods is relegated largely to the various academic disciplines (epidemiology, economics, psychology, etc.) in which substantive scientific issues arise.

When dealing with the subject of bias, on the other hand, neatly severing theory from practice is not feasible. Certain general principles can be abstracted from the scientific context, but the motivation for these ideas cannot be grasped fully in the abstract. The relevant intellectual framework for thinking about bias has evolved primarily in the natural course of scientific research, and secondarily in generalizations made by philosophers of science observing this research. In particular, the concept of *causality* will be central to our discussion of bias throughout this book. Causal analysis would ideally be grounded in extensive background knowledge. For example, two anthropologists arguing about the effect of a certain cultural practice in various societies would share a foundation of theory and information necessary for a meaningful interchange. Obviously, we cannot hope to approach such a breadth and depth of contextual understanding. On the other hand, a theory of bias divorced completely from concrete scientific issues in the real world would be hopelessly sterile.

To partially address this conundrum, we present in this chapter a set of case histories. Each of these describes an actual study, or set of studies, to which we can refer later when discussing various sources of bias. The narratives offered here are necessarily somewhat sketchy. We attempt to highlight the main aspects of the research sufficiently to provide the reader with a tangible feel for the methodological challenges in making causal inferences. In selecting these case histories, several criteria have been taken into account. Each example pertains to an issue that was (or still is) considered to be important and subject to substantial uncertainty and disagreement. The range of these cases in terms of research design and subject matter is quite broad. Epidemiology, clinical trials, and social science research are all represented. Each of these narratives highlights a particular pivotal article or report that was central to the controversy. Most important, this set of studies allows us to illustrate a wide range of biases that either were or could have been considered by the investigators.

Throughout the book, we will draw upon these studies to provide context for various points, often introducing hypothetical elements. For instance, we may suggest a possible distorting influence that could theoretically have affected the study, even though there is no actual evidence to indicate that such a source of bias actually existed. When such liberties have been taken with the facts, the fictitious aspects will be noted. More substantive discussion of possible conclusions regarding the issues will be deferred to the final chapter.

The narratives presented here are intended in part to illustrate why statistical methods fail to address fully the range of methodological concerns related to bias and causation. Classical statistical methods are designed to provide answers to specific questions. Is this new medication superior to the standard treatment for a certain disease? Will this new educational approach improve academic performance? But the questions of practical interest are often much more particular, subtle, and complex. A practitioner may need to decide whether to try a new drug on her patient. A teacher may need to decide whether the new educational approach will work in her classroom. These practitioners may be interested in what is known about the possible causal effects (both beneficial and adverse) of these interventions for different kinds of individuals. Their decisions ultimately must be based on all the available evidence, both statistical and nonstatistical, that they can bring to bear, filtered through years of practical experience. For this purpose, they need information that is quantitative and rigorous, but also open-ended enough to connect with their richer base of qualitative knowledge in fruitful ways. To develop such information may call for a new analytic paradigm that is both more tolerant of ambiguity and more respectful of subject-matter knowledge.

Recently, there has been an explosion of interest in causal analysis within the field of statistics. This represents a very positive development, but there is a danger that causal inference will be reduced to just another mathematical–statistical technology. It would be regrettable if causal models were judged narrowly by their ability to solve statistical problems in the ways such problems have conventionally been formulated.

> But paradigm debates are not really about relative problem-solving ability, though for good reasons they are usually couched in those terms. Instead, the issue is which paradigm should in the future guide research on problems many of which neither competitor can yet claim to resolve completely. A decision between alternate ways of practicing science is called for, and in the circumstances that decision must be based less on past achievement than on future promise. The man who embraces a new paradigm at an early stage must often do so in defiance of the evidence provided by problem-solving. He must, that is, have faith that the new paradigm will succeed with the many large problems that confront it, knowing only that the older paradigm has failed with a few. A decision of that kind can only be made on faith. (Kuhn, 1962, 156–157)

This book was motivated by faith that a new paradigm for dealing with bias is possible, based on a deeper understanding of causality. This new paradigm will not reject the existing paradigm, but will define its limits, identifying the research questions for which it is not applicable. This new paradigm may provide some improved solutions to conventional problems, but its larger value will be in daring to pose and address novel questions. For example, traditional approaches concentrate almost exclusively on average effects, ignoring a largely unmet need to tailor effective interventions to individual characteristics and conditions. Causal theory has the potential to offer a proper language within which useful answers can be articulated, although the form of these answers may appear unfamiliar—and lacking the mathematical precision to which statisticians have become accustomed.

Evaluating the Efficacy of Antityphoid Vaccine

In a classic article, epidemiologist Mervyn Susser (1977) discussed the need for sound judgment grounded in subject-matter expertise to augment statistical analysis. The article referred to several historical examples, including one of the earliest uses of statistical analysis to evaluate the efficacy of a new medical treatment. The story begins in England in 1896, when Dr. Almroth Wright developed a vaccine to prevent typhoid fever. After several tests of the new vaccine among volunteers in the British Army, the Medical Advisory Office to the War Office was attempting to decide whether the army ought to adopt routine vaccination as a general policy. To aid in making this decision, the available data were submitted to Karl Pearson, the most eminent statistical expert of his day.

Pearson's analysis of the data led him to conclude that efficacy had not been firmly established; his published report suggested that more research was needed (Pearson, 1904). Wright took issue with Pearson's report in an accompanying editorial, and a heated debate in the pages of the *British Medical Journal* ensued. For a variety of reasons that went beyond the purely scientific issues, Wright "won" the debate and a policy of vaccination was adopted. However, a program of continued research was implemented, as Pearson had recommended. The results were summarized by Leishman, a colleague of Wright who directed the follow-up program (Leishman, 1910). The analysis appeared to provide strong support for the decision to implement inoculation in all units being sent overseas. Shortly after Leishman's data were released, the vaccine was adopted for routine use by the British, French, and American militaries. An article appearing in the *New York Times* hailed the success chronicled by Leishman as a triumph of modern medical science and its heroic practitioners:

> Trained scientists have labored weary hours without number in their laboratories bending over their microscopes and watching their test tubes to attain the golden truth. The result has been victory, a new triumph in the domain of medicine. It has not only been proved, say its champions that typhoid fever can be prevented by vaccination by anti-typhoid serum, but they claim immunity already has been conferred upon thousands and thousands of persons—soldiers chiefly—in this and other lands. (*New York Times*, June 5, 1910)

The data upon which the dispute between Pearson and Wright was based are presented in Tables 1.1 and 1.2, which we have adapted from Susser's summary (Susser, 1977). Table 1.1 speaks to the possible prophylactic effect of the vaccine. In each of the cohorts, the rate at which typhoid fever was contracted was lower among those inoculated with the vaccine than among those who were not. However, the magnitude of the rates and the difference between the two groups varied widely. Table 1.2 pertains to the question of whether the vaccine lowered mortality among those who contracted the disease. Here again, the mortality rates vary across cohorts. With one exception (Ladysmith garrison) the rates are lower in the inoculated group. Leishman's data are not presented here but show a similar and even stronger pattern of apparent effectiveness.

Table 1.1 Prophylactic Effect of Antityphoid Vaccine[a]

Cohort	Inoculated		Not Inoculated	
	N	Rate	N	Rate
Hospital staffs	297	10.8%	279	26.9%
Ladysmith garrison	1,705	2.1%	10,529	14.1%
Methuen's column	2,535	1.0%	10,981	2.3%
Single regiments	1,207	6.0%	1,285	6.4%
Army in India	15,389	0.8%	136,360	1.6%

[a]Adapted from Susser (1977).

Table 1.2 Effect on Mortality of Antityphoid Vaccine[a]

Cohort	Inoculated		Not Inoculated	
	N	Rate	N	Rate
Hospital staffs	32	6.3%	75	16.0%
Ladysmith garrison	35	22.9%	1489	22.1%
Single regiments	72	12.5%	82	25.6%
Special hospitals	1174	7.3%	4991	10.8%
Various military hospitals	764	8.2%	3374	10.8%
Army in India	84	13.1%	1475	28.7%

[a]Adapted from Susser (1977).

Our main purpose in presenting this data will be to consider possible sources of bias and the extent to which these biases may have compromised the studies. One obvious concern was that the various substudies all relied on data from soldiers who had volunteered for the experimental inoculation. Another pertained to the potential lack of reliability in diagnosis of typhoid fever at that time. Furthermore, the medical officers assigned to monitor the results were aware of whether a patient was or was not inoculated. In addition, the specificity of the treatment was subject to uncertainty about inoculation histories and lack of quality control in manufacturing the vaccine. Also, because soldiers were often transferred into or out of units, obtaining a valid count to use for a denominator in calculating the rates in particular units was complicated. Moreover, recording of the duration of exposure was also untrustworthy, because exposure status (inoculated or not) was recorded as of the end of the observation period. Finally, it is possible that the apparent effectiveness of the vaccine was attributable to other changes in personal hygiene or the water supply that were occurring at the same time (Cockburn, 1955).

With all these potential problems, most of which were recognized at the time, the true value of Wright's typhoid vaccine was far from certain. However, the public seemed to regard the matter as case closed. Major F. F. Russell of the Medical Corps of the U.S. Army speaking at Johns Hopkins was quoted at length in the *New York Times* article:

> Among the exposed regiments who had been inoculated with the vaccine in use at present there were 3.7 cases per 1,000 against 32.8 per 1,000 among the untreated. … The observation of this group of 12,000 men covers a period of over three years, and no more perfect or convincing statistics are needed to show the value of this method of prophylaxis. (As quoted in the *New York Times*, June 5, 1910)

A more sober and professional statistical analysis several years later came to a similar conclusion, while recognizing the methodological limitations of the existing data (Greenwood and Yule, 1915).

Despite whatever lingering doubts may have existed in the scientific community, Wright's antityphoid vaccine with various refinements remained in use without benefit of a controlled clinical trial for five decades. Considerable observational data accumulated attesting to reductions in typhoid incidence throughout the world that appeared to result from vaccination. Then in the 1950s the discovery of the antibiotic chloromycetin made possible a randomized test of typhoid vaccine, because those assigned to the control group who contracted typhoid fever could be cured. Fortunately, the vaccine was able to satisfy the more rigorous testing needed to receive the stamp of modern scientific validation (Cvjetanovic, 1957).

Racial Disparities in Death Sentencing

The death penalty is one of the most controversial issues related to the U.S. criminal justice system. In *Furman v. Georgia*, decided in 1972, the U.S. Supreme Court ruled essentially that the death penalty was being administered in a way that was arbitrary, capricious, and based on impermissible factors. Although not rejecting its use globally, the Court effectively set a higher standard for the manner in which death penalties could be imposed. The *Furman* decision led to reforms by many states aimed at avoiding the completely unstructured sentencing statutes that the Supreme Court had ruled unconstitutional (Baldus et al., 1990).

Since the *Furman* decision, lawyers seeking to overturn death penalty convictions have often argued that the decisions were disproportionate and/or discriminatory. Disproportionate would mean that the severity of the sentence was out of proportion to that received by other similarly situated defendants. Discriminatory would mean that the conviction and/or sentencing were tainted by impermissible factors, such as race or socioeconomic status. Many critics of the death penalty believed that judicial systems post-*Furman* remained permeated by racial discrimination and lack of proportionality in sentencing. Against this backdrop, David Baldus and his colleagues undertook two major interrelated studies of capital punishment in Georgia during the years 1973–1980.

Their main purposes were to estimate the extent of disproportionality and of racial discrimination in death-penalty decision making. These studies have been described in detail in a book titled *Equal Justice and the Death Penalty* (Baldus et al., 1990).

The second and larger of the studies was called the Charging and Sentencing Study (CSS). The CSS included 1066 cases from both the pre-*Furman* and post-*Furman* periods. These cases comprised a stratified random sample from a total of 2484 defendants "arrested and charged with homicide who were subsequently convicted of murder or voluntary manslaughter." Among the cases sampled, 127 resulted in a death penalty. For each of the 1066 cases, a wide array of variables was collected pertaining to five stages of the charging and sentencing process:

- Grand-jury indictment decisions
- Prosecutorial plea-bargaining decisions
- Jury guilt-trial decisions
- Prosecutorial decisions to seek a death penalty after conviction
- Jury penalty-trial sentencing decisions

Broadly speaking, the degree of discretion exercised by the decision-makers becomes more structured and constrained as a case moves through the process.

If the guilt trial results in a conviction for capital murder, the prosecutor must decide whether to seek the death penalty. Statutory criteria for potential death-eligibility are spelled out in general terms, but they must be interpreted by the prosecutor. If she believes that the death penalty is warranted, a second and entirely separate penalty trial will be held. The sole issue is to determine whether a death penalty should be imposed. To reach this decision, the penalty-trial jury is instructed to weigh specific aggravating and mitigating circumstances. To impose the death penalty, the jury must find at least one of the statutory aggravating factors. However, the jury is also permitted to consider any potentially mitigating factors.

One motive for undertaking the CSS was its potential use by attorneys representing convicted killer Warren McCleskey. McCleskey's death sentence had been imposed after his conviction for murdering a police officer named Frank Schlatt. In 1980, when the CSS study was first being considered, McCleskey's appeal was working its way through the Georgia legal system. An important basis for the appeal was McCleskey's assertion that the decision was tainted by racial discrimination; he was black and the victim white. Previous research in Georgia and elsewhere had suggested that both the race of the defendant and the race of the victim might play a role in death-sentencing decisions. A main goal of the CSS was to establish the extent to which death-sentencing decisions in Georgia had been influenced by race.

Based on their extensive database of cases in Georgia, Baldus and his team performed a variety of statistical analyses aimed at assessing possible

Table 1.3   Sentencing by Race of Defendant and Victim[a]

Category	Defendants	Death Sentences	Rate
Black on white	233	50	21.5%
White on white	748	58	7.8%
Black on black	1443	18	1.2%
White on black	60	2	3.3%
All cases	2484	128	5.2%

[a]Adapted from Baldus et al. (1990).

discrimination. Their book presents the data and analyses, along with details of their presentation in federal district court, and eventually to the U.S. Supreme Court. The book also deals extensively with various methodological issues raised during the appeals process and provides the authors' views on the validity of various criticisms. Rarely has a statistical study been subjected to so much scrutiny and with so much potentially at stake. Because issues of potential bias were dissected in great depth from both a statistical and a legal perspective, this case is highly instructive.

The basic data at issue can be summarized very simply. Table 1.3 shows the results of sentencing decisions in Georgia for cases in the post-*Furman* period (Baldus et al., 1990, 315). The unadjusted rates reveal some striking racial disparities, especially with respect to the victim's race. These rates were then adjusted in a variety of ways to account for the circumstances of the cases, especially as these pertained to the "moral culpability" of the defendant. One statistical model that was highlighted in court relied on a logistic regression that included 39 independent variables. This "core model" contained variables that both statistically and theoretically "appeared to exercise the greatest influence in determining which defendants indicted for murder would actually receive a death sentence." The coefficient (odds ratio) for the race-of-victim variable in this model was 4.3 and had a p-value of 0.005 (Baldus et al., 1990, 326).

As powerful as this statistical evidence appears to be, it did not carry the day. Eventually, the U.S. Supreme Court on April 22, 1987, in a 5–4 decision failed to overturn McCleskey's sentence. A part of the reasoning articulated by several of the justices was related to the appropriateness of *any* statistical argument. However, many specific criticisms of the methodology were also raised. In two long methodological appendices, the CSS investigators thoughtfully addressed these and other issues.

In terms of bias, there were three main areas of potential concern. One area related to the way that cases in the CSS had been selected. Only defendants convicted of voluntary murder were included, leaving out other potential death-penalty candidates whose cases reached other dispositions. Therefore, the potential for selection bias in estimating racial effects existed. A second problem related to the measurement of the culpability measures being used as covariates in the model. There were several aspects of the judicial system that made accurate and consistent data collection difficult. By far the most complex issue, however, related to the adequacy of the covariates collected, extensive as they were, to rule out other confounding factors. Was the observed difference in sentences truly the result of the victim's race, or alternatively of some other factors that were not measured but were correlated with race?

In Mr. McCleskey's case, further appeals on his behalf were put forward based on nonstatistical evidentiary grounds. To settle these, the case was eventually heard again by the Supreme Court. In the end, on September 26, 1991, Warren McCleskey was executed. In a *New York Times* editorial run 3 days later, the fourfold disparity estimated by David Baldus and his colleagues was prominently mentioned (*New York Times*, 1991).

Evaluation of Employment Training Programs

In the United States during the 1970s political support for government-sponsored social interventions to eliminate poverty and social inequity was strong. A number of major experimental educational and social programs were initiated, and the methodology of program evaluation became a major preoccupation of social scientists. The great majority of such government-sponsored efforts were observational (i.e., did not involve random assignment to different types of programs). Rather, subjects were assigned either to the innovative program being evaluated or to a more conventional control program according to some known criteria. For example, the program might be offered to those satisfying some needs-based eligibility criterion. Because this assignment mechanism was deliberate rather than random, the groups assigned to the different programs might be different in important respects.

In general, randomized experiments were not considered feasible in social program evaluation for a variety of ethical and practical reasons. However, a fortuitous exception to this limitation occurred in the area of worker training programs. The National Supported Work Demonstration (NSW) aimed to assist disadvantaged workers to enter the labor market successfully by providing work experience and counseling in a sheltered work environment (Dickinson and Maynard, 1981; Masters and Maynard, 1981). The target population for the NSW was composed of two main subgroups: women in the Aid to Families with Dependent Children (AFDC) program, and men who were high-school dropouts and often had a background that included drug addiction and criminal activity. Unlike most other government programs, individuals were selected for the available slots *randomly* from among a pool of qualified applicants, and the candidates who were not chosen became the controls.

Data related to annual income and various related socioeconomic and demographic individual characteristics were collected at baseline and at three follow-up points over 36 months. The postprogram income after 36 months was the primary outcome variable for the NSW. The official report of the study's findings found a very small impact of the NSW on the male participants and a fairly substantial improvement in earnings for the AFDC women (Manpower Demonstration Research Corporation, 1983). Because of the randomized design, these results were widely viewed as authoritative. Although there were some issues related to retention and compliance of participants that needed to be addressed, the statistical analysis was relatively straightforward. In contrast to the situation with other employment and training programs, no complex adjustments to deal with potential differences between the subjects who received the intervention and those who did not were necessary.

The existence of such a "gold standard" was viewed as a golden opportunity by econometrician Robert LaLonde. He wondered what would have been concluded if the NSW had relied on the more common quasi-experimental approach. To simulate such observational results, he created several different comparison groups based on available survey data. He then applied several alternative statistical techniques to obtain estimates based on these comparisons. Each of these statistical adjustments was based on a somewhat different mathematical model. LaLonde was primarily interested in whether any of the nonexperimental approaches could faithfully reproduce the "true" experimental findings. In addition, LaLonde asked whether it would be possible to discern bias in an observational study based on its own data, without reference to the gold standard. Specifically, would some violation of the adjustment model's assumptions be apparent to tip the researcher off that a problem existed? LaLonde's answers to these questions were published in an article that shook the econometric world:

> This study shows that many of the econometric procedures and comparison groups used to evaluate employment and training programs would not have yielded accurate or precise estimates of the impact of the National Supported Work Program. The econometric estimates often differ significantly from the experimental results. Moreover, even when the econometric estimates pass conventional specification tests, they still fail to replicate the experimentally determined results. Even though I was unable to evaluate all nonexperimental methods, this evidence suggests that policymakers should be aware that the available nonexperimental evaluations of employment and training programs may contain large and unknown biases from specification errors. (LaLonde, 1986, 617)

LaLonde's results, along with similar findings by Fraker and Maynard (1987), were hailed by methodologists who were strong advocates of randomized experiments (Burtless and Orr, 1986; Barnow, 1987). On the other hand, some econometricians and statisticians continued to defend nonexperimental studies as both necessary and viable (see Heckman and Hotz, 1989, with discussion; Heckman and Smith, 1995). In particular, James Heckman and his colleagues performed their own reanalyses of the NSW data and came up with estimates closer to those in the original randomized study. Their statistical models were selected based on the ability to pass certain tests of the model's assumptions. According to Heckman and Hotz, models that were not ruled out by these "specification tests" tended to perform quite well.

The debate over the ability of such specification tests to identify valid estimates in observational studies continues to this day. For example, Dehejia and Wahba (1999) have attacked the NSW problem from the perspective of propensity-score analysis (Rosenbaum and Rubin, 1983b). The merits of this relatively new approach have been debated (Smith and Todd, 2005a,b; Dehejia, 2005). How this controversy will ultimately play out is uncertain. What seems clear, however, is that proponents of observational research have been put on the defensive by those who argue that only true randomized experiments can yield reliable evidence of causality.

Phenylpropanolamine and Hemorrhagic Stroke

Phenylpropanolamine (PPA) was a component of many popular cold and cough medicines available over-the-counter prior to the year 2000. It was also used as an appetite suppressant. In several decades of use, few serious side effects had been observed. However, starting in the 1980s sporadic reports of hemorrhagic stroke (bleeding in the brain), particularly in young women, began to emerge. Epidemiologic evidence of a causal relationship was quite tenuous, but concerns persisted. By 1992, the Food and Drug Administration (FDA) decided to commission a large-scale case–control study to determine whether PPA was in fact implicated as a risk factor for hemorrhagic stroke.

The study, conducted by a team of researchers at Yale University, began in 1994 and was expected to require four years to complete. Results of the Yale Hemorrhagic Stroke Project (HSP) were reported to the FDA on May 10, 2000, and eventually published on December 21, 2000, in the *New England Journal of Medicine* (Kernan et al., 2000). The authors concluded that "phenylpropanolamine in appetite suppressants, and possibly in cold and cough remedies, is an independent risk factor for hemorrhagic stroke in women." The FDA indicated an intention to reclassify the drug as being unsafe for over-the-counter use and urged manufacturers to withdraw it from the market. All of the PPA producers agreed to this FDA request and made plans to transition to other medications for use in cold and cough products. Unfortunately for the manufacturers, the story did not end there, as a flood of lawsuits followed, brought by people who suffered strokes they believed were caused by PPA.

During the ensuing litigation, the HSP study's findings were hotly contested by the contending parties. Medical and epidemiological experts were retained by both sides. Experts brought in by the plaintiffs extolled the virtues of the HSP study as a model of scientific rigor. Experts for the defendant companies raised many serious methodological criticisms. As a result, the court records contain a wealth of information that sheds light on potential biases. To understand why the study was so controversial, it will be useful to describe briefly the study's design and the specific results obtained.

Because hemorrhagic stroke is such a rare event in young adults, it would be difficult to study its occurrence prospectively, waiting for a sufficient volume of cases to accumulate. So, the Yale investigators decided to conduct a "case–control" study. A case–control study is essentially run in reverse. The study starts by collecting a group of patients who have experienced the outcome event (e.g., hemorrhagic stroke) and then looks backward to identify factors that appear to be responsible. Loosely speaking, the method proceeds by comparing the rate of exposure to a particular risk factor (e.g., PPA) among the cases and among a group of noncases. If exposure is more prevalent among cases than among noncases, the risk factor may be a cause of the event. A more rigorous discussion of the case–control methodology will be presented in Chapter 4.

The HSP study enrolled 702 men and women 18–49 years of age who were recruited at 43 U.S. hospitals and had experienced a hemorrhagic stroke within 30 days prior to enrollment. For each of these cases, two control subjects were identified through the use of random-digit telephone dialing. Each of the two controls was matched with the corresponding case based on telephone exchange, race, sex, and age. For each case, the "focal time" was defined as the calendar day and time believed to mark the onset of the stroke-related symptoms. A focal time for each control was defined as the same day-of-week and time-of-day as the focal time of the matched case. Interviews of controls were conducted within 7 days of this focal time. Case and control interviews employed a structured questionnaire to obtain demographic, clinical, behavioral, and pharmaceutical data. Exposure to PPA was defined as use of a product containing PPA on the day of the stroke (prior to the event) or on any of the previous 3 days.

The analyses were performed using a technique called conditional logistic model for matched sets. This approach attempted to adjust for several other variables, in addition to the matching variables. The final model included an adjustment for hypertension, smoking status, and education. These factors were considered because each was believed to be associated with the occurrence of hemorrhagic stroke, either as a direct cause (hypertension, smoking) or indirectly as a surrogate for other unknown causal factors (education). Therefore, the statistical relationship between exposure and being a case of hemorrhagic stroke might be related to these "confounding factors" rather than a causal effect of PPA.

The analysis performed on the HSP data resulted in an estimated odds ratio (OR) of 1.49, with a p-value of 0.17. An odds ratio is a measure of effect that is roughly equivalent to the ratio of event rates with and without exposure. Thus, the study estimated a 49% increase in the frequency of hemorrhagic strokes attributable to PPA. However, the significance level of 0.17 was above the conventional 0.05 criterion commonly applied. The estimated odds ratio of 1.98 for women only was barely significant ($p = 0.05$), and the OR of 0.62 for men was not significant ($p = 0.41$).

For use of PPA in appetite suppressants, however, the HSP reported a whopping OR value of 16.58 ($p = 0.02$), among women (there was no male exposure to appetite suppressants), but based on only six exposed cases vs. one exposed control. Furthermore, a secondary analysis based on "first use" of PPA resulted in an odds ratio of 3.13 for women ($p = 0.08$). Here first use was defined as use of PPA within 24 hours of focal time, but no prior use within the past 2 weeks.

The HSP investigators read these results to suggest a causal association between PPA and hemorrhagic stroke in young women. Their report also acknowledged several possible sources of bias, and discussed measures taken that were believed to have minimized any problems. Confounding could have affected the estimated OR values, despite attempts to identify and correct for important confounding variables. Publicity about PPA might have influenced referral and diagnosis patterns: physicians could have preferentially identified as cases those who were thought to have consumed PPA (selection bias); case subjects could have had either clearer or less accurate memories of events just prior to the index date (recall bias). Finally, the report mentioned *temporal-precedence bias*, which can occur "when exposure is counted although the exposure occurs after the onset of the disease under study, often in response to disease symptoms" (Kernan et al., 2000). This concern was raised by awareness of a phenomenon known as *sentinel headaches* in which a transient headache may herald the onset of a stroke that is not recognized for hours, or even days. As a result, an individual who had used PPA after the sentinel headache but before the index date would be incorrectly regarded as exposed.

In the course of the litigation that followed, biostatistical experts retained by the PPA manufacturers raised these and several other potential biases as reasons to doubt the HSP conclusions (e.g., Weisberg, 2004). Some of their arguments are discussed later in this book to illustrate how various sources of bias can arise in a case–control study. In the majority of trials, the drug companies prevailed, deterring some plaintiffs from pursuing cases and motivating many others to settle for relatively modest amounts (Frankel, 2006). In the end, we will probably never know whether PPA really was responsible for causing strokes, as its removal from the market has made this question moot.

Postmenopausal Hormone Replacement Therapy and Cardiovascular Risk

Prior to 2002, estrogen supplementation was being used routinely by millions of postmenopausal women to control vasomotor symptoms (hot flashes, night sweats) and by many in the hope of reaping a variety of health benefits. The most common formulation of replacement hormones used in the United States consisted of conjugated equine estrogen, possibly in conjunction with progestin. Simplistically, because declining hormone levels were a natural concomitant of aging, replacement of the lost estrogen seemed to many women a logical step to help retain health and vitality. During the 1980s and 1990s a large number of observational studies appeared to confirm that hormone replacement therapy (HRT) did indeed provide a number of health benefits, in addition to generally effective relief of vasomotor symptoms. This "wonder drug" seemed to reduce the risk of osteoporosis, fractures, and cardiovascular disease and possibly even to slow progression toward dementia; the only known serious adverse effect was a possible slight increase in breast cancer. On balance, the profile of risks and benefits was generally considered quite favorable by hard-headed scientists as well as more subjective enthusiasts.

The only fly in this promising ointment was the lack of definitive evidence from randomized controlled trials. Skeptics argued that the observational data reflected potentially serious methodological weaknesses. Most significant was the suspected lack of comparability between women who were using HRT and those who were not; the HRT users appeared to be generally healthier and better educated than nonusers. Attempts were made in various ways to control for this healthy-user effect, but the success of these statistical adjustments was uncertain. To obtain more definitive answers, several large-scale randomized controlled trials were implemented during the 1990s. The focus of these efforts was on a range of health endpoints thought to be influenced by hormone levels. Of particular interest were cardiovascular outcomes, considered to be a major potential benefit of HRT.

In 1998, the results of a major clinical trial with a primary focus on cardiovascular disease were published (Hulley et al., 1998). The Heart and Estrogen/Progestin Replacement Study (HERS) was intended to evaluate HRT for secondary prevention in a cohort of postmenopausal women who had previous coronary heart disease (CHD). The main endpoint was occurrence of a serious CHD event (myocardial infarction or sudden death). The results were disappointing, as no overall difference between the treated and untreated groups emerged. Then in 2002, a much-anticipated randomized study of HRT for primary prevention of CHD yielded even more disturbing news. The Women's Health Initiative (WHI) study showed that in a large cohort of healthy postmenopausal women, HRT was associated with a modest *increase* in CHD events (relative risk of 1.29 overall), as well as elevated risk of breast cancer and stroke (Writing Group for the Women's Health Initiative Investigators, 2002). Although the study demonstrated that some beneficial effects accrued for other endpoints, the apparent harm caused by HRT for these serious adverse events clearly tilted the risk–benefit balance against routine use of HRT.

Other randomized studies, including a multiyear extension of HERS called HERS-II (Hulley et al., 2002) seemed generally to confirm the results of HERS and WHI that suggest either a neutral (e.g., Grady et al., 2002; Pentti et al., 2006) or harmful (Vickers et al., 2007) effect of HRT on CHD. The 180-degree turn between the observational studies and randomized experiments has created confusion among researchers and distress among women and their physicians. Hormone replacement therapy is still recommended for short-term relief of vasomotor symptoms, but not as a long-term regimen to promote good health. For the scientific community, it has been especially unsettling that observational studies seemed so convincing a few years ago, but apparently got the story completely wrong! Or did they? A variety of possible explanations have been offered by biostatisticians and clinical researchers.

Most of this methodological soul-searching accepts that the observational designs were flawed and tries to understand exactly why in order to avoid similar mistakes in the future. However, a substantial minority of methodologists refuse to accept the results of the randomized trials as gospel (e.g., Machens and Schmidt-Gollwitzer, 2003; Naftolin et al., 2004). These skeptics point to a number of methodological problems with the clinical trials. In particular, the possibility of selection bias has been suggested, based on the inclusion in both HERS and WHI of primarily older women who had not previously used HRT and were many years beyond menopause (Naftolin et al., 2004; van der Schouw and Grobbee, 2005). Other aspects of the eligibility criteria and screening process for entry into the trial may also have resulted in an unusual study population (Michels, 2003). Furthermore, it has been suggested that the particular hormone regimen (type and dose) utilized in HERS and WHI may not have been optimal, at least not for all women, and possibly different from that usually received in routine practice (Grodstein et al., 2003; Hoffman and Zup, 2003; Garbe and Suissa, 2004). So, it is possible that the restrictions imposed by the trials in order to enhance internal validity may have engendered a lack of external validity.

Although at present the pendulum has swung strongly away from long-term use of HRT, especially for cardioprotective purposes, there remains much uncertainty. Many still believe that HRT in some form can play a valuable role for some women under certain circumstances. The biological processes leading to CHD are complex, and the impact of hormonal supplementation may be highly variable across different individuals. If so, the challenge is not to determine simply whether or not to use HRT but when and for whom, and in what manner to apply this approach. A recent reanalysis of the WHI represents a potentially important step in this direction. This study found that, for relatively younger women within 10 years of menopause, the risk of CHD events was actually reduced (Roussouw et al., 2007). This finding reinforces the idea that HRT is safe for short-term use by newly menopausal women to relieve vasomotor symptoms.

Much of the recent research and controversy about HRT concerns the extent to which specific HRT formulations can safely provide health benefits, as well as alleviation of discomfort, to specific subgroups of women. For example, it has been hypothesized that women with more severe menopausal complaints may be those for whom HRT would tend to be most beneficial (van der Schouw and Grobbee, 2005). Identifying this or other markers of substantial benefit and low risk would be extremely helpful in practice and might reconcile apparently conflicting results of observational and controlled studies. Clinical research is also progressing with respect to various novel preparations that may provide the benefits of HRT without the alleged side effects of the conventional estrogen and estrogen/progestin regimens. For example, the synthetic steroid tibolone (Tib) that is used in Europe but not approved in the United States has shown promise in small-scale trials (Koh et al., 2005).

Finally, the idea of so-called *bioidentical* hormones has great appeal to many women. Bioidentical preparations are derived from plant extracts that have been chemically modified to be indistinguishable from hormones produced naturally in the body. These products are generally compounded by pharmacists who are not subject to FDA manufacturing regulations. Therefore, practices employed in compounding can vary widely. Some pharmacists customize prescriptions based on saliva tests or blood serum levels. The use of bioidentical hormone therapy is controversial. Advocates are swayed by the rationale that these products are "natural" and tuned to individual characteristics. They push for additional research, while being encouraged by the very limited scientific evidence available (e.g., Moskowitz, 2006). The medical research community, on the other hand, seems generally much more skeptical. Scientists tend to emphasize the lack of controlled trials, as well as the essential similarity of bioidentical and "synthetic" hormonal products (Fugh-Berman and Bythrow, 2007).

The hormonal changes that occur during and after menopause have profound and complex implications, but it has become clear that modifying or regulating these changes safely is not a simple matter. For biostatisticians, the efforts to understand when, how, and for whom HRT can be beneficial will continue to shed valuable light on the relative strengths and weaknesses of controlled trials and observational studies. From a methodological perspective, the impact of the WHI study on epidemiologists was similar to the impact of LaLonde's study on econometricians and other social scientists. The apparent reversal of what seemed a well-established body of knowledge shook the faith of many in the reliability of observational studies.

Antidepressants and Adolescent Suicide

During the 1990s a new generation of medications became widely available to treat major depression (MD) and anxiety disorders. Most of these new antidepressant drugs were in a class known as selective serotonin reuptake inhibitors (SSRIs). Initially, these new drugs were considered effective and safe, leading to rapidly expanding use that was thought by some to have played a role in observed decreases in population suicide rates (Olfson et al., 2003). However, some concerns began to surface in case reports and one clinical trial that these drugs might actually prompt suicidal thoughts and behavior in some patients, particularly adolescents. The existence of such an effect would be ironic, as suicidal tendencies can be a concomitant of MD that antidepressants are intended to treat. However, the emerging evidence was deemed sufficient by late 2003 to result in warnings by several European regulatory agencies. Then in October 2004 the U.S. Food and Drug Administration (FDA) delivered a *coup de grace* by ordering pharmaceutical companies to add a "black box" warning regarding possible risk of suicidality to the labeling of all antidepressants prescribed for pediatric use.

The FDA action was based primarily on a meta-analysis conducted to summarize the available evidence from randomized placebo-controlled trials on the risk of suicidality in adolescents who used modern antidepressants (U.S.

Food and Drug Administration, 2006; Hammad et al. 2006a,b). A meta-analysis is a type of study that produces an overall estimate of treatment effect by combining the results of several individual studies. Based on 24 clinical trials, the FDA meta-analysis found an approximate doubling of risk apparently attributable to the use of new-generation antidepressants. The regulatory actions by both U.S. and European authorities during 2003 and 2004 precipitated a rapid decline in prescriptions for these medications (Wheeler et al., 2009). The impact of this decrease on suicide rates is not yet clear, and the meaning of the available data is being debated (e.g., Leslie et al., 2005; Dubicka et al., 2006; Bridge et al., 2007). Psychiatric professionals are uncertain about the true balance of risks and benefits associated with the use of antidepressants. All agree that additional research is needed to better understand the circumstances, if any, when antidepressants may do more harm than good.

Much of the uncertainty derives from the limitations of the studies upon which the regulatory bodies based their decisions. In particular, the FDA meta-analysis has had great influence and been subject to much discussion. The major strength of this study is that it is based on placebo-controlled randomized trials, generally considered to be the gold standard of clinical research. However, a number of methodological problems have been pointed out by those who remain unconvinced that SSRIs and other antidepressants increase suicide risk. Some of these problems pertain to the paucity of relevant data in the clinical trials. Fortunately, there were no actual suicides in any of these study populations. So, the analyses of "suicidality" were based on indirect measures of relatively rare serious adverse events that reflected "suicidal behavior or ideation" as judged by a panel of experts (Hammad et al., 2006a,b). However, most of the trials were of short duration (4–16 weeks), so that even using this indirect proxy endpoint, very few events occurred.

Besides the small numbers of events in the clinical trials, there were two major potential sources of bias. First, the relationship between suicidality as measured in the studies and actual potential for self-harm is unclear. There was potential for inter-rater disagreement among the expert ratings of the adverse events. Second, the evidence contained in adverse-event reports might not have been adequate to allow accurate prediction of real suicidal intent.

An even more vexing issue pertains to the selection of study samples. Nearly all of the clinical trials on antidepressant use for adolescents attempted to exclude individuals who appeared at high prior risk for suicide. Several reviewers have noted that such screening could have affected the generalizability of the results (Dubicka and Goodyer, 2005; Greenhouse et al., 2008; Weisberg et al., 2009). Specifically, it is plausible that the observed relative risk was inflated by excluding some of those most likely to benefit from treatment with antidepressants. If so, it is conceivable that the regulatory actions may have been counterproductive by discouraging use of products that, properly monitored, could exert a net beneficial effect. From a methodological perspective, this situation highlights the difficult realities that arise in evaluating many complex interventions. Randomized controlled trials are geared primarily to establishing an overall or average treatment effect. When the effect on individuals can vary, perhaps even in direction as well as size, this overall effect can be misleading. Understanding when, how, and for whom antidepressants should be prescribed will require years to unravel. As in the HRT situation, there is much grist here for the methodological mills to grind.

GUIDEPOST 1

This chapter has introduced the topic of bias in comparative studies and presented several case studies that illustrate both the importance and the difficulties inherent in causal inference. These examples were presented against a backdrop of introductory ideas that emphasized the limitations of classical statistical theory for causal inference. We suggested that methods for dealing with bias must be built upon a deep understanding of the real problems posed by attempting to estimate causal effects. Out of such understanding, a new paradigm may emerge that draws on statistical theory but expands beyond its borders to better connect with subject-matter knowledge and clinical insight.

In the next chapter, we explain in more detail the theoretical basis for recent developments related to analysis of causation in comparative studies. The basic concepts of counterfactuals and potential outcomes are defined. The central idea of viewing human populations as collections of "response patterns" is introduced and illustrated with a simple hypothetical example. This idea then leads to a notion of "exchangeability" that, at least conceptually, solves the "apples and oranges" dilemma.

CHAPTER 2

Causality and Comparative Studies

Virtually any statistical analysis of practical importance is potentially susceptible to various forms of bias. Therefore, it might be supposed that biased
comparisons would be of central interest to statisticians and other research
methodologists. In fact, the field of statistics has traditionally treated bias more
as a minor annoyance than a topic worthy of serious academic consideration.
In scientific articles, discussion of possible bias is often relegated to a few
paragraphs on the "limitations" of the study. These afterthoughts often have
a perfunctory quality; they seem intended more to deflect criticism than to
invite a serious evaluation of the study's validity.

Why has the subject of bias received so little attention by statisticians?
Perhaps the main reason is that any meaningful discussion of bias in comparative studies cannot avoid dealing with the core concept of *causality*. Unlike
ordinary statistical concepts, bias pertains explicitly to the causal nature of
things. The existence of bias cannot be determined by mere calculation; it
depends as well on the interpretation of the data in light of some causal theory.
Thus, bias has become an "inconvenient truth" for the almost strictly mathematical discipline that statistics has become.

2.1 BIAS AND CAUSATION

As we suggested in Chapter 1, the field of statistics has long regarded causality
as lying outside the bounds of its scientific mandate. Statistics deals with
empirical quantities and with the strengths of interrelationships among measurable variables. Thus, measures of *association*, such as correlation and regression coefficients, are of interest. The only exceptions apply when experimental
control of treatment allocation is feasible, as in a randomized experiment.
However, we will see that these techniques do not really deal with the issue

of causation directly. Rather, they cleverly sidestep this issue by creating special situations in which bias can be avoided. In this way, a causal effect can be estimated without having to worry about what this effect really means.

To assess whether an observed effect reflects causation or mere coincidence entails considerations beyond the purview of purely mathematical analysis. External information derived from background knowledge, intuition, or theory may be required to identify relationships that a scientist would normally describe as causal. Because these sources of information possess varying degrees of credibility and subjectivity, statisticians are generally reluctant to consider them, preferring to rely exclusively on observed data within the study at hand (and if possible on experimental manipulation). Many statisticians regard causality to be a quasi-scientific notion that is not amenable to rigorous mathematical analysis.

> Throughout the 19th and most of the 20th centuries, the mathematics and concep-
> tual apparatus of probability and statistics became ever richer, while the causal ideas
> that drove much of that development remained tacit, informal, and obscure.
> Kolmogorov's axioms may be seen in retrospect as the completion of that tendency;
> all connection between probability and causal ideas is lost. No wonder, perhaps, that
> many statisticians since have treated ideas of causation as an embarrassing meta-
> physical entanglement, even while routinely practicing causal analysis without
> naming it. (Glymour, 1998, 1513)

Rejecting causality as a useful concept within mainstream statistical theory can be viewed as part of a broader phenomenon. The modern statistical para-digm came to maturity during an era in which *logical positivism* dominated the philosophy of science. Positivistic thought focuses on the description of observable, and presumably objective, facts and patterns, distrusting all meta-physical speculation. Classical statistical methodology, with its emphasis on observable distributions and stable relationships evolved naturally out of this worldview.

The straitjacket imposed on scientific thought by positivism has left statisti-cians struggling to improvise methods for analysis of causation that do not explicitly deal with the concept of causality. These attempts have foundered on the essential limitation of positivistic thought:

> There is a curious misconception that somehow the mathematical mysteries of
> Statistics help Positivism to evade its proper limitation to the observed past. But
> statistics tell you nothing about the future unless you make the assumption of the
> permanence of statistical form. For example, in order to use statistics for prediction,
> assumptions are wanted as to the stability of the mean, the mode, the probable error,
> and the symmetry or skewness of the statistical expression of functional correlation.
> (Whitehead, 1933, 126)

As a result, statistical thinking about causal questions has until recent decades been stunted. In particular, the concept of bias within statistics has been reduced to a mathematical formalism that is entirely devoid of causal content. Yet the obvious question for any comparative study remains: what does an observed difference between two groups really mean?

The answer to such a question usually depends on considerations that reach beyond the available data. The existence of bias must be judged in the context of a broader conceptualization of reality, and "metaphysical entanglements" are thus inevitable. Unlike error related to random variability, bias cannot be assessed without external knowledge of the world. Reliance on this imperfect knowledge may have risks, but failure to apply the knowledge can have even greater risks.

> The point is that speculative extension beyond direct observation spells some trust in metaphysics, however vaguely these metaphysical notions may be entertained in explicit thought. Our metaphysical knowledge is slight, superficial, incomplete. Thus errors creep in. But, such as it is, metaphysical understanding guides imagination and justifies purpose. (Whitehead, 1933, 128)

Pure empiricism tries to avoid all risk of serious error by focusing on what can be analyzed with mathematical precision. Thus, sources of bias that cannot be reduced to mathematical expression pose a dilemma: admitting their existence would be an acknowledgment of failure. The goal of research design and analysis is understood to be the complete elimination of bias. This objective can sometimes be attained through experimental design but can never be absolutely assured when experimental control is absent. So in the vast majority of studies, there is no practical way to justify validity, other than to rationalize that bias is negligible. Is it really surprising that so little academic attention has been paid to understanding exactly how bias can and often does affect causal inferences?

2.2 CAUSALITY AND COUNTERFACTUALS

What exactly do we mean by a cause? The meaning of causality has been an object of philosophical inquiry for centuries. Although it is not necessary for a practicing researcher to wade deeply into these philosophical waters, a basic conceptual framework for causal thinking is essential. Statistics has been applied to help elicit causal relationships in many areas of research. At some level, the ideas presented in this book are applicable to all areas of scientific research in which statistical methods are used. However, the focus of this book is on research that involves human subjects. The examples used to illustrate various types of bias are drawn from the biomedical and social sciences. The types of bias considered are primarily those that affect studies of human health and behavior.

In recent years a new paradigm for thinking about causal effects in the biomedical and social sciences has emerged. Attempts to deal with causality in these areas have led to the development of a widely accepted conceptual framework for causal analysis. This approach is generally known as *counterfactual* analysis, or sometimes as the theory of *potential outcomes* (Rubin, 1974). Throughout this book, the term "causal" will implicitly refer to this counterfactual–potential outcomes framework. For example, a causal model will mean a statistical model based on counterfactual concepts.

The philosophical roots of counterfactual analysis have been traced back to the eighteenth century. David Hume focused attention on the alternative events that would have occurred in the absence of a specific cause. He is famously quoted as proposing that:

> We may define a cause to be an object, followed by another ... where, if the first object had not been, the second had never existed. (Hume, 1748, 115)

In other words, an event A is a cause of an event B if "occurrence of A was necessary for occurrence of B under the observed background circumstances" (Greenland, Robins, and Pearl, 1999). This conception envisions a hypothetical scenario that is identical in all relevant respects to what took place, except for the occurrence of A. Let us suppose that, in this alternate reality, B would not have occurred. Instead, some other outcome, which we can call not-B, would have occurred. Then, because the only difference between the actual and alternate realities was that A in fact *did* occur, A can be considered a cause of B. The event not-B is a *counterfactual* because it is contrary to what actually happened. It is a *potential outcome* in the sense that it *would* have occurred if event A had not.

Note that we did not say that A was *the* cause of B, because there may have been other events present in the "observed background circumstances" that were also required to bring about event B. For example, advanced age is evidently a cause of cardiovascular disease. However, there are many other contributing factors (e.g., smoking history, cholesterol levels, hypertension, obesity) that together determine whether and when a heart attack or stroke will actually occur. This book focuses on issues of bias and causation for a single factor. We will often refer to this factor of interest as a risk factor, although the outcome event is not necessarily something undesirable. In the biological and social sciences, it is almost never possible to completely isolate this risk factor from the context in which it is embedded. Researchers must often pretend that it is possible to "hold everything else constant" while observing the effect of modifying a specified factor. But it is dangerous to forget that ceteris paribus is at best a useful approximation.

John Stuart Mill is often cited as the first to propose a concept of causation that recognizes the role of multiple factors working in concert to produce a given outcome (Mill, 1874). Nearly a century later, philosopher J. L. Mackie introduced a refinement of these ideas that has become very influential among social scientists (Mackie, 1965; 1974). Mackie stated that a causal factor is a condition that is an *insufficient* but *necessary* component of an *unnecessary*

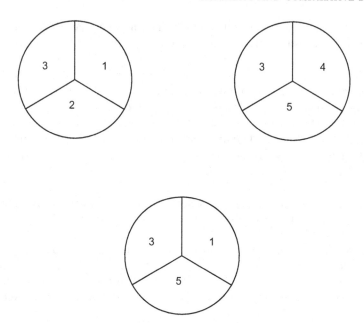

Figure 2.1. Illustration of Rothman's SCC model: three sufficient causes, each with three component causes.

but *sufficient* condition. He abbreviated this formulation to the acronym INUS. An INUS is one of a constellation of conditions that together always result in the outcome event. Because the INUS is necessary but does not *by itself* bring about the event, it is insufficient but necessary (IN-). Because the constellation need not be unique, it is unnecessary but sufficient (-US).

Epidemiologist Kenneth Rothman independently derived essentially the same concept in the context of epidemiology (Rothman, 1976). Rothman represented graphically the various conditions that could jointly precipitate an event (e.g., disease) in a specific individual as sectors of a pie chart. Each of these conditions he called a *component cause* of the event. The combined effect of all the component causes is a *sufficient cause*. Consequently Rothman's model is usually called the sufficient component cause (SCC) model. Figure 2.1 is a schematic diagram similar in essence to that in Rothman's original exposition of the SCC model.

In this diagram, each number denotes a component cause necessary for one or more of the three sufficient causes. A sufficient cause need not be unique; there could be multiple pie charts that would represent combinations of conditions giving rise to the event. Furthermore, a particular component cause might be a constituent of more than one sufficient cause. If a component cause appears in all possible sufficient causes, then it can be considered a *necessary cause* for the event to occur. In the illustration, component 3 would be necessary, because it is a part of all three sufficient causes.

As a (grossly oversimplified) concrete example, suppose that five factors determine whether a particular individual dies from an influenza epidemic:

Factor 1: Not inoculated with seasonal flu vaccine
Factor 2: Age over 75
Factor 3: Exposed to the virus causing the epidemic
Factor 4: No access to medical care
Factor 5: Weakened immune system

Suppose that Fig. 2.1 represents this imaginary situation. Then there would be three possible scenarios in which this person would die. All three sufficient causes require exposure to the flu virus (Factor 3). However, there are three possible pairs of additional component causes that could occur in conjunction with Factor 3 to form a sufficient cause. For example, the first sufficient cause would include being over 75 and not being inoculated.

Note that each pie chart represents a set of circumstances that pertain to a particular individual. Thus, Rothman's SCC model explicitly represented causation for an individual. The component causes are conceived as individual characteristics that could *potentially* be switched on or off. The values at which these switches are set determine whether or not the outcome event actually occurs. Such an idea is impossible to represent within the traditional statistical framework, because only the individual's realized outcome is considered. The concept of latent causal factors can illuminate many causal issues that are obscured within the traditional statistical framework (Rothman and Greenland, 2005).

For our purposes, it will be convenient to adapt the SCC model, so that it can be more easily connected to counterfactual ideas. We will first define the *state of the world* heuristically as the complete set of all conditions that can potentially influence the value of an outcome event. This state reflects the values of all factors that together determine the outcome value. In the SCC model, a factor is a component cause that is either present or absent. Similarly, the outcome either occurs or does not occur. For illustrative purposes, we will retain these simplifications. However, the basic logic extends to situations in which the factors and outcomes assume many possible values. A factor is said to potentially influence the value of the outcome if altering its value would change the outcome value under at least one possible state of the world. Such a factor can be called a *relevant* factor.

Now, let us assume that the sectors of a single pie chart can stand for the entire state of the world, with all the relevant factors explicitly represented. Note that these factors could be aspects of the society or the physical environment, as well as individual characteristics. In effect, we collapse the component causes included in different sufficient causes into a single pie chart. For simplicity, we will always include five sectors in our diagrams. Furthermore, instead of assuming that all factors in the chart are present, as in Fig. 2.1, we allow

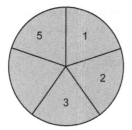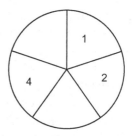

Figure 2.2. Causal model: Two individuals, each with a causal configuration comprising the values of five causal factors. The outcome event occurs for one of them (indicated by shading) but not for the other.

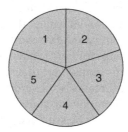

Figure 2.3. An individual with all five causal factors present. This particular causal configuration causes the outcome event to occur, as indicated by the shading.

each such factor to either be present or absent. If the factor is present, we will display its identifying number in the corresponding sector of the pie chart. If the component is absent, we will leave the corresponding sector blank. In this conception, the outcome is determined by the pattern of presence and absence of the relevant conditions, which we will term the *causal configuration*. Finally, if a particular configuration of factors causes the outcome event, then the pie chart will be shaded, but if it does not, the pie chart will remain unshaded.

Figure 2.2 displays two possible configurations. The pie chart on the left represents a situation in which Factors 1, 2, 3, and 5 are present, but Factor 4 is absent. The shading indicates that the outcome event occurs. In this case, both the first and third sufficient causes shown in Fig. 2.1 would be fulfilled. In terms of our artificial influenza epidemic, this person would not be vaccinated and would be over age 75, be exposed to the virus, and have weakened immunity. On the other hand, the right-hand pie chart is not shaded, because none of the three sufficient causes is present.

To illustrate the counterfactual concept of causality, suppose in our hypothetical example that all five relevant factors (1, 2, 3, 4, and 5) are present for a particular person. This causal configuration causes the event to occur, as represented in Fig. 2.3 Now suppose we ask the following counterfactual question: what would have happened with Factor 1 absent, but everything else in this scenario remaining fixed? By "everything else" we mean the values of all

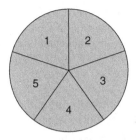 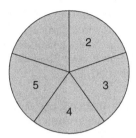

Figure 2.4. An individual for whom the outcome event occurs (indicated by shading) whether or not Factor 1 is present. Therefore, Factor 1 has no causal effect for this individual.

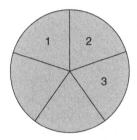 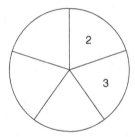

Figure 2.5. An individual for whom the outcome event occurs if and only if Factor 1 occurs. Therefore, Factor 1 is a cause of the event for this individual.

other factors besides Factor 1. The constellation of factor values besides the causal factor(s) of interest will be called the *causal context*. With the causal context portrayed in Fig. 2.3, the outcome event would have occurred even without Factor 1, because Factors 3, 4, and 5 were present. This situation is illustrated in Fig. 2.4. Because the event would have resulted anyway, Factor 1 would not be construed as a "cause" of the event. More precisely, the causal effect of Factor 1 would have been the null effect (no difference).

Alternatively, suppose that the causal context for Factor 1 consisted of Factors 2 and 3 only. In this case, without Factor 1 the event would *not* have happened. So, given this causal context (i.e., presence of Factors 2 and 3, but not 4 and 5) the outcome event would have occurred if and only if Factor 1 also occurred. In this sense, Factor 1 was truly a cause of the event. This situation is portrayed in Fig. 2.5.

Two important points emerge from this discussion. First, the causal effect of a given factor on a particular individual *depends on her particular background conditions*. In general, it is not meaningful to define a causal effect in the abstract, without reference to the causal context. Second, the causal effect of a given factor *can vary across individuals*, because the causal context depends on individual characteristics and circumstances.

Some methodologists have gone so far as to assert that estimating a general causal effect is nearly always a gross simplification and thus misleading:

> I consider it pointless to speak of causes when all that can be validly meant by reference to a cause in a particular instance is that, on one trial of a partially specified manipulation *t* under conditions A, B, and C along with other conditions not named, Phenomenon P was observed. To introduce the word *cause* seems pointless. (Cronbach, 1982, 137)

In context, Cronbach's extreme position was apparently intended as a counterweight to a widespread view among methodologists that evaluations of educational and social programs should focus primarily on "main effects" and that "interactions" with background conditions were less likely to occur (e.g., Campbell, 1969). He argued cogently that, in the setting of educational program evaluation, there is usually no logical or empirical justification for assuming that individuals and background circumstances are homogeneous. Consequently, causal effects are more likely to be highly variable.

2.3 WHY COUNTERFACTUALS?

The counterfactual approach provides a clear and coherent way to think about a variety of important concepts related to causation. In particular, a counterfactual framework will allow us to distinguish between the *causal effect* of some factor and an *empirical effect* that can potentially be measured. The empirical effect is the effect that would be observed in a particular study if an effectively infinite sample could be obtained. We distinguish the empirical effect from the *observed effect* to emphasize that the issues discussed relate exclusively to systematic error, as opposed to the random error that can arise because of finite samples. The observed effect reflects the impact of both random variation and bias.

The goal of a comparative study is to estimate an empirical effect that faithfully represents an underlying causal effect. Methodological bias is the discrepancy between such an empirical effect and the causal effect of interest. Our definition of bias as the difference between an empirical effect and the corresponding causal effect is deliberately narrow. Contrast this definition with that given in *A Dictionary of Epidemiology*:

> Deviation of results or inferences from the truth, or processes leading to such deviation. Any trend in the collection, analysis, interpretation, publication or review of data that can lead to conclusions that are systematically different from the truth. (Last, 2001, 14)

A general definition like this would include many aspects of research design and analysis that extend beyond our core concerns about causal validity. The main purpose of this book is to help researchers understand, identify, and deal with the various kinds of bias that can undermine causal inferences. To accomplish this objective, we will employ counterfactual analysis primarily as a *conceptual* tool to sharpen our thinking about bias.

The presentation in this book tries to steer a middle course between two methodological extremes. On the one hand, a counterfactual perspective sometimes refers to a general philosophy or language for scientific inquiry. This "softer" application of counterfactual methodology tends to be more conceptual and qualitative, a form of "scientific common sense" (Phillips and Goodman, 2006). On the other hand, a more quantitative form of counterfactual analysis may apply sophisticated mathematical techniques, such as structural equation models, to work out the details of a complex causal nexus among variables of interest. When it comes to understanding bias in comparative studies, counterfactual thinking offers much more than common sense, but usually much less than precise mathematical answers.

The approach adopted throughout this book is to utilize basic algebra and statistical concepts as devices to help methodologists formulate (and sometimes answer) questions that arise in practice. For example, every researcher is aware that allowing individuals to choose whether to participate in a treatment group or a control group is a recipe for bias. But exactly why and in what sense is this true? Are there circumstances in which such a design will not produce bias? When can the extent of possible bias be estimated? Counterfactual concepts provide a useful "language" in which such questions can be discussed.

The counterfactual perspective is not a unique conceptual framework for addressing such questions meaningfully. There are certainly others. Although counterfactual thinking has become extremely popular recently, it is by no means universally accepted by philosophers of science or research methodologists. Energetic debate about the merits of the approach persists (Holland, 1986, with discussion; Dawid, 2000, with discussion; Maldonado and Greenland, 2002, with discussion). It might well have been possible to write a book about methodological bias grounded in a different conceptualization of causal effects (Shafer, 1996; Dawid, 2000). But so far at least, counterfactual analysis is distinguished by the extent to which it has been usefully applied in biomedical and social sciences. For the problems posed by comparative studies, counterfactual thinking has made major contributions.

2.4 CAUSAL EFFECTS

This book deals with studies in which statistical analysis can be used to estimate causal relationships. Such studies attempt to measure the changes in a specific outcome that would be caused by manipulating (perhaps hypothetically) some preceding factor. For example, if elderly nursing-home residents are inoculated with a newly developed vaccine, how much will the rate of influenza decrease? To study this kind of question, a researcher must find a way to *compare* what happens to groups of different individuals (or the same individuals at different times) under alternative specified conditions. If we design this comparison appropriately, it can reveal something useful about the causal effect of interest. If we design the comparison badly, the resulting empirical effect will be misleading. Suppose we decide to compare the influenza rate for a group of elderly patients given the vaccine with the rate for a similar group who were not. Under what conditions will this comparative study yield meaningful and useful results? To what extent might the results of this comparison be misleading? How could we conduct analyses that adjust for any existing bias? Trying to address such questions is essentially the subject matter of this book.

We have defined bias as the difference between an empirical effect and a causal effect. To be more precise, we introduce some standard terminology to describe comparative studies. For simplicity, we assume that a study is concerned with the causal effect of a single factor (treatment, risk factor, program, intervention) and a single well-specified variable potentially affected by this factor (outcome, response, endpoint). Our objective is to determine the causal effect of this factor on the outcome. If the factor of interest is operative for a certain individual, we will say that she has been *exposed* to the factor. More generally, exposure can be framed in terms of a scale, with varying levels of exposure, possibly even on a continuum. For simplicity, we will consider the exposure status for any individual to have only two possible values: *exposed* and *unexposed*.

In many contexts, an exposed individual is said to experience the *treatment* and an unexposed to experience a *control* condition. The terms "treatment" and "control" derive from the early development of statistical analysis in the context of controlled experimentation on agricultural treatments (e.g., insecticides, herbicides, seed types, fertilizers). Alternatively, the exposed group may receive the treatment and the unexposed group an alternative treatment, or *comparator*. Sometimes, particularly in a medical context, the groups being compared are said to receive alternative treatment *modalities*. In the social sciences, the terms *intervention* and *program* often take the place of treatment. Context-specific terminology can be awkward when discussing general methodological issues that cut across different areas of application. The more neutral term "exposure" will usually be adopted throughout this book. Furthermore, we will generally refer to the two groups being compared as the *study groups*.

What exactly do we mean by the causal effect of an exposure? The traditional statistical perspective offers no simple answer. Statisticians deal with measures of association in a specified population. Is lung cancer associated with smoking cigarettes? An observed relationship of this kind may be accorded a causal interpretation only if certain conditions are believed to hold. Generally speaking, the association is said to be causal if alternative (noncausal) explanations can be dismissed from consideration. The process of adducing evidence to rule out all other possible reasons for the observed association can be complex and contentious, as it was for many years with smoking and cancer.

From this traditional statistical perspective, the concept of causation is conceptually secondary to that of association. (The terms *correlation* and *dependence* are also commonly used to describe statistical association, although each has a more specific technical meaning as well.) Modern statistical theory, since the early days of Karl Pearson and his disciples, has nearly expunged the word "cause" from its vocabulary (Pearson, 1900). Causation is discussed elliptically if at all; statisticians typically employ circumlocutions such as "independent risk factor" or "explanatory variable" to avoid causal language. One important consequence is that the causal effect applicable to a particular *individual* cannot be formalized. From the usual statistical perspective, the primary realities are variables and populations.

In contrast, counterfactual analysis posits a meaningful causal effect for each individual. Before a research study has been implemented, each individual in the study population is imagined to possess two potential outcomes. One of these will actually be realized in the study; the other will be counterfactual. For example, imagine that we are evaluating a new program of tutoring for the mathematics portion of the Scholastic Aptitude Test (SAT). Jane Smith's score if she is assigned to the intervention (New Program) is one potential outcome (say 640) and her score if assigned to the traditional approach (Old Program) is another potential outcome (say 600). This pair of potential outcomes has been termed the *response pattern* (Greenland and Poole, 1988). Jane's response pattern can be written as (640, 600).

The New Program received by Jane can be considered a causal factor contributing to the determination of her outcome score. However, we would have no easy way to represent graphically the entire range of values that can be assumed by potential outcomes. (For the SAT there are 61 possible scores, ranging from 200 to 800 in increments of 10.) In principle, we could extend the model to accommodate such a numerical outcome by using different gradations of shading and/or color to represent various outcome scores. Obviously, such an approach would not be practical. Therefore, when considering a numerical outcome, we will utilize algebra rather than a graphic visualization. Furthermore, we will for now suppress the possible dependence of the potential outcomes on the causal context.

Let V_{iE} designate the potential outcome value when subject i in a study is exposed, and V_{iU} the outcome if she is unexposed. For individual i, the exposure has no causal effect if $V_{iE} = V_{iU}$. However, if $V_{iE} \neq V_{iU}$ under the study conditions, then exposure *causes* a change in the observed outcome value from V_{iU} to V_{iE}. It might be tempting to jump to the conclusion that the causal effect for an individual is $V_{iE} - V_{iU}$. Indeed, this is one way to define an individual causal effect, but we could also define it as some other mathematical function, such as V_{iE}/V_{iU} or $\ln(V_{iE}) - \ln(V_{iU})$. Most generally, we define the causal effect as the response pattern (V_{iE}, V_{iU}). A particular mathematical "difference" between the values of V_{iE} and V_{iU} is just one manifestation of this underlying causal effect.

From the counterfactual perspective, the response pattern for a given individual contains all the information that is relevant to any conceivable causal inference. The distribution of response patterns for the individuals in a given population represents the full *causal effect* of the exposure on that population. If we knew the response patterns, we would "see" what would happen to each individual under alternative scenarios, such as being exposed and unexposed. We would then be in a position akin to that experienced by the three-dimensional being who visits the two-dimensional world of Flatland in Edwin Abbott's popular allegory (Abbott, 1884). Stepping out of ordinary reality into the extra dimension of "potentiality" would allow direct observation of potential outcomes, and therefore of causation. Of course, we have no means of witnessing the response patterns. How then can causal effects be made visible?

Mainstream statistical theory has almost nothing to say about individual causal effects. Classical statistical theory shifts attention away from individuals and onto a population (actual or hypothetical) of which the individual is assumed to be representative. Probability distributions of variables that describe important individual characteristics are of central importance. Any individual is typically regarded as a "random" observation whose value for any variable is a "deviation" from, or "error" around, the average in the population.

Concentrating on parameters of probability distributions was undeniably a huge step forward in the development of modern statistical inference. Within this framework, the elaborate machinery of statistical methods for managing random error grew to maturity. Meanwhile, causal inference withered on the vine. Causation became viewed as an outmoded relic of a prescientific era. Ignored in this evolution of statistical theory was the irony that the "hypothetical infinite" populations commonly assumed by statisticians as the source of observed samples are no less imaginary or mystical than the potential outcomes for which causal models are frequently criticized.

Potential outcomes are "real" in the same sense as hypothetical infinite populations. Both are valuable conceptual devices. Imagining that a set of observations was drawn at random from a probability distribution, such as the *normal distribution*, places the data in a useful perspective. A particular data set is seen as one of many that *could* potentially have arisen. Thus, the assumed probability distribution can be viewed as a counterfactual model for analyzing *random variability*. In a similar way, potential outcomes facilitate thinking about *systematic variability*. Both statistical inference and causal analysis require consideration of what might have been but was not. Hypothetical infinite populations describe counterfactuals under random sampling from a fixed population under stable circumstances. Potential outcomes describe counterfactuals for a fixed individual when one or more specified circumstances change. Data analysis must ideally take proper account of both these important sources of variability.

Formulating problems in terms of potential outcomes facilitates clear thinking about research design and analysis. Especially important, we can discuss causation at the level of the individual. Ideally, research would allow us to measure the actual response pattern that is applicable to each individual. For example, a clinical trial of some new medication might yield a set of detailed instructions on how to tailor the timing and dosage of a particular drug to a patient's specific characteristics and circumstances. Eventually, the detailed understanding of the human genome may lead to advances in personalized medicine that approach this ideal.

For now, such fine-grained inferences remain in the realm of science fiction. Moreover, individualization in the social sciences may seem an even more remote possibility. In practice, most studies currently aim to generate an unbiased estimate of an aggregate causal effect. Does the Head Start program increase future academic achievement in school for disadvantaged children? Does treatment with a statin drug reduce the incidence of myocardial infarction in middle-aged women with elevated lipids? Can exposure to second-hand smoke cause asthma in young children?

In effect, these studies attempt to estimate some useful measure of the typical (often average) effect in a target population. As mentioned previously, the ideal object of inquiry is the entire distribution of response patterns in the population. Knowledge of the response-pattern distribution would allow us to calculate any possible *measure of causal effect*. For example, we could determine what proportion of individuals with any specified characteristics do better when exposed than when unexposed, and by how much. For concreteness, let us consider a class (Class A) of 10 high school students about to begin the hypothetical new tutoring program (New Program) to improve their SAT math scores. Their potential outcomes are shown in Table 2.1.

Table 2.1 displays three possible measures of causal effect. Note that the average score for the students in Class A with the New Program is 637, and the average with the Old Program would be 570. So, one way to express the causal effect would be as a 67-point average increase. However, an alternative measure would be the 60-point increase in the median, or perhaps the increase of 20% in the number of students who achieve a score of at least 600 points.

Table 2.1 SAT Tutoring Experiment: Class A

Student	New Program	Old Program
A1	570	520
A2	480	480
A3	800	600
A4	640	600
A5	680	700
A6	780	660
A7	600	550
A8	560	480
A9	590	540
A10	670	570
Mean	637	570
Median	620	560
% 600+	60%	40%

There are limitless additional possibilities, each of which represents a specific way of summarizing the underlying distribution of response patterns for the students in Class A.

In general, let us define V_E to be a random variable defined over the study population. This variable assumes the value V_{iE} if individual i is exposed (E). Let us define V_U as the corresponding random variable if he is unexposed (U). For simplicity of presentation, we will in some contexts use V_E also to represent the *distribution* of the variable V_E. In our example, V_E stands for both the vector of ten scores in the first column of Table 2.1, and the variable that has these values. In general, we define a causal effect as a "difference" denoted by $D(V_E, V_U)$.

For example, suppose we want to know the median improvement in SAT score attributable to the New Program vs. the Old Program. If we could somehow determine the full distribution of response patterns, this parameter would be computable. One possible approach would be to find matched pairs of students who appear very similar on relevant characteristics. Then one of the students in each pair could receive the New Program and the other could receive the Old Program. The observed difference between the two after the tutoring programs would be an estimate of the program's effect for each individual in the New Program. Of course, such a study might be criticized, because there is no guarantee that the two paired students really were similar. There are usually many possible differences between individuals, so that their causal contexts are far from identical. An observed difference between outcomes for two *different* individuals can almost never be safely assumed to represent the difference for the *same* individual under two different conditions. As a result, measuring an effect such as the median improvement, which depends on observing individual effects, is rarely feasible.

Alternatively, suppose we define $D(V_E, V_U)$ as the difference of the medians, rather than the median of individual differences. That is, we would calculate the median of V_E and the median of V_U and find the difference between them. For example, the difference of median SAT scores shown in Table 2.1 would be 60 points. This approach depends only on the *marginal distributions* of V_E and V_U. It is not necessary to know the individual response patterns, which depend on the more complete *joint distributions*. As we will see, this is a major advantage. Indeed, effect measures that depend only on the marginal distributions of V_E and V_U are virtually the only ones ever employed. By far the most common effect measure is the difference of averages. We will denote the average (mean) of V_E by μ_{VE} and the average of V_U by μ_{VU}. In this case, $D(V_E, V_U)$ becomes simply $\mu_{VE} - \mu_{VU}$.

2.5 EMPIRICAL EFFECTS

So far, we have formalized various concepts related to the true, but unattainable, causal effect. Now we venture into the world of actual research. In the real world, we cannot simultaneously assign both E and U to every member of the study population. For each subject in a study, only one of the two potential outcomes can be observed. This methodological conundrum has been called the *fundamental problem of causal inference* (Holland, 1986). On the one hand, therefore, we can think of a true causal effect as a platonic ideal that is never actualized. On the other hand, scientists are often able to estimate causal effects by applying principles of experimental design and analysis to fashion appropriate comparative studies.

Suppose we calculate the mean (or some other parameter) for a group of subjects who are exposed and a "control" group who are unexposed. How can we be certain that the observed difference is attributable to the exposure of interest rather than to some preexisting differences between the two study groups? In an intuitive sense, we want assurance that the two groups being compared are *equivalent in all relevant respects*. In that way, any observed difference will depend only on the different modalities assigned to the two groups. However, the phrase "equivalent in all relevant respects" is too vague to be very helpful. We have expressed the causal effect as $D(V_E, V_U)$, the true effect of the program in the population under study. Because, for each individual, only one of V_{iE} and V_{iU} is ever observable, it is not obvious how $D(V_E, V_U)$ might be estimated. It seems sensible to assume that a comparison of the exposed and unexposed groups could somehow be useful, but we require a logical connection between our empirical study data and the unobservable causal effect. Understanding the nature of this potential link is the key to solving the fundamental problem of causal inference.

In our tutoring example, suppose that we decide to compare the results for Class A with those obtained by another Class B that receives the Old Program. Table 2.2 displays the hypothetical set of potential outcomes for Class B. The

Table 2.2   SAT Tutoring Experiment: Class B

Student	New Program	Old Program
B1	610	500
B2	670	520
B3	540	570
B4	690	600
B5	720	700
B6	740	670
B7	770	620
B8	650	560
B9	690	610
B10	800	770
Mean	688	612
Median	690	605
% 600+	90%	60%

Table 2.3   SAT Tutoring Experiment: A vs. B

	New Program	Old Program
	570	500
	480	520
	800	570
	640	600
	680	700
	780	670
	600	620
	560	560
	590	610
	670	770
Mean	637	612
Median	620	605
% 600+	60%	60%

study will compare the actual outcomes in the New Program column of Table 2.1 with the actual outcomes in the Old Program column of Table 2.2. The results of this comparison are displayed in Table 2.3. The observed difference of the means is 25 points. However, from Table 2.1 we know that the causal effect for the students in Class A is really 67 points. In an actual study, the observed difference would be affected by both random and systematic variability. As explained previously, we are pretending that the random error does not exist, so this observed effect is interpreted as the true difference between the two classes being compared. This difference is what we have termed the *empirical effect*.

In our example, the empirical effect is biased because the potential outcomes for Class B under the Old Program did not result in a mean of 570. If they had, the mean for Class B could have been substituted for the unobserved mean of Class A under the Old Program. In this special situation we would say that Class B is *partially exchangeable* with Class A (Greenland and Robins, 1986). For the two groups to be *completely exchangeable*, we would need to be able to reverse the process as well, so that the potential outcomes for Class A under the Old Program could serve as a substitute for the potential outcomes of Class B under the New Program. In that way, we could estimate the causal effects for both classes.

Note that if we had complete exchangeability, the mean values for the potential outcomes of the classes would be the same. Therefore, the causal effects (as measured by the difference of means) would be the same for the two classes. Also, under complete exchangeability it would not matter which class actually received the New Program and which received the Old Program. We could exchange the roles of the two groups without affecting the resulting empirical effect. This situation is illustrated in Tables 2.4 and 2.5. Exchangeability in this sense is one way to express what we mean by saying that the

Table 2.4   SAT Tutoring Experiment: Class B1

Student	New Program	Old Program
B1	580	550
B2	500	520
B3	800	640
B4	620	600
B5	680	680
B6	780	660
B7	590	580
B8	570	430
B9	540	470
B10	710	570
Mean	637	570
Median	605	575
% 600+	50%	40%

Table 2.5   SAT Tutoring Experiment: A vs. B1

	New Program	Old Program
	570	550
	480	520
	800	640
	640	600
	680	680
	780	660
	600	580
	560	430
	590	470
	670	570
Mean	637	570
Median	620	575
% 600+	60%	40%

study groups are "equivalent in all relevant respects." But it is not at all obvious how we would know whether such exchangeability can be assumed. Furthermore, this definition of exchangeability only expresses the idea of equivalence between groups in a very limited sense of *mean-exchangeability*. For example, Table 2.5 shows that the difference of medians is 45 instead of the corresponding causal effect of 60.

A more intuitive notion of equivalence between study groups might be reflected by the situation exhibited in Tables 2.6 and 2.7. Here the entire (marginal) distribution of potential outcomes under either tutoring program in Class B2 is identical to the corresponding distribution in Class A. In this

Table 2.6   SAT Tutoring Experiment: Class B2

Student	New Program	Old Program
B1	640	600
B2	560	570
B3	480	480
B4	600	700
B5	670	480
B6	570	520
B7	680	550
B8	800	540
B9	590	600
B10	780	660
Mean	637	570
Median	620	560
% 600+	60%	40%

Table 2.7   SAT Tutoring Experiment: A vs. B2

	New Program	Old Program
	570	600
	480	570
	800	480
	640	700
	680	480
	780	520
	600	550
	560	540
	590	600
	670	660
Mean	637	570
Median	620	560
% 600+	60%	40%

case, *any* empirical effect that can be measured will be identical to the corresponding causal effect. For example, the difference of means, difference of medians, and difference of percentages above 600 points all have the correct (causal) values. Of course, as explained in the previous section, we could not estimate the mean individual effect. That would require a way to observe the distribution of response patterns (i.e., the joint distribution of the potential outcomes).

Although exchangeability in the sense exemplified by either Class B1 or Class B2 is theoretically appealing, how could we obtain information to decide whether such exchangeability is plausible? Any intuition or knowledge

Table 2.8   SAT Tutoring Experiment: Class B3

Student	New Program	Old Program
B1	590	540
B2	780	660
B3	680	700
B4	480	480
B5	670	570
B6	600	550
B7	640	600
B8	570	520
B9	800	600
B10	560	480
Mean	637	570
Median	620	560
% 600+	60%	40%

we may possess about group formation pertains to which types of individuals are assigned to each study group. But we have noted that with respect to causal effects, the only relevant characteristics of individuals are their response patterns. Therefore, the critical question becomes whether the assignment process is related to response patterns in such a way that the response-pattern distributions are identical (apart from sampling variation) in the two groups.

If these distributions differ in some way, then there exists a *potential* for bias. As seen in the tutoring example, whether bias actually occurs depends on the specific nature of the difference between the groups and on the particular effect measure chosen. It is certainly possible that the empirical effect will fortuitously have the correct value even if the response-pattern distributions in the treatment groups happen to be different. But it is hard to imagine how we could ever discern that such measure-specific exchangeability is present. In the broadest sense, *two study groups can be deemed exchangeable if and only if they have identical distributions of response patterns.* This concept of exchangeability has been tremendously influential since its introduction (in a somewhat different form and terminology) by statisticians Paul Rosenbaum and Donald Rubin (Rosenbaum and Rubin, 1983a). Unless otherwise indicated, exchangeability throughout this book will refer to this definition.

Tables 2.8 and 2.9 illustrate exchangeability in this sense of identical response-pattern distributions. Note that in terms of response patterns, the students in Class B3 are identical *as a group* to those in Class A. Unfortunately, there is no way to pair a student in Class A (say A1) to her counterpart in Class B3 (B8). So, we still cannot identify the causal effects for any individual students. More generally, let (W_E, W_U) represent the response-pattern distribution in the group of individuals assigned to E and (Z_E, Z_U) the corresponding

Table 2.9   SAT Tutoring Experiment: A vs. B3

	New Program	Old Program
	570	600
	480	570
	800	480
	640	700
	680	480
	780	520
	600	550
	560	540
	590	600
	670	660
Mean	637	570
Median	620	560
% 600+	60%	40%

distribution in the group assigned to U. These distributions depend on exactly how the two study groups are selected. In the study, W_E and Z_U are observable, but Z_E and W_U are not. If exchangeability holds, then the distributions represented by (W_E, W_U) and (Z_E, Z_U) are identical to each other and therefore also to the overall distribution (V_E, V_U) in the study population. Thus, we can utilize W_E and Z_U as substitutes to be exchanged for V_E and V_U, respectively.

How can we guarantee that such exchangeability holds? Sir R. A. Fisher, widely regarded as the founder of modern statistical theory, recognized that the only surefire method was to assign subjects to study groups *randomly* (Fisher, 1925). *Randomization* was originally implemented by methods similar to tossing a coin. Later, tables of random number sequences were generated and published for use by study designers, and today "pseudorandom" numbers can be easily generated using computer algorithms. Although randomization is a powerful tool, it is only applicable when experimental manipulation of subjects is feasible. We cannot, for example, test the effect of cigarette smoking by forcing a random subset of potential experimental subjects to take up smoking. Furthermore, randomization only assures exchangeability in a probabilistic sense, an ideal that is approached in very large samples.

Comparative studies that do not involve control of assignment to comparison groups by randomization are usually termed *observational* studies. If the observational study involves an intervention that is manipulated by the researcher, it is sometimes called a *quasi-experiment*. This terminology highlights the idea that a quasi-experiment incorporates some elements of statistical control but falls short of the degree of validity provided by a randomized experiment. An observational study in which the exposure is a factor that is not governed by the researcher is sometimes called a *correlational study* or *natural experiment* (Shadish et al., 2002). In the social sciences, methodological issues related to quasi-experiments have been analyzed extensively in the context of *program evaluation*. In social science research, different types of bias are often described as *threats to validity* (Campbell and Stanley, 1963; Cook and Campbell, 1979).

In an observational study, the study groups being compared are generally not exchangeable. Observational studies are therefore much more susceptible to various forms of bias than are randomized studies. Observational studies have traditionally been regarded by research methodologists as much inferior to randomized experiments. In the absence of random assignment, the burden of proof about the exchangeability assumption is on the data analyst. Several kinds of evidence can be brought to bear in reaching a conclusion regarding exchangeability. Sometimes information about the actual process of forming the study groups is available. More often, judgments about exchangeability are based on analyses of differences in the distributions of various measurable characteristics (covariates) between the groups.

To appraise the magnitude of possible bias requires more than a general assessment of whether exchangeability holds. Informed speculation about the specific ways in which the study groups may depart from exchangeability is necessary. In some cases, it will be possible to compensate for lack of exchangeability through analytic strategies. When multiple studies of the same intervention or risk factor are available, the pattern of results across studies may help to reveal the nature and extent of bias and suggest possible corrections. The counterfactual perspective can be helpful in clarifying the methodological issues that must be addressed and suggesting fruitful approaches.

Assessing the potential for bias in an actual study involves familiarity with the various *sources of bias* that may apply. There are many specific aspects of a study's context, design, or implementation that can lead to bias. For example, allowing volunteers to self-select for participation in a social intervention is an obvious source of bias. Discussions of many specific sources appear scattered throughout the methodological literature. The definitions and terminology used to describe these various sources are inconsistent and often confusing. In this book, the counterfactual framework is used to help organize and understand these sources of bias according to their essential features.

This book draws heavily on a base of knowledge that has evolved over approximately the past 30 years. The basic concepts of counterfactual analysis have been widely disseminated throughout the academic literature. These theoretical developments in causal analysis provide a powerful lens through which to view bias, especially in observational research. But this conceptual clarification has so far had little practical impact on how most methodologists and research practitioners deal with bias. One purpose of this book is to help link these new concepts to the realities "on the ground" faced by research practitioners and professionals who apply the results of comparative studies.

GUIDEPOST 2

This chapter has introduced several fundamental themes that will reverberate throughout this book. An explicit framework for addressing causation is central to the understanding of bias in comparative studies. The counterfactual perspective, especially as expressed in the idea of potential outcomes, has proven useful as a basis for theoretical discussions of causal effects. These concepts lead naturally to the centrality of the individual response pattern. A causal effect can be defined as a summary of the distribution of response patterns. Bias is defined as the difference between the empirical effect that would actually be approached by the study in large samples, and the causal effect. A critical requirement for unbiased causal inference is that the study groups be exchangeable, in the sense of having the same distributions of response patterns.

In the next chapter, we consider a number of major issues that pertain to bias. These issues have been discussed extensively in the methodological literature but have given rise to much confusion. The purpose of Chapter 3 is to describe the general nature of these issues, to show why clarification is needed, and to hint at how an explicit causal model can help to address these issues more appropriately.

CHAPTER 3

Estimating Causal Effects

The main purpose of a comparative study is to estimate a causal effect. We have defined a causal effect most generally as a distribution of individual response patterns in the study population. The response-pattern distribution can be viewed as the raw material from which any specific causal effect can be forged. This distribution could tell us everything we need to know about causal effects, but it cannot be observed directly. It is highly informative on a conceptual level, but is not directly useful for actually estimating an effect.

As explained in Chapter 2, the causal effects of interest typically take the form of a difference between (or ratio of) the values of two summary statistics. One of these statistics is a summary of the potential-outcome distribution for a group of exposed individuals; the other is the corresponding summary for the unexposed group. The difference of means and the ratio of proportions are by far the two most common summary measures of effect. The adoption of a particular measure of effect seems so natural and obvious that it is easy to overlook its significance. Implications that flow from the mathematical properties of a particular measure are sometimes interpreted incorrectly as properties of causal inference in general.

The empirical effect has the obvious advantage that its value can be calculated from potentially observable data. However, the empirical effect may have only an indirect relationship to the actual causal effect, which is based on the underlying response-pattern distribution. Any specific aspect of a study that creates a discrepancy between the empirical effect and the causal effect is a *source of bias*. Most attention by methodologists has focused on sources that lead to a lack of equivalence between study groups. We have defined this concept in counterfactual terms as a lack of exchangeability between the groups. However, we will see that such a lack of exchangeability is not the only way that bias can occur.

Bias and Causation: Models and Judgment for Valid Comparisons, By Herbert I. Weisberg
Copyright © 2010 John Wiley & Sons, Inc.

A classic example of a biased comparison is the use of volunteers to test a new program. Lack of exchangeability between the volunteers and nonvolunteers would imply that *any* empirical effect might be distorted. However, the existence and extent of bias would depend also on the particular mathematical form of the empirical effect and on the specific way in which the property of exchangeability has been violated. Consider again the imaginary tutoring experiment discussed in Chapter 2. The magnitude of bias when comparing Class A with Class B1 was zero for the difference of means, but 15 points for the difference of medians (see Tables 2.1 and 2.5). Was the study biased? It depends on the preferred measure of effect.

The point of these examples is that assessment of possible bias requires counterfactual thinking to understand exactly how the particular empirical effect might differ from the causal effect of interest. In particular, informed judgment about the likelihood and nature of nonexchangeability is critical. Moreover, in some situations, counterfactual analysis can suggest ways to obtain empirical evidence about the extent of bias.

3.1 EXTERNAL VALIDITY

We have so far defined bias as the difference between the empirical effect and the causal effect in a study population. Exchangeability as defined previously allows a valid causal inference for the population that is actually being studied. However, this population is not necessarily the *target population* of interest. For example, suppose our study of tutoring was performed on private-school students. If the new program is found to be more effective than the traditional one, we cannot know whether this result will generalize to public-school students as well. Social scientist Donald Campbell first formalized this distinction by coining the terms *internal validity* and *external validity* (Campbell, 1957; Campbell and Stanley, 1963):

> *Internal validity* is the basic minimum without which any experiment is uninterpretable: Did in fact the experimental treatments make a difference in this specific experimental instance? *External validity* asks the question of generalizability: To what populations, settings, treatment variables and measurement variables can this effect be generalized? Both types of criteria are obviously important, even though they are frequently at odds in that features increasing one may jeopardize the other. While *internal validity* is the sine qua non, and while the question of external validity, like the question of inductive inference, is never completely answerable, the selection of designs strong in both types of validity is obviously our ideal. (Campbell and Stanley, 1963, 5)

Campbell and Stanley originally included generalizing to "treatment variables and measurement variables" as part of their conception of external validity. In later writings, Campbell narrowed this definition by separating out generalization regarding treatment variables and measurement variables (Cook and

Campbell, 1979). Apart from that one modification, this quotation remains a clear and relevant statement of several key themes that will reverberate throughout this book.

Should bias be defined as the total difference between the empirical effect in a study population and the causal effect in the target population, or only as the "internal bias" within the study population itself? This question will be addressed in Chapter 5 in the context of a theory of selection bias. For now, it suffices to say that, although internal validity is usually the primary concern, there are circumstances in which external validity should be encompassed within the assessment of bias. In counterfactual terms, we must consider whether the study population is exchangeable with the target population. However, information about the selection of the study population is often very hard to obtain. It may not be feasible to measure relevant characteristics of the entire target population, or even a random sample.

To appreciate the difficulty, consider a situation that is in many respects ideal: a randomized controlled trial (RCT) to compare a new drug with a standard comparison drug. It is common to recruit experimental subjects who appear to meet certain initial screening criteria. Prospective subjects may respond to advertisements, be referred by their physicians, or learn of the RCT by word of mouth. The potential subjects are then interviewed and tested to determine if they meet the study's eligibility criteria. Those deemed eligible are given further information and required to provide informed consent. During the orientation process, some may decide not to participate for a variety of reasons. The pool of study subjects who eventually enter the RCT after surviving this complex gauntlet of requirements may not be representative of future patients for whom the new drug would be prescribed.

During the selection process, the emphasis of the investigators is on enrolling an eligible group, not on collecting detailed and systematic data about those who were screened out. Consequently, even in the most tightly controlled studies, there may be scant data regarding the selection mechanism. In an observational study exchangeability may be less problematic, because the individuals are studied under more "natural" conditions. However, there are still many ways in which one or both of the study groups can differ from the target population.

Among researchers, possible lack of external validity tends to be downplayed. Part of the reason for this attitude is the priority usually accorded internal validity concerns, as reflected in the preceding Campbell and Stanley quote. Another part stems from the belief that scientific research is intended to disclose causal relationships that are as close to universal as possible. Therefore, an effect that can vary materially across individuals and circumstances may be deemed a troubling exception to the more general rule of uniform effects. This perspective is even stronger in the biomedical sciences than in the social sciences. Consider this statement in a popular epidemiology text:

Epidemiologic study designs are usually stronger if subject selection is guided by
the need to make a valid comparison, which may call for severe restriction of admis-
sible subjects to a narrow range of characteristics, rather than by an attempt to make
the subjects representative, in a survey sampling sense, of the potential target populations. ... Selection of study groups that are representative of larger populations in
the statistical sense will generally not enhance the ability to abstract universal statements from observations, but selection of study groups for characteristics that
enable a study to distinguish effectively between competing scientific hypotheses
will do so. (Rothman and Greenland, 1998, 133–134)

Recently, there has been increasing recognition that the pendulum may have
swung too far away from concerns about external validity. Explicit models of
the selection process are needed in order to address these concerns. Such
models may help to restore a better balance between internal and external
validity in the design and analysis of comparative studies.

Whether the context be medical or educational or social, the norm seems to be that
people vary in their response to treatment. ... A planner choosing a treatment rule
for a heterogeneous population cannot easily extrapolate findings on treatment
effects from an easily studied subpopulation to its complement, as optimal treatments in the two may differ. Hence the planner should value sample data that is
representative of the entire population of interest. Indeed, the planner may judge
an observational study of the entire population to be more useful than an ideally
executed experiment performed on a subpopulation. (Manski, 1999, 280)

3.2 MEASURES OF EMPIRICAL EFFECTS

To specify the mathematical form of an effect measure, we must first consider
whether the outcome of interest is on a *numerical* scale or is simply a *dichotomy*. An SAT test score is an example of a numerical value, whereas occurrence or nonoccurrence of a myocardial infarction is a dichotomy. Of course,
if we designate the two values of a dichotomy as 0 and 1, the outcome is technically numerical. However, the use of a number (0 or 1) in order to create a
binary variable is merely a mathematical convenience. A binary variable
defined in this way is sometimes called an *indicator variable*, because it indicates whether or not the event of interest occurs.

   If the outcome is numerical, it is usual to define the effect as either an
arithmetic difference or a ratio between a summary measure of W_E and the
comparable measure of Z_U. (Recall from Chapter 2 that W_E represents the
potential outcome variable for the exposed group under the condition of
exposure, and Z_U represents the potential outcome variable for the unexposed
group under nonexposure.) Other types of mathematical functions are sometimes useful, but not often enough to warrant discussion here. The summary
measure for each study group could be any standard statistical parameter, such
as the mean, median, or variance. If we denote this parameter by π, then we

can express the difference version of the empirical effect as $\pi_{WE} - \pi_{ZU}$ and the ratio version as π_{WE}/π_{ZU}. By far the most commonly used measure of effect is the difference $\mu_{WE} - \mu_{ZU}$ between the mean values.

For any measure of an empirical effect, there is a corresponding measure of causal effect. The causal effect has the same mathematical form as the empirical effect, except that π_{VE} replaces π_{WE}, and π_{VU} replaces π_{ZU}. Therefore, most causal effects have the form $\pi_{VE} - \pi_{VU}$ or π_{VE}/π_{VU}. (Recall that V_E and V_U represent the potential outcome variables for the entire study population under the two alternative exposure conditions.) We have defined bias as the difference between the empirical and causal effects. If the causal effect is a simple arithmetic difference, then it is usual to express bias also as a difference:

$$\text{Bias} = (\pi_{WE} - \pi_{ZU}) - (\pi_{VE} - \pi_{VU}) \tag{3.1}$$

As mentioned, the difference of averages is by far the most common measure when the outcome is numerical. We will therefore often be assuming that

$$\text{Bias} = (\mu_{WE} - \mu_{ZU}) - (\mu_{VE} - \mu_{VU}) \tag{3.2}$$

When the outcome is dichotomous, attention invariably focuses on the *proportion* of individuals in each study group who experience the outcome event. Let us denote this proportion as q_{WE} for the exposed group and q_{ZU} for the unexposed group. The results of any study can be expressed as a 2×2 *contingency table*, sometimes called a *fourfold table*, as illustrated by Table 3.1.

The 2×2 table contains all the basic data necessary to calculate any possible empirical effect. In the social sciences the *difference of proportions* $q_{WE} - q_{ZU}$ is often used. This measure is fairly common when the proportions are not very small, and it is sometimes called the *risk difference*, denoted by RD. The bias of the RD can be expressed as

$$\text{Bias} = (q_{WE} - q_{ZU}) - (q_{VE} - q_{VU}) \tag{3.3}$$

When the outcome has a small probability of occurrence, the relative frequency q_{WE}/q_{ZU} is a more typical choice than the absolute difference of proportions. The quantity q_{WE}/q_{ZU} is called the *risk ratio* or the *relative risk*. In either case, these measures are often written as RR. The word "risk" in these terms derives from the idea that the event of interest is usually something

Table 3.1   Standard 2 × 2 Table

	Event	No Event
Exposed	$q_{WE}N_E$	$(1 - q_{WE})N_E$
Unexposed	$q_{ZU}N_U$	$(1 - q_{ZU})N_U$

undesirable, such as the onset of a certain disease. In this case the bias can be represented either as a difference or as a ratio:

$$\text{Bias}_{\text{diff}} = \frac{q_{WE}}{q_{ZU}} - \frac{q_{VE}}{q_{VU}} \qquad (3.4)$$

$$\text{Bias}_{\text{ratio}} = \frac{(q_{WE}/q_{ZU})}{(q_{VE}/q_{VU})} \qquad (3.5)$$

Ratio bias is simply the RR_{emp} divided by RR_{caus}. This ratio is sometimes expressed as a percentage increase or decrease. For instance, if the empirical RR is 1.8 and the causal RR is 1.2, we might say there is a 50% bias, or that the causal effect is overestimated by 50%.

3.3   DIFFERENCE OF MEANS

The empirical effect and its corresponding causal effect are aggregate statistics that are defined at the population level. In particular, the difference of means is a summary measure of effect for a study population. However, it is often inferred that aggregate effects apply at least approximately to each individual in the population. For example, suppose that our new tutoring program claims to produce SAT scores that are 50 points higher on average than those attributable to the traditional approach. A mother evaluating options for her daughter might anticipate approximately this advantage for her child. Under what conditions would this expectation be justified?

Let us assume that the exposed and unexposed groups are exchangeable, so that the empirical effect can be interpreted as a causal effect. In this section we consider the question of how this effect applies to specific individuals. From a traditional statistical perspective, the concept of an individual treatment effect is not explicitly defined. So questions about individual effects are rarely entertained. In his classic introductory text on experimental design, statistician D. R. Cox advises researchers to assume

> ... that either only average treatment effects are required or that the treatment effects are constant. Special precautions should be taken if the treatment effects are expected to depend in an important way on the value of some supplementary observation, or to be different for different groups of units. (Cox, 1958, 22)

In other words, the issue of effect variability across individuals is generally finessed in one of three different ways:

- Focus on the average causal effect.
- Assume a uniform (constant) effect.
- Allow for effect modification.

Each of these approaches has its own advantages and limitations.

The individual effect is typically formulated as a difference (additive effect) or a ratio (multiplicative effect), but most multiplicative effects are usually converted to additive effects by using a logarithmic scale. For simplicity, we consider an additive effect, $V_{iE} - V_{iU}$. Suppose there are N subjects in the study population. Then the *average causal effect* (ACE) can be expressed as

$$\frac{1}{N}\sum_{i=1}^{N}(V_{iE} - V_{iU}) = \left(\frac{1}{N}\sum_{i=1}^{N}V_{iE} - \frac{1}{N}\sum_{i=1}^{N}V_{iU}\right) = \mu_{VE} - \mu_{VU} \qquad (3.6)$$

So the difference of means is also the average of the individual effects. Furthermore, because the empirical effect $\mu_{WE} - \mu_{ZU}$ is an estimate of the causal effect, it is also an estimate of the ACE for the population.

Focusing on the ACE is mathematically convenient, because we can calculate the quantity $\mu_{WE} - \mu_{ZU}$ even though the individual effects are unobservable. Note that such a trick would not be possible for other measures, such as the median individual effect. However, when there is substantial effect variability, the ACE may be misleading when applied to many individuals in the population. Thus, the technical advantages of adopting the ACE as a measure of effect are obtained at the cost of overlooking individual differences.

When the ACE is used to measure the causal effect, it is not uncommon to go further and assume that the causal effect is uniform in the population, that is, has a constant value for all individuals. If pressed, most statisticians will admit that they do not literally believe in constant effects. The assumption may be viewed as a mathematically convenient approximation when there is no particular reason to expect substantial variability across individuals. This pragmatic view is understandable, but it lacks any real justification. Moreover, the assumption of uniformity is often made implicitly, without conscious consideration of its implications. This occurs because many statistical methods intended to adjust for confounding are based on mathematical models that assume a constant effect. The validity of the results produced by these methods can depend on this assumption; if there is effect variability, the results may be difficult to interpret.

The third approach to analyzing individual causal effects depends on being able to identify measurable factors that can *modify* the effect. Let X be some *ancillary variable*, or *covariate*, measured for each individual in the study population. Then in general we can say that X modifies the causal effect if the value of the effect varies with the value of X. In our SAT tutoring example, suppose that the new program is actually 75 points better on average than the traditional program for boys, but only 25 points better for girls. Then breaking down the overall effect in this way would certainly be relevant to our hypothetical mom's decision regarding her daughter.

In statistical terms, variation of an aggregate effect according to the value of some covariate X is termed an *interaction*: there is an interaction between the causal effect and X. Statistical models for interaction effects generally assume the causal effect is constant within strata defined by given values of X (boys and girls in our example). Specifying interaction effects thus allows for a limited degree of individual causal-effect variation. A statistical interaction of this kind is to some extent dependent on the particular measure of effect (Rothman and Greenland, 1998, 330). For example, the difference of means might vary between boys and girls in the tutoring example, but the difference of medians might not vary.

A statistical interaction can arise in two possible ways. One possibility is that there is causal "synergy" operating at the level of the underlying biological or social processes giving rise to the data. For example, the combined effect of smoking and radiation may be greater than the sum of the individual effects. An extreme form of causal synergy is portrayed in Fig. 3.1. The causal context for a certain risk factor F for a particular individual includes four other factors. Assume that Factors 3 and 4 are present, but Factor 2 is absent for this individual. As for Factor F and Factor 1, there are four possibilities, as shown. For each possible situation, the event may or may not occur.

Suppose that the event occurs if and only if both of these factors are present. In this illustration, we have not shaded the three pie charts that represent scenarios in which either F or Factor 1 is absent. Observe that F causes the

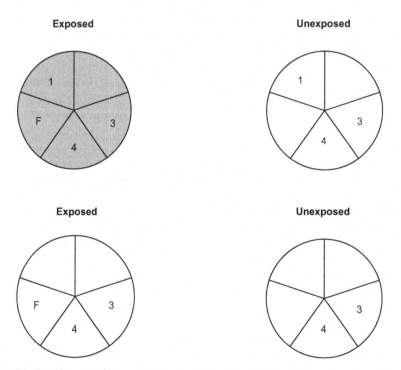

Figure 3.1. Causal synergy between two factors, Factor F and Factor 1. The outcome event occurs if and only if both causal factors are present.

event when Factor 1 is present (conditional on the other fixed factors as well). However, F does not cause the event if Factor 1 is absent. When the causal effect of a certain factor such as F can depend on the value of another factor, such as Factor 1, we will say that the second factor is a *causal modifier* of the first. Put differently, the response pattern for F depends on the value of Factor 1. More generally, causal modification means that the response pattern for an individual depends on the value of one or more variables in his causal context.

The second way that a statistical interaction can arise may be unrelated to causal modification. That is, there will be effect modification but not causal synergy. Suppose we decide to divide the total study population into two subpopulations (strata) on the basis of a dichotomous variable X that is not a causal factor. Then the causal effect *for each individual* within the two different strata will not depend on X, but the *measure of causal effect* may still vary across the strata. This can occur if the distributions of response patterns differ in the two strata. For example, the two strata might be two different schools in a community, such as a public school and a private school. The causal effect of an educational program might differ in the two schools because of the different populations attending these schools, even if the type of school does not influence the effect for any individual student. The relationship between effect modification and causal synergy will be explained in more detail in Chapter 7.

3.4 RISK DIFFERENCE AND RISK RATIO

When the outcome is dichotomous, the empirical effects are usually framed as either a risk difference or a risk ratio. To help understand the problems inherent in the interpretation of empirical effects, we introduce a simple model. Let us represent the two possible outcome values as D (for disease) and H (for healthy). Then the potential outcomes V_{iE} and V_{iU} can each be either D or H, and there are only four possible response patterns. These response patterns define four types of individuals, as shown in Table 3.2.

The basic mathematical model and the terminology to describe the four types were presented by epidemiologists Sander Greenland and James Robins in a seminal paper that introduced several counterfactual concepts, including exchangeability, to epidemiologists (Greenland and Robins, 1986). A *doomed*

Table 3.2 Four Possible Response Patterns

Response Pattern	Exposed	Unexposed	Proportion
1: Doomed	D	D	P_1
2: Causal	D	H	P_2
3: Preventive	H	D	P_3
4: Immune	H	H	P_4

individual will experience the event of interest regardless of exposure status, whereas an *immune* individual will always be spared. A *causal* individual experiences the event only if exposed, and a *preventive* individual only if unexposed. Note that individuals are *causal* or *preventive* in this model not in an absolute sense, but relative to the alternative exposure state under consideration. For example, a subject in a drug trial might be *preventive* with respect to a placebo comparison, because the disease would occur if there were no treatment. However, the same subject would be *immune* if the controls received an alternative drug that would also protect him from the outcome disease.

Throughout this book, the letter j will be used to index the possible response patterns that can occur in a given situation. With a dichotomous outcome, there are four possible response patterns, and P_j will represent the proportion of individuals with response pattern j in the target population ($j = 1,2,3,4$). (Later we will use a symbol Q_j to represent the corresponding proportion in the study population.) The four proportions summarize the entire distribution of response patterns and thus determine the value of any causal effect. To understand the nature of a causal effect, let us assume the exposed and unexposed groups are exchangeable, so that the values of P_1, P_2, P_3, and P_4 pertain to the response patterns in both groups. Then the (expected) values in the four cells of the 2×2 table can be represented as Table 3.3. For example, a person in the exposed group will experience the outcome event if she is either *doomed* or *causal*.

Based on Table 3.3 we can express the causal risk difference RD_{caus} as

$$RD_{caus} = P_1 + P_2 - (P_1 + P_3) = P_2 - P_3 \tag{3.7}$$

This expression reveals that the causal effect depends on the relative proportions of *causal* and *preventive* individuals in the study population. Thus the value of RD_{caus} reflects the net change in the population, but is not applicable to any individual. Indeed, the causal effect can be zero ($P_2 = P_3$) even if there are many *causal* and *preventive* individuals. Of course, this balancing out of effects could not occur if P_3 (or P_2) were known to be zero. An effect that can only operate in one direction (e.g., $P_2 = 0$) has been called *monotonic*. Under monotonicity the RD_{caus} value can be interpreted straightforwardly as the proportion of individuals who are affected by the exposure.

As mentioned in Section 3.2, the risk ratio RR_{caus} is often utilized as a measure of the causal effect. From Table 3.3 it is evident that we can write

Table 3.3 Standard 2×2 Table: Causal Model

	Disease	Healthy
Exposed	$(P_1 + P_2)N_E$	$(P_3 + P_4)N_E$
Unexposed	$(P_1 + P_3)N_U$	$(P_2 + P_4)N_U$

$$RR_{caus} = \frac{P_1 + P_2}{P_1 + P_3} \tag{3.8}$$

Once again, the value of RR_{caus} depends on the relative values of P_2 and P_3 and can have a null value of 1.0 only when $P_2 = P_3$, but now the value of P_1 does not cancel out of the expression. Therefore, the risk ratio is a complex measure that depends both on the balance between *causal* and *preventive* proportions, *and* on the proportion of *doomed* individuals in the population (Maldonado and Greenland, 2002). Again, under the assumption of monotonicity, the interpretation simplifies. For example, if exposure can never be beneficial, the causal risk ratio becomes a measure of how many individuals experience D because of exposure (*causal*), relative to the number who would have experienced D whether or not exposed (*doomed*).

In an actual study we observe the empirical effect, which is meant to estimate the unobservable causal effect. Now matters become more complicated, because the exposed and unexposed groups are not necessarily exchangeable. The empirical effect (risk difference or risk ratio) depends on the nature of possible nonexchangeability. The resulting bias depends on how the lack of exchangeability affects the particular measure of empirical effect. The traditional statistical model does not distinguish explicitly between what can be observed (empirical effect) and the underlying causal structure. Lack of an explicit causal model can lead to confusion about the meaning of an empirical effect.

For example, some researchers appear to believe that the value of a risk ratio reflects some sort of absolute truth about an underlying causal relationship. If the risk ratios obtained in several studies differ substantially, questions may arise about which study has the "right" answer. But the variation across studies can involve issues of internal validity, external validity, and the properties of the measure of effect. Distinguishing these interrelated but conceptually separate sources of variation among studies is necessary to interpret correctly the pattern of results across all the studies. Therefore, to expect strongly consistent values of a causal effect across studies performed in various circumstances may be setting the bar too high (Susser, 1988). The counterfactual perspective clarifies why a certain amount of variability across studies is the norm (Maldonado and Greenland, 2002).

3.5   POTENTIAL OUTCOMES

We have defined the potential outcomes as the values of an outcome for an individual that would occur under different exposure states, holding all else equal. The most common formulation of counterfactual analysis assumes that the potential outcomes for an individual are *deterministic*. Probability theory enters statistical models only through the variability of the response patterns across the individuals in a specified population. An alternative counterfactual

approach would define the potential outcomes for an individual to be probabilistic, or in statistical terminology *stochastic*. The values V_{iE} and V_{iU} can be
regarded as probability distributions, instead of single values. Under this
approach, the distributions of potential outcomes in the study population (V_E
and V_U) would be composed of two components: interindividual variation and
intraindividual variation.

From a practical standpoint, a fully stochastic counterfactual model would
become prohibitively complex. One way to simplify the model is to assume
that we are primarily interested in the mean values of the potential outcomes.
The intraindividual variation can be viewed as random "noise" around the
"true" mean value. From this perspective, the individual mean value simply
replaces the fixed value in most applications of the counterfactual model. If
the outcome is numerical, then a numerical mean replaces a numerical fixed
value. So there is no additional mathematical complexity under this type of
probabilistic interpretation. The counterfactual model retains exactly the same
form, although the theoretical interpretation is somewhat different. However,
under certain conditions, as when multiple measurements of outcomes for the
same individual are possible, the stochastic model may expand the range of
useful inferences that can be made (Steyer, 2005).

If the outcome is dichotomous, the variable that assumes one of two values
(D, H) would be replaced by a variable representing the "potential probability" of D if the individual is exposed and if not exposed. Because a probability
can assume any value between 0 and 1, potential outcomes are thus converted
from a dichotomous to a numerical scale (albeit with a restricted range). In
principle, this would permit us to define a uniform causal effect, a concept that
is not meaningful under the deterministic scheme. For example, we might
hypothesize that a new drug doubles the probability of a particular adverse
event, such as a myocardial infarction, for every patient. However, even with
a stochastic counterfactual model, interpreting an estimate of relative risk can
be quite complicated (Greenland, 1987b). For example, a risk ratio is not
simply an average of the individual causal effects.

In terms of our causal model, we assume that only a few of potentially many
causal factors are known. For example, we may know there exist other relevant
conditions that determine whether the event occurs, but do not have information on the values of these factors.

When causal components remain unknown, one may be inclined to assign an equal
risk to all individuals whose status for some components is known and identical.
Thus, men who are heavy cigarette smokers are said to have a 10% lifetime risk of
developing lung cancer. Some interpret this statement to mean that all men would
be subject to a 10% probability of lung cancer if they were to become heavy
smokers, as if the outcome, aside from smoking, were purely a matter of chance. In
contrast, we view the assignment of equal risks as reflecting nothing more than
assigning to everyone within a category, in this case male heavy smokers, the average
of the individual risks for people in that category ... assigning the average value
to everyone in the category reflects nothing more than our ignorance about the

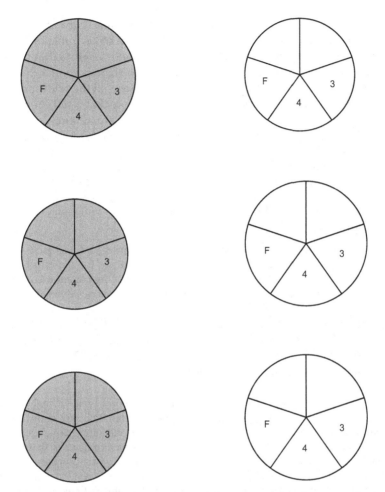

Figure 3.2. A situation in which the occurrence of the outcome event is not completely determined by the causal factors included in the causal model. The result is a probability of occurrence. In this example, the probability is 0.5, because the event occurs in half of the individuals with the particular causal configuration.

determinants of lung cancer that interact with cigarette smoke. (Rothman and Greenland, 1998, 8–9)

A pie chart depicting this type of situation is presented in Fig. 3.2. Here we have six individuals, each exposed to Factor *F* and with identical patterns of all *known* factors. The outcome event occurs in three of the individuals and fails to occur in three. It is certainly possible to conceptualize this result as implying a potential probability of 0.5 for the event's occurrence under exposure to *F*. Like Rothman and Greenland, we will interpret this situation to mean that some additional hidden factor(s) must be present in 50% of the population.

The stochastic model perspective effectively assumes that individuals are inherently indistinguishable, except possibly for a few measurable factors. The deterministic counterfactual model, on the other hand, assumes that individuals are in principle distinguishable with respect to their potential outcomes, even though we lack the ability to specify and measure all the relevant factors. However, this statement should not be taken to imply that a full set of causal factors exists as a well-defined objective reality. There are infinitely many ways in which these factors and their interrelationships might be formulated.

> A person develops lung cancer as a result of years of smoking. It is true that if he had not smoked he would not have developed lung cancer. It is also true that he would not have developed lung cancer if he had not possessed lungs, or even if he had not been born. But it is absurd to think his possession of lungs or even his birth caused his lung cancer.

> Common sense draws a crucial distinction between causes and background conditions, so disqualifying them from being difference-makers for the effect. Several philosophers of causation have stressed the importance of this commonsense distinction in connection with the view of causes as difference-makers. (Menzies, 2004, 143)

What we have termed the causal context includes all the relevant factors *as conceptualized within some scientific theory*. This theory is an intellectual abstraction carved out of reality to serve certain purposes. Thus, causal thinking inevitably entails metaphysical aspects.

> It is a platitude—but one worth repeating—that the world is exceedingly complex in its causal structure. Within any spatiotemporal region, there are many levels of causation, and within each level many cross-cutting and intersecting causal processes. To determine the structure of these processes, we are necessarily forced, by the finitude of our minds, to focus selectively on some aspects of what is going on and to ignore others or place them in the background. The causal schemas by which we interpret the world are irremediably permeated by abstractions that enable this selective focusing. (Menzies, 2004, 154)

When we say that the potential outcomes are the counterfactuals that would occur while holding all else equal, we must keep in the back of our minds that this condition is relative to some ultimately subjective causal schema.

3.6   TIME-DEPENDENT OUTCOMES

The counterfactual models we have discussed so far are completely static; the description of potential outcomes has not made reference to a time dimension. In some contexts, however, the outcome of interest is observed not at a single point in time, but throughout some relevant time period. If the outcome is numerical, we might record its value continuously, or at two or more discrete points in time. For example, studies of human growth and development track the changes over time in some characteristic of interest, such as height or language skills. In principle there is a continuum of potential outcomes $V_{iE}(t)$ and $V_{iU}(t)$ that could be measured for all time points t during some interval. These values trace out a *growth curve* for each individual. Studies that involve a time-dependent, or dynamic, outcome are more complex than those with a static outcome for two related reasons.

First, the outcome itself is some mathematical function of time whose form must be specified. For example, the simplest form of growth model would be linear:

$$V_{iE} = \alpha_{iE} + \beta_{iE}t$$
$$V_{iU} = \alpha_{iU} + \beta_{iU}t \qquad\qquad (3.9)$$

Even if a linear model is assumed, there are now four parameters that characterize the potential outcomes for a single individual, rather than two values. If a more complex model is specified, there may be several more parameters. So the mathematical analysis becomes much more complicated. Second, to analyze change or growth, at least two measurements of the outcome value are necessary. With multiple measurements for each individual, the statistical analysis of resulting data can become very involved (e.g., Rogosa et al., 1982; Strenio et al., 1983; Bryk and Raudenbush, 1992).

Comparative studies that examine changes over time play a major role in the social sciences. Research often focuses on how a particular intervention increases or decreases some aspect of human behavior or performance. Typically, the characteristic of interest has a value prior to the study but can change during the study. For example, an educational program might be aimed at improving the child's level of literacy. Certain forms of bias are specifically related to the dynamic nature of these situations, and these types of bias figure prominently in discussions of social science methodology (e.g., Cook and Campbell, 1979; Shadish et al., 2002).

Complex dynamic outcomes can also be important in epidemiology and clinical research. However, in the biomedical arena, attention is often concentrated on a dichotomous outcome, such as the occurrence of disease. So far, we have considered studies of *whether* a particular factor (program, exposure, etc.) causes a particular event to occur. Now, suppose we are also interested in *when* the outcome event may occur. Then we can define the growth curve for each individual as a simple step function. The values of $V_{iE}(t)$ and $V_{iU}(t)$ are each 0 up to the point when (and if) the event occurs. This special case of a dichotomous time-dependent outcome variable is of great practical importance, because we are often interested in time-to-event data. Such data are also called *survival data*, because the focus is typically on how long a person remains alive or remains free of a particular disease. There are many textbooks devoted exclusively to survival analysis, which has become a mainstay of biomedical research (e.g., Cox and Oakes, 1984; Hosmer and Lemeshow, 2000).

When the outcome is time-to-event, we could apply the approach suggested in Section 3.4 to specify the empirical effect. For each individual, we could simply note whether or not the event occurred during the study. Then the relative risk RR_{emp} would serve as the measure of effect. In some circumstances, this is a practical and appropriate strategy. However, essentially ignoring the time-dependence in this manner can be problematic if the study duration varies across individuals, as when some subjects drop out of the study for reasons unrelated to occurrence of the outcome of interest. For example, suppose that subjects in the exposed group tend to remain in the study for a longer duration than subjects who are not exposed. Such differential attrition can lead to bias if we depend on the risk ratio as the empirical effect. To correct for this bias, we need to adjust for the longer average period of observation enjoyed by the exposed subjects.

One simple way to accomplish this adjustment is by using an alternative measure of causal effect. In Section 3.4 we denoted the proportions of individuals in the study population who experience the event of interest as q_{VE} for the exposed group and q_{VU} for the unexposed group. We defined the risk ratio as q_{VE}/q_{VU}. Similarly, we can define the event occurrence *rates* per unit of time as r_{VE} and r_{VU}. If these rates are assumed to be constant over time, the usual causal effect of interest is the *rate ratio*, or *incidence rate ratio* (IRR). We can express the relative risk in this context as r_{VE}/r_{VU}. The corresponding empirical effect is r_{WE}/r_{ZU}, where r_{WE} and r_{ZU} are the observed event rates in the two study groups.

To estimate the empirical rate ratio is relatively straightforward. Let us define the normal time-on-study for each individual as his total follow-up until the planned completion of the study. For an individual without an event who completes the study, the duration is the normal time-on-study. For an individual who experiences the outcome event during the study, duration is defined as the time elapsed prior to the event. For someone who drops out, the duration is said to be *censored* and is defined as the time until censoring.

Let n_{WE} represent the number of events that occur in the exposed group, and let d_{WE} represent the sum of the individual durations for all subjects in the exposed group. Let n_{ZU} and d_{ZU} be the corresponding quantities in the unexposed group. The quantities d_{WE} and d_{ZU} are called the *total time at risk*, or sometimes the number of *person-years* (or *person-weeks*) of observation, in the two groups. Now we can define the statistical estimates of r_{WE} and r_{ZU} as n_{WE}/d_{WE} and n_{ZU}/d_{ZU}, respectively. Under the assumption that r_{WE} and r_{ZU} are constant over time, these estimates will converge to the true event rates. Consequently, the ratio of the estimated rates (observed effect) will be close to the rate ratio (empirical effect) that is being measured by the study's methodology.

If the event rates are not constant, the situation becomes more complex. The rates must now be expressed as functions of time. Using methods of calculus, we can define an instantaneous rate of occurrence per unit time at each value of time. This instantaneous rate is called the *hazard function*. The ratio of the hazard rates is called the *hazard ratio* and represents a more general time-dependent version of relative risk than the rate ratio. Sophisticated statistical methods for estimating hazard ratios fall generally under the heading of Cox regression analyses (Cox, 1972; Cox and Oakes, 1984). Anderson et al. (1980) present a basic introduction to the use of Cox regression and other methods for survival analysis in the context of comparative studies.

We will not explicitly consider the time dimension further. Although there is inevitably a time delay between initiation of exposure and occurrence of the outcome, the nature of any time-dependence will remain implicit. Consequently, counterfactual analyses of causal effects related to growth curves and hazard functions are not presented. In general, the conceptual issues that pertain to such time-dependent causal effects are extensions of the main ideas that are discussed. However, the specific application of these ideas to time-dependent outcomes entails complexities that cannot be covered adequately within the confines of this book.

3.7   INTERMEDIATE VARIABLES

We mentioned in Section 2.1 that our focus is on the effect of a single factor (exposure), effectively holding constant other related causal factors (component causes). For the most part, we do not deal with situations in which a more complex causal network involving several interconnected variables is of interest. Even in this simplified situation, however, the effect of the exposure can be *mediated* by other factors that must be considered. In terms of the causal model, the effect of interest is conditioned on the values of some causal modifiers. In general, these causal modifiers may occur at different points in time. In particular, certain components may come into play at some time after the exposure of interest. These other factors may be at least partially caused by the exposure.

Suppose we are interested in the question of whether eliminating alcohol consumption contributes to longevity. We carry out a randomized study that somehow motivates the subjects in an intervention group to give up drinking, and we monitor the mortality of the study participants for 20 years. The result is a slight reduction in the mortality rate for the "quitting" group relative to a nonquitting comparison group. What would such a study tell us about the effect of abstinence from alcohol? The modest improvement in all-cause mortality is a *total effect* that includes possible *indirect effects* of the treatment. For instance, some ex-drinkers might have become more physically fit, leading to participation in dangerous activities such as motorcycle racing or mountain climbing. Should the indirect effects of mortality from such activities be chalked up to avoidance of alcohol?

The answer depends in part on the real question of practical interest. For example, a biologist might be seeking to understand a *direct effect* of giving up alcohol, such as the impact on liver function. However, a public health specialist might be more concerned about the broader implications, both direct and indirect, that affect the bottom line of mortality. This issue is particularly salient when the outcome develops over a lengthy period of time. As time passes, the exposure has ample opportunity to interact with other factors in ways that might affect the outcome. If these unplanned effects are essentially similar across the study groups, these intermediate effects may approximately balance out. If they differ across groups, exchangeability could be affected and bias result. Furthermore, even if internal validity is preserved, the nature of the treatment could be significantly modified. In effect, the treatment as implemented and realized over time could be quite different from the treatment as planned.

We can think of a direct effect of some factor as the effect under the assumption that all other factors are in fact held fixed. The total effect, on the other hand, is the effect of this factor "when all other variables are permitted to run their natural course" (Pearl, 2000, 164). Total effects are generally much easier to measure, because the intermediate factors may be difficult to control. However, even if the total effect is of interest, as in a public health context, there is no guarantee that uncontrolled variables will run their course in a way that generalizes to other situations. So the relevance of the study's results may be in doubt. Conversely, if we attempt to rigidly control other factors deemed important, we may end up distorting the "natural" situation so much that the results become virtually meaningless. In clinical research, this dilemma is sometimes framed in terms of the concepts of *efficacy* and *effectiveness*. Efficacy is the effect of some intervention "under ideal conditions," whereas effectiveness refers to the effect "when deployed in the field in routine circumstances" (Last, 2001, 57–58).

One of the most common problems related to intermediate variables pertains to the exposure itself. The existence or degree of exposure might vary over time in ways that are related to other time-varying factors. A very common and potentially problematic source of exposure variability is related to noncompliance. There can be many reasons certain subjects fail to adhere consistently or diligently to their assigned treatment modality. For example, noncompliance may be induced by the exposure itself (e.g., unpleasant side effects) or by exogenous forces (e.g., changed medical or economic circumstances). A widely accepted approach to address this problem is to analyze the data according to the *intention-to-treat* (ITT) principle. The outcomes for all subjects who are initially randomized are included, regardless of how well subjects comply with the requirements imposed by the study design. This ITT approach preserves exchangeability at the cost of "diluting" the treatment; the actual treatment modality received by a noncompliant subject will often lack the full potency of the intended intervention.

3.8 MEASUREMENT OF EXPOSURE

So far, we have implicitly assumed that the exposure state for every subject as observed and recorded is accurate. Suppose, however, that some individuals are incorrectly identified as having been exposed when they actually were not. Conversely, some in the unexposed group may be wrongly categorized as exposed. These types of mistakes are called *misclassification errors*. As a consequence of exposure misclassification, the actual study groups being compared do not correspond to the real study groups of interest. We end up comparing *apparently* exposed and unexposed individuals rather than truly exposed and unexposed.

Misclassification of exposure is a subtle phenomenon that can lead to bias in essentially two different ways. First, such misclassification can sometimes affect the exchangeability of the two study groups. Imagine that in our tutoring experiment the outcome was the score on an actual math SAT administered just after program completion. Suppose that Jane Smith was a student in Class A, but during the analysis of outcome results she was for some reason mixed up with Joan Smith, a student in Class B3. Then the distributions of response patterns in the two groups *as measured* would include Jane's score in Class A and Joan's in Class B3. For instance, if Joan's potential outcomes were much lower than Jane's, then the observed Class A would have a less favorable distribution of response patterns than the observed Class B3. The two classes of truly exposed students were exchangeable based on the study design, but misclassification has resulted in two "classes" that are no longer exchangeable.

Suppose, though, that the two study groups remain exchangeable, despite the mis-classifications. We would still face a second problem. The apparently exposed group is partially unexposed, because it contains some proportion of individuals who are actually unexposed. Similarly, the "unexposed" group is partially exposed. Thus, the causal effect being estimated is the effect of *observed exposure* rather than the effect of true exposure. If the study groups being compared are exchangeable, this comparison can yield an unbiased causal effect. However, this causal effect would pertain to a "treatment" that may be difficult to define, either conceptually or in practice. This observed treatment is *variable* across study subjects, with some receiving the true exposure of interest and some remaining unexposed. In Chapter 9 we propose a model for describing exposure misclassification that clarifies the nature of the resulting problems.

As a practical matter, exposure misclassification occurs primarily in retrospective epidemiological studies, when exposure status is obtained after the fact. In the social sciences, measurement error issues pertain primarily to the outcome and will be discussed in the following section. In the social sciences, measuring the exposure in a narrow technical sense usually poses few difficulties. However, attention may be given to a more subtle issue that is also related to the assessment of exposure. Suppose that exposure status is ascertained with perfect accuracy, but the nature of the exposure has been misconstrued. For example, the new tutoring program is supposed to involve only a change in curriculum, but in fact the students in Class A begin to meet and work together outside the classroom. Then the responses under the treatment are actually being driven by a set of activities that extend beyond a simple change in the curriculum.

More complex and subtle issues frequently occur in psychological and social research. For example, a psychologist might be interested in the influence of "marital stress" on the propensity for aggressive behavior. The general phenomenon of marital stress may be realized in a variety of specific ways. If research on marital stress is to have any generality, the realizations in particular circumstances (and studies) must reflect some common underlying *construct*. Ideally, this construct must be a parameter of scientific interest whose interpretation is unambiguous. The notion of *construct validity* was introduced to describe this desirable property of a variable (Cronbach and Meehl, 1955). For meaningful causal inference, the exposure specification must possess construct validity in the sense of representing the true causal factor of scientific interest.

Construct validity is closely akin to external validity in the sense that both pertain to the potential applicability of an effect beyond the bounds of the particular study (Cook and Campbell, 1979; Shadish et al., 2002). However, lack of construct validity does not necessarily lead to bias in our narrow sense of causing a deviation between the empirical effect and the causal effect. It is perfectly possible to estimate a causal effect for the intervention as operationalized, regardless of whether it truly mirrors the "real" construct of scientific interest.

In addition to assuming adequate construct validity for the outcome, we will also assume that the exposure is *stable* in a particular sense. Imagine that we could hypothetically repeat the tutoring study, possibly even altering its design. If Jane Smith ended up receiving the same program in both instances, we would expect her response patterns to be exactly the same. Suppose, however, that Jane's response pattern (and those of others) could vary across different studies. Then the resulting causal effect might also be variable. But why might the response pattern for a given student vary? This could occur if the intervention being studied did not remain stable, so that each student was not really receiving an identical modality in the different replications. In other words, the treatment modalities as experienced might be variable and not well defined.

To address this potential problem, Donald Rubin has proposed that a necessary condition for meaningful causal inference is that the potential outcome for any particular unit (person) i following treatment t "would take the same value for all other treatment allocations such that unit i receives treatment t" (Rubin, 1990a). In other words, an individual's response pattern will not depend on the particular way in which treatments have been assigned to individuals in the study. Rubin has called this the *stable unit–treatment value assumption* (SUTVA):

> There are two components to SUTVA. The first is that there is only one version of a specific treatment: all individuals assigned to "treatment" receive the same active

treatment, and all individuals assigned to "control" receive the same control treatment. The second component of SUTVA is that there is no interference between units: the value of each unit's potential outcome does not depend on the treatment assignments of other units. (Rubin et al., 2004, 108)

Rubin regards SUTVA as an important criterion for "deciding which questions are formulated well enough to have causal answers" (Rubin, 1986).

The first component of SUTVA envisions the possibility of various subtypes or gradations of the treatment. Suppose there is a mix of treatment versions being received. Then not only would the distribution of response patterns in the study groups depend on which individuals were assigned to each group, but also on which versions of the treatments were assigned. For example, the new SAT tutoring program might include both a classroom component and individual tutoring with one of five different tutors. Then the *combination* of the two components might actually comprise the effective intervention. Each student's performance in the New Program might result in part from the particular version (tutor). Therefore, the distributions of response patterns in the two groups could depend on exactly how the tutors were assigned. For example, Class A might happen to get lucky and be assigned tutors who were more experienced or sensitive to individual learning styles. Therefore the New Program would appear more effective than it would have with a less fortuitous allocation of tutors. In a replication of the experiment, a less favorable allocation might occur.

The second component of SUTVA implies that the impact of a treatment on someone must depend on her own treatment alone and not on the treatment received by anyone else. When SUTVA is satisfied, there is no contamination, diffusion, or other form of interference by which one individual's treatment could influence the treatment actually received by another. For example, suppose that Jane Smith's best friend, Sarah Smart, is assigned to Class B3 and receives traditional tutoring. Then Jane might pass on some of the knowledge she has gained in her program to Sarah and vice versa. Such diffusion of the treatment could undermine the study's integrity by changing the nature of the two programs.

In the social sciences, certain types of interference have been recognized under the heading of threats to internal validity (Cook and Campbell, 1979, 54–55). For example, the *compensatory equalization of treatments* can occur when there is reluctance to withhold a desirable intervention, such as increased funding or other resources, from a control group. As a result, unintended ways of channeling additional resources to the "nontreated" controls could materialize. Here the structure of the study results in contamination of the control condition, which is exposed to some aspects of the treatment. Another potential threat to validity is *compensatory rivalry*, in which a less favored group might be motivated to try especially hard to succeed. This response would be an indirect effect on the control group of being in a "competition" against the study group perceived to be favored. In either of these examples, the treatments are distorted in ways that make it difficult to interpret the study's results.

3.9 MEASUREMENT OF THE OUTCOME VALUE

In measurement of the outcome variable, a critical concept is the reliability of the measurement process. Loosely speaking, reliability is the extent to which the observed measurement agrees with the actual value. Psychometricians have developed many approaches to estimation of reliability (test–retest reliability, inter-rater reliability, parallel-forms reliability, internal consistency reliability, etc.) applicable in different situations (e.g., Lord and Novick, 1968). Psychometric theory of measurement applies almost exclusively to numerical variables. The central idea is that an observed outcome score comprises two components: a "true score" and "measurement error." Reliability concerns the extent to which the observed scores agree with the true scores and are not distorted by the measurement process.

Classical psychometric theory is based on certain assumptions about the mechanism that generates the observed outcome values. These assumptions are reasonable in certain contexts, but not realistic in others. In essence, the measurement errors are envisioned as random "noise" that is added to the underlying true scores. The main impact of this random error is to reduce the precision of statistical estimates, but not to induce bias. Within the classical statistical framework, our ability to conceptualize the nature of measurement error and its impact on empirical effects is quite limited. In Chapter 9, we propose an alternative framework for viewing measurement error. The resulting causal model suggests that the implications of outcome measurement error may be more complex than implied by the classical measurement model.

In the biomedical sciences, the outcomes are often dichotomous. Consequently, outcome measurement error takes the form of misclassification. The issue of reliability boils down to two key parameters: sensitivity and specificity. These are, respectively, the probabilities of correctly classifying the outcome when the event of interest occurs and when it does not occur. The implications of this model are elaborated in Chapter 9, and an alternative causal model is proposed.

The concept of construct validity, discussed in the context of exposure measurement, is also important when considering the observed outcome. Analyzing the "wrong" outcome does not necessarily result in bias. It is still possible to interpret an estimated effect as causal, even if the effect is exerted on a variable that does not correspond precisely to the construct of interest. The major concern is that the study's results may not be applicable in real-world situations. Defining the outcome variable tends to be somewhat more problematic in the social sciences than in biomedical research. The social sciences are concerned primarily with complex human behavior, from which it may be quite difficult to distill simple constructs. Just as we would need an operational concept of marital stress in studying its effect on propensity for aggressive behavior, we would also need some way to specify how we would identify unambiguously the kinds of behavior that are considered to be aggressive.

In biomedical research, we may be more interested in outcomes that are relatively well defined, such as the occurrence of a particular disease. Therefore, much less attention has traditionally been paid to the construct validity of outcomes by epidemiologists and clinical researchers than by social scientists. However, there are significant exceptions to this generalization. Increasingly, disease entities being investigated are complex both in their etiology and clinical manifestations. Alzheimer's disease and other forms of dementia exemplify the difficulties that are posed by long-term chronic illnesses, especially in the neurological domain. Diagnosing the existence and progression of these conditions can be extremely challenging and based largely on indirect evidence. Therefore, the principles developed by social scientists for achieving construct validity may well be relevant.

Paul Meehl thought deeply about issues of construct validity in the development of disease entities (Meehl, 1977). As a psychologist, he was motivated largely by an interest in psychopathological conditions. However, his ideas also apply to such physical conditions as cancer and heart disease. Meehl described the identification of a disease entity as a complex and dynamic process. The process starts with tentative recognition of a rather loosely related set of symptoms and signs that seem to cluster together and are thought to reflect some underlying "specific etiology" that is yet to be elucidated. Eventually, as more specific clinical and laboratory findings accumulate, a coherent disease entity may emerge.

Much current research is focused on long-term chronic diseases related to the aging process (e.g., heart disease, cancer, arthritis, diabetes, dementia). Both the causes and manifestations of these illnesses are extremely complex and multifactorial, involving a delicate balance of biochemical forces that can be tipped in one direction or another (Vineis, 2003). New pharmacological treatments for such conditions may have multiple impacts on various inter-related biological processes. Rather than focusing exclusively on narrowly defined disease events, the endpoints in clinical trials and epidemiological research may include composite measures of morbidity or all-cause mortality. Furthermore, there may be increasing emphasis on developing surrogate markers that can be observed much earlier or more easily than the ultimate outcomes of clinical interest. For example, a study of cardiovascular disease may rely on measurement of carotid artery thickness rather than waiting for occurrences of actual thrombotic events. Finally, the recent explosion of knowledge about the human genome has fueled the hope that truly personalized medical treatment will become the norm in the coming decades (President's Council of Advisors on Science and Technology, 2008).

In this complex and rapidly evolving scientific environment, issues related to construct validity may become increasingly salient. For the most part, these issues do not pertain to bias directly. However, a construct that reflects a complex biological or social system is unlikely to be affected by an exposure in a way that is uniform across individuals. Consequently, the possibility of substantial variability of individual effects will often need to be considered quite seriously.

3.10 CONFOUNDING BIAS

One particular category of bias holds a preeminent place in any discussion of threats to validity. A recent search of the PubMed® database of medical literature using the search term *confounding* resulted in nearly 25,000 citations! Confounding is a term that is widely understood in a general sense but very difficult to define unambiguously, especially within the traditional statistical framework. The intuition that gives rise to the idea of confounding is straightforward. If the study groups are not "equivalent in all relevant respects," there must be factors that "explain" the difference. The causal effects of these other factors are entangled, or *confounded*, with the causal effect we wish to isolate.

If we can manage to somehow "control for" these confounding variables, an unbiased estimate of the causal effect will emerge. Randomization is the ideal method for controlling confounding because it assures balance between the study groups for all potential confounding factors, whether measured or not. In an observational study, on the other hand, we must attempt to control for confounding explicitly by stratifying on values of the confounding variables or by some other more sophisticated approach. The main practical problem is that we can never be sure that all important confounders have been included in the analysis.

From the counterfactual viewpoint, confounding is a source of bias if it can potentially distort the distributions of response patterns so that the groups are no longer exchangeable. As a simple example, suppose we are studying the relationship between smoking and coronary heart disease (CHD). We plan to follow a cohort of people for several years and record the occurrence of CHD events. Suppose that the smokers in our cohort tend to be substantially older than the nonsmokers. Intuitively, this would suggest that the smokers would have higher CHD rates even if smoking were benign. So, a higher observed rate for the smokers could not unambiguously be attributed to smoking.

From the counterfactual perspective, we will see that the essence of the confounding problem lies in the lack of exchangeability between study groups. The problem is that we cannot observe the response patterns directly in order to assess possible nonexchangeability. Observable differences on a variable such as age provide an indirect indication of nonexchangeability. Stratifying on the basis of such variables may arguably reduce, or perhaps even eliminate, confounding bias by creating subgroups (strata) within which exchangeability exists.

GUIDEPOST 3

Chapter 3 has introduced a number of issues that pertain to the estimation of causal effects in the face of biases that may affect a comparative study. These issues encompass the core set of concerns that research methodologists have wrestled with for over a century. However, for the most part traditional discussions of each issue have not penetrated to the heart of the matter. We have tried to outline clearly the nature of each issue and to suggest how a counterfactual perspective may lead to a deeper understanding.

In Chapter 4, we will survey the treatment of bias in the methodological literature. This treatment has been rather ad hoc, primarily because causality in general, and bias in particular, have not until recently been part of mainstream statistical thinking. Chapter 4 is meant to provide a link to the main concepts and terminology that many readers may have encountered in previous studies. It begins with a basic review of research design approaches, because some types of bias are specific to particular designs. Then the main typologies of bias within the social sciences and biomedical sciences are reviewed. We suggest that it would be more useful to think in terms of various *sources* of bias, rather than the types of bias that have traditionally been discussed. Finally, a preliminary taxonomy of the main sources of bias is suggested.

CHAPTER 4

Varieties of Bias

In reporting the results of a comparative study, investigators often mention the possibility of bias, even if statistical adjustments have been made. A few specific types of bias may be highlighted and discussed briefly. Most commonly, a rationale is offered to justify a low level of concern about such residual bias. Less frequently, some actual evidence is provided to shed light on whether certain types of bias are plausible. Even less frequently, a sophisticated mathematical analysis may be performed to account for the uncertainty about residual bias. In the end, the reader may feel vaguely dissatisfied. It may be unclear whether *all* potential sources of bias have been considered. Are there other varieties of bias that could have threatened the validity of the study's findings? Equally important, the nature and implications of possible biases are unclear.

Discussions of bias have been hampered by the lack of an explicit model for causation. Without such a model, it may be difficult to specify the circumstances in which bias seriously threatens the study's conclusions. In this chapter, we describe the conventional ways of thinking systematically about varieties of bias. The fundamental problems faced in the biomedical and social sciences are quite similar, but the language used to describe these biases tends to differ. The difference is partly a function of the kinds of research that characterize these fields.

Epidemiologists and clinical trialists are typically concerned about causal effects of specific factors that are usually quite stable and fairly well defined. Furthermore, the physiological and biochemical science underlying many causal effects of interest may facilitate relatively precise definitions of the relevant concepts. The primary emphasis is on estimating the causal effect of a specific exposure or treatment modality in the face of possible alternative explanations. Because most epidemiological research is based on observational data, the groups being compared are typically not exchangeable. In

Bias and Causation: Models and Judgment for Valid Comparisons, By Herbert I. Weisberg
Copyright © 2010 John Wiley & Sons, Inc.

various contexts, critics of particular studies have discovered a wide variety of specific ways that exchangeability might be violated. Some of these have become well known, while others lie buried in specialized academic journals. The result has been a somewhat disorganized proliferation of disparate concepts.

Social scientists tend to be concerned with aspects of human behavior that are often embedded in complex systems. Identifying stable and meaningful constructs that can be generalized is very challenging. Furthermore, human behavior is extremely diverse and context-dependent. Research design must take into account existing social organizations and networks of interpersonal relationships. Unlike their biomedical counterparts, social scientists cannot ordinarily draw upon the "hard sciences" to ground their theories. In this environment, issues in defining concepts (construct validity) and measuring outcomes (reliability) are pervasive, along with the consequences of social organization (selection). Threats to validity, including what we have been calling bias, tend to be addressed at a higher level of generality.

So far in this book we have said little about the way in which the study groups are actually formed. The details of assembling subjects and collecting data lie at the heart of the *research design*. To be specific about the potential threats to validity in a particular study, we need to understand the study's design. One of the reasons for confusion in many discussions of bias is that both the potential for bias and the nature of the bias can depend on the particular research design. Attempts to enumerate the various types of bias often fail to make clear that certain biases apply only, or at least primarily, to particular research designs.

4.1 RESEARCH DESIGNS AND BIAS

Throughout this book, we will discuss bias in the context of two main types of studies: *randomized experiments* and *observational studies*. The basic distinction between a randomized and an observational study was introduced in Chapter 2. In a randomized experiment, the alternative treatment modalities are assigned to subjects by a random mechanism, ensuring exchangeability of the study groups. In an observational study, the study groups are formed in some other manner, so exchangeability usually cannot be assumed.

We noted that a randomized experiment, or *randomized controlled trial* (RCT), is generally regarded as the gold standard for causal inference. However, the practical conditions necessary to conduct an RCT are often not satisfied. Consequently, most studies of human health and behavior are observational studies. Among these, *cohort studies* are by far most common, encompassing nearly all studies in the social sciences and many in the biomedical sciences. These studies compare outcomes for two or more subgroups of the individuals in a defined population (cohort). For certain epidemiological questions, it is impractical to monitor a cohort straightfor-wardly until a sufficient number of outcome events have occurred, which could entail many years of follow-up. Therefore, epidemiologists have devel-oped an alternative research design known as the *case–control* methodology, which will be explained later in this chapter. The basic idea of a case–control design is to work backward by comparing a group of individuals who have already experienced the outcome event with a group who have not experi-enced it.

Randomized Experiments

In its simplest form, a randomized study uses the computerized equivalent of tossing a coin to decide which modality (e.g., active treatment vs. placebo) will be administered to each subject. More complex designs involving randomiza-tion within pairs or blocks of subjects are also possible and have technical advantages in certain situations. For example, subjects might be organized into pairs who share certain critical characteristics (i.e., suspected important causal factors). Then one of the two subjects in each pair is exposed and the other unexposed. In that way, the outcome difference for each pair "controls" per-fectly for the variables on which the matching is based. The advantages and disadvantages of various schemes involving matching and blocking are covered in textbooks on experimental design (e.g., Cochran and Cox, 1957; Cox, 1958; Box et al.; 1978) and will not be considered in this book.

In theory, randomization produces study groups that are equivalent in all relevant respects (ignoring random variability). In practice, there are various circumstances in which the randomization process can be subverted, either deliberately or unconsciously (Berger, 2005). For example, if the researcher is aware in advance that the next patient to enter a clinical trial will be allocated to the control group, he could find pretexts to exclude this person based on eligibility criteria. For the most part, such situations can be avoided by ensur-ing that the allocation schedule determined by the randomizing mechanism is hidden from the investigators. This precaution is often known as *allocation concealment* (e.g., Schulz, 1995; Schulz and Grimes, 2002). Although it is clear that implementation of randomization may be imperfect, we will not consider this source of bias further in this book.

Of greater practical importance are biases that result from the failure of subjects to adhere to the study requirements. There are a number of reasons certain subjects may deviate from the planned treatment modality, either by choice or necessity. So, even if the randomization process has been perfect, bias can creep in during the course of study follow-up. Some mea-sures can be taken to minimize these problems. For example, the technique of "blinding" the subjects and investigators to the specific treatment modality received by each individual can be important in some studies. However, in practice it is often difficult to maintain complete integrity of the randomized design.

Observational Cohort Studies

In a *cohort study*, we designate a group of "subjects" whose experience we intend to observe. This group is the cohort. For example, a cohort might consist of all women born in Boston during the "baby boom" years 1946 through 1964. Within this cohort, some of the members are exposed to the treatment or risk factor and some are not. The basic idea in a cohort study is straightforward: we compare the outcomes for the two (or more) study groups.

In planning a comparative study, one basic decision concerns whether the data will be collected prospectively or retrospectively. In a *prospective study*, subjects are selected for observation before the outcome has occurred. Then the outcome for each subject is subsequently observed and recorded. In a *retrospective* study, on the other hand, the outcomes for the study subjects have already occurred. Therefore, the data on exposure and outcome are gleaned from records that have been generated previously. These records are typically maintained for administrative purposes or are based on the recollections of study participants. In general, prospective studies allow for greater methodological rigor, because the data are obtained by observing and recording events as they unfold by trained research professionals. However, in some contexts a lengthy wait to obtain results is the necessary price. Retrospective studies can sometimes be performed much more quickly and efficiently.

In most studies, the exposed and unexposed groups are monitored over the same time period, whether prospectively or retrospectively. However, sometimes a *historical control* consisting of observations taken during an earlier time-period is used. Ideally, the historical control group consists of very similar individuals, or even the same individuals, prior to exposure. Monitoring the experience of the same group at two or more time points can be a very useful design element, especially in certain quasi-experimental research. For example, the *regression discontinuity* design tracks the trend in an outcome measured at several points in time before and after an intervention (Campbell and Stanley, 1963).

Assume for now that a cohort of individuals has been selected from the target population and allocated in some unspecified manner to an exposed and an unexposed group. If the study is prospective, these subjects are then followed for some appropriate period of time and their outcomes observed. If the study is retrospective, the outcome value and exposure state for each individual were determined prior to the study and subsequently recorded by the researchers. In either case, the study would be a *longitudinal* study, because the outcomes occurred at some point in time *after* the exposure to the treatment condition. On the other hand, suppose we measure both the exposure state and outcome value of a sample of individuals at a particular point in time. For example, we might perform a survey in which educational level and income are ascertained as of a certain date. Such a study is said to be a *cross-sectional* study.

In effect, the temporal relationship between exposure and outcome is collapsed in a cross-sectional study. The survey just described will not necessarily provide evidence about the effect of education on subsequent income. The cross-sectional approach offers potential advantages in terms of cost and convenience. However, in a cross-sectional study the relationship between current exposure (e.g., education) and current outcome status (e.g., income) may depend on when the exposure and outcome actually occurred for the individuals in the study. We may not even know whether the outcome for a given individual occurred before or after the exposure. A cross-sectional study can thus be regarded as a degenerate form of a longitudinal study in which a critical piece of information is missing, namely the temporal relationship of exposure and outcome. For this and related reasons, cross-sectional studies are considered much weaker for making causal inferences than longitudinal studies.

Case–Control Studies

In a cohort study we follow the natural approach of comparing the distributions of outcomes in two specified groups. However, when we are concerned with an outcome that is dichotomous and relatively rare, waiting for a sufficient number of events to occur can require a very large cohort and a long duration of study. Primarily to avoid these problems, epidemiologists devised the approach of starting with a set of observed events and working backward to obtain the data on previous exposure (Lane-Claypon, 1926).

Throughout this book, our discussion will apply most directly to prospective longitudinal cohort studies, both randomized and observational. Such research designs are predominant in both the biomedical and social sciences, but retrospective studies, especially with a case–control design, are not uncommon in epidemiology. Although case–control studies are affected by the same types of potential biases as longitudinal cohort studies, the "reverse logic" of the case–control approach introduces a number of additional complications that can lead to some unique forms of bias. The details of biases that are specific to case–control studies will not be considered here. However, to appreciate the general nature of the special challenges posed by case–control methodology, it is useful to explain its workings in some more detail.

In a *case–control study*, an attempt is made to ascertain all instances of the event in question that occur within a particular geographic or administrative area and period of time. Each such event is called a *case*. In addition, a set of *controls* (individuals who did not experience the event) is chosen. Note that this terminology can (and often does) cause confusion, because the "controls" here are not the subjects assigned to the control (i.e., unexposed) group. Indeed, the method of estimating the causal effect from a case–control study is often described as comparing the proportion who are exposed (i.e., in the "treatment group") among the cases vs. the proportion who are exposed (i.e., in the "treatment group") among the controls. To avoid confusion, the term *case–referent* study is preferred by some methodologists. Although case–referent is a clearer description of the technique, case–control has unfortunately become the firmly established terminology.

The advantages and disadvantages of case–control studies have been elaborated in several epidemiology texts and numerous expository articles (e.g., Schlesselman, 1982; Wacholder, McLaughlin, et al., 1992; Wacholder, Silverman, et al. 1992a,b). Although the essential logic of the case–control method was initially considered to be fundamentally different from that of a cohort study, this design is now seen as drawing on the same principles of design but using a special method of sampling the subjects. By harvesting a large number of outcome events (cases), this sampling approach is very efficient when the event is infrequent. In addition, because the exposure status of a case or control is not fixed in advance, as in a prospective cohort study, it is possible to assess the effects of several different exposures within the same study. However, the critical challenge lies in being able to select cases and controls in such a way that an unbiased estimate of the risk ratio can be derived.

To understand the basic trade-off between cohort and case–control studies, let us consider how the empirical risk ratio RR_{emp} defined in Section 3.2 might be estimated in each type of study. For a cohort study, we could simply calculate the ratio of observed proportions experiencing the outcome in the exposed and unexposed groups. To consider a case–control study, we reproduce Table 3.1 here as Table 4.1. Ideally, we could draw the cases and controls from a large cohort of people for whom both exposure and outcomes have previously occurred and have been recorded. Such a study is called a *nested case–control* study because the case–control sample is nested within a larger cohort (e.g., Rothman and Greenland, 1998, 94). Let us assume for simplicity that we are dealing with such a situation.

If the event D is rare, the study may actually capture all of the cases, or at least a large fraction of them, say f_D. In addition, a much smaller proportion (f_H) of the healthy (H) controls will be sampled. The resulting situation is represented in Table 4.2. If we attempt to calculate the RR_{emp} using the data in Table 4.2, then our estimate based on the restricted case–control sample will be

$$RR_{emp} = \left(\frac{q_{WE}}{q_{ZU}} \right) \frac{q_{ZU} f_D + (1 - q_{ZU}) f_H}{q_{WE} f_D + (1 - q_{WE}) f_H} \qquad (4.1)$$

Table 4.1   Standard 2 × 2 Table: Cohort Study

	Disease	Healthy
Exposed	$q_{WE} N_E$	$(1 - q_{WE}) N_E$
Unexposed	$q_{ZU} N_U$	$(1 - q_{ZU}) N_U$

Table 4.2 Standard 2 × 2 Table: Nested Case–Control Study

	Disease	Healthy
Exposed	$q_{WE}f_D N_E$	$(1 - q_{WE})f_H N_E$
Unexposed	$q_{ZU}f_D N_U$	$(1 - q_{ZU})f_H N_U$

Table 4.3 Example of Cohort Study Data

	Disease	Healthy
Exposed	200	9800
Unexposed	100	9900

Table 4.4 Example of Case–Control Study Data

	Disease	Healthy
Exposed	200	196
Unexposed	100	198

According to Equation 4.1, unless f_D and f_H are equal, which would defeat the purpose of the case–control methodology, the value of RR_{emp} will be distorted from its "true" value of q_{WE}/q_{ZU}. A simple numerical example is presented in Tables 4.3 and 4.4 to illustrate this situation. Table 4.3 shows the data that would be obtained if we could observe the entire cohort. From Table 4.3, the risk ratio that would be obtained in this cohort study is 2.0. However, the data from a case–control study with 2% of the controls sampled would yield Table 4.4. If we attempt to calculate RR_{emp} from the case–control study's data in Table 4.4, a value of 1.51 is obtained.

Fortunately, there is an alternative method of calculating the risk ratio (to a close approximation) when q_{WE} and q_{ZU} are small. This approach is based on calculating the *odds ratio* OR_{emp} instead of the risk ratio RR_{emp}. The odds associated with an event can be defined as the probability of occurrence divided by 1 minus this probability. For example, if the probability were 0.8, the odds of occurrence would be 4 to 1 in favor of the event. Thus, the odds ratio is obtained by dividing the odds under exposure by the odds under nonexposure:

$$OR_{emp} = \frac{q_{WE}f_D(1-q_{ZU})f_H}{q_{ZU}f_D(1-q_{WE})f_H} = \frac{q_{WE}(1-q_{ZU})}{q_{ZU}(1-q_{WE})} \tag{4.2}$$

Equation 4.2 indicates that the empirical odds ratio calculated from the case–control subpopulation will be identical to the odds ratio in the full study population. That is, the odds ratio (unlike the risk ratio) is unaffected by selection of (random) subsets of the cases and controls. However, because we are assuming q_{WE} and q_{ZU} are small, the odds ratio in the study population is approximately equal to q_{WE}/q_{ZU}, the risk ratio in the study population. Thus, the odds ratio calculated from a case–control study turns out to be an approximation to the risk ratio that would have been observed in the corresponding cohort study (Cornfield, 1951). In our numerical example displayed in Table 4.4, the odds ratio is 2.02, which is very close to the actual risk ratio of 2.0.

Although this description suggests that a case–control study can be virtually equivalent to the corresponding cohort study, there is one very important caveat. In practice, the samples of cases and controls are *not* randomly selected. In most applications of case–control methodology there is no convenient population available that contains all the potential cases and controls. Rather, a series of cases is typically acquired through surveillance of some population over time and recruitment of individuals who experience the outcome event. Usually, the goal is to include *all* cases over a certain period of time and within a specified geographic area. For example, in the Yale Hemorrhagic Stroke Project (HSP) study introduced in Section 1.3, the investigators recruited hemorrhagic stroke victims who were seen in one of the 43 designated hospitals during a specified period. For serious health-related outcomes such as hemorrhagic stroke, hospital-based sampling of cases is very common. When a *case series* is assembled via some purposive selection process, the question of identifying an appropriate sample of controls becomes critical.

Biostatistician Olli Miettinen described the definition of the study population in this situation as being *secondary* to the case definition (Miettinen, 1985). In contrast with the *primary* definition of the population in the nested case–control situation, a secondary definition of the population entails many potential complications. Miettinen coined the term *study base* to describe the hypothetical study population defined *implicitly* by the manner in which cases have been selected. In the HSP study, the cases treated at the 43 hospitals were drawn from a much larger population of individuals who would have been seen in these hospitals if they had experienced a hemorrhagic stroke. More generally Miettinen explained that:

> Given that the case series is the totality of cases in a secondary base, such a base must be *the* population experience (the entirety of it) in which *each* potential case, had it *occurred*, *would* have been included in the case series. (Miettinen, 1985, 545)

Miettinen proposed that the study base should represent the source of appropriate controls to include in the study. Therefore, the main goal of study design is to determine a series of controls that represents essentially the same population as the study base. Such a control series is *coherent* with the case series. The concept of coherence has been widely adopted as a conceptual guide in planning the selection of controls. Much has been written on the advantages and disadvantages of various approaches for achieving coherence (e.g., Schlesselman, 1982; Wacholder, Silverman, et al., 1992a; Savitz, 2003). There are two especially common methods for selecting a group of controls: *community controls* and *hospital controls*. Community controls, also sometimes called *neighborhood controls*, are usually proximate in place and time to the cases. Hospital controls are individuals who came to the same hospitals (or other facilities) as the cases during the time frame of the study, but for other reasons.

In the HSP study, the theoretical study base consisted of all hemorrhagic stroke victims between the ages of 18 and 49 who would have been treated at one of the 43 designated hospitals. The HSP employed the community-control approach; random-digit dialing was used to identify a sample of controls similar to the general population of potential cases. This community-based sample was intended to mimic the theoretical study base. An alternative option would have been to select patients who actually entered the same 43 hospitals for some condition known to be unrelated to phenylpropanolamine (PPA) use.

A possible choice of controls could have been patients who were treated at the same hospitals for *ischemic strokes* (caused by a blood clot in the brain). Ischemic stroke is thought to be unrelated to the use of PPA but to have symptoms virtually identical to those of the outcome event, *hemorrhagic stroke* (caused by bleeding in the brain). Consequently, ischemic stroke victims might have been regarded as an ideal group of hospital controls. However, as a practical matter, the rarity of ischemic strokes in relatively young adults might have severely limited the number of controls who could have been recruited for the study.

Although the notion of coherence between the cases and controls is helpful conceptually, its precise meaning is unclear. In what sense are the control population and the case population (study base) required to be the same? One possible response is that the control population and the study base should be exchangeable. That is, the distribution of response patterns in the population from which controls are sampled should be identical to the corresponding distribution in the study base. However, since the study base is really a hypothetical construct, finding evidence of coherence in this sense may be difficult.

Furthermore, the criterion of exchangeability by itself would not be sufficient. It would still be possible for the rates of exposure in the two populations to differ. Suppose that the total numbers of exposed and unexposed individuals among the controls were M_E and M_U, respectively. Then the data would appear as displayed in Table 4.5. As a result, the empirical odds ratio would be distorted:

$$\text{OR}_{\text{emp}} = \frac{q_{WE}(1 - q_{ZU})N_E M_U}{q_{ZU}(1 - q_{WE})M_E N_U} \tag{4.3}$$

Table 4.5   Standard 2 × 2 Table: Case–Control Study

	Disease	Healthy
Exposed	$q_{WE}f_D N_E$	$(1 - q_{WE})f_H M_E$
Unexposed	$q_{ZU}f_D N_U$	$(1 - q_{ZU})f_H M_U$

The issues related to possible lack of coherence make case–control studies especially susceptible to bias. Lack of coherence can compound the threats to validity that can affect any observational study. Therefore, case–control studies are often viewed as more appropriate for exploratory research to uncover possible causal effects than as a tool to reach definitive conclusions.

4.2   BIAS IN BIOMEDICAL RESEARCH

There have been several attempts to develop a comprehensive glossary of the various types of bias. Some of these include issues that extend beyond the boundaries of our narrow definition of bias as a difference between an empirical effect and a causal effect. Our brief review in this section will be restricted to bias in this narrower sense. For example, we exclude such issues as systematic errors related to improper use of statistical analyses, misinterpretations of findings attributable to researchers' preconceptions, and selective publication of results (publication bias).

For biomedical studies, David Sackett's much-cited article appears to contain the first attempt at a catalog of biases (Sackett, 1979). Sackett listed 56 different forms of bias, with a brief description of each. Sackett's roster of biases appeared in a paper written for a conference on methodological issues pertaining to case–control studies. He discussed in considerable detail nine types of bias thought to be of particular importance in case–control studies. Sackett viewed this effort as preliminary, and he encouraged other methodologists to contribute additional varieties of bias. He also suggested the need to study the "empiric elucidation of the dynamics and results" of biases and he chided methodologists for failing to "measure the occurrence and extent of bias" in actual research.

Sackett also classified his 56 biases according to the stage of the research process at which each was most likely to occur. He identified seven stages of research, which we can paraphrase as follows:

1. Review of relevant literature
2. Specification and selection of the study sample
3. Implementation of the study design
4. Measurement of exposures and outcomes

5. Data analysis
6. Interpretation of study results
7. Publication of results

In this book, we are primarily concerned with sources of bias that arise during stages 2 through 4. Sackett identified a total of 41 specific biases in these two categories.

Around the same time that Sackett's article was published, a high-level taxonomy of bias was starting to gain acceptance among epidemiologists. This approach to categorizing biases, originated by Olli Miettinen, identified three main sources of bias: the *selection* of subjects, the *information* collected, and *confounding* between the risk factor and other possible factors (Miettinen and Cook, 1981). This terminology has now become widely accepted among epidemiologists and clinical trialists.

Selection bias derives from the process employed to choose study subjects from the target population. If the actual selection process favors individuals with certain characteristics for inclusion in the study, bias is possible. In any study, a selection process can lead to external bias if the selected study participants are unrepresentative. In a cohort study, internal validity would be threatened if the study groups are not exchangeable. In a case–control study, internal validity can additionally be affected by the manner in which the cases and controls are sampled.

A simple example of selection bias is *volunteer bias*. This bias can arise because volunteers "may exhibit exposures or outcomes (they tend to be healthier) which differ from those of non-volunteers" (Sackett, 1979, 61). In a cohort study, the subjects are recruited prior to the occurrence of the outcome. Suppose, like Almroth Wright and his colleagues a century ago, we were testing a new vaccine, and individuals were allowed to decide whether to be vaccinated or not. Then it is possible that those who volunteered to be inoculated might have been healthier (or perhaps less healthy) than those who did not.

Information bias pertains to the collection of data about the exposure or the outcome. Various forms of information bias derive from errors in measuring accurately these key components of a study. As we discussed in Sections 3.8 and 3.9, the consequence of such errors when the variables are categorical is *misclassification* of the affected variable. For example, *diagnostic suspicion bias* occurs when knowledge of the subject's exposure state influences the diagnostic process (Sackett, 1979). In a study of the relationship between passive smoking and asthma, an asthmatic child whose family included no smokers might be misdiagnosed more frequently than an asthmatic child whose parents smoked.

In a cohort study, such misdiagnosis could result in information bias. Each study participant would be followed and his outcome would be recorded, but the occurrence of an asthmatic episode would be missed more often among children of nonsmokers than among those of smokers. Therefore, the estimated effect of smoking would be exaggerated. In a case–control study, diagnostic suspicion could also have another kind of effect. Suppose that in the hypothetical study just described, we attempted to include all cases of asthma in a particular time period and geographic area. If some cases were overlooked, and all of them were for children not exposed to passive smoking, these cases would be excluded rather than misclassified. In this situation, underdiagnosis linked to lack of exposure could cause selection bias instead of information bias. This form of selection bias is sometimes called *case ascertainment bias*. This example illustrates that a single *source of bias*, in this case increased diagnostic suspicion related to exposure, can sometimes lead to two different types of bias.

Confounding was introduced in Section 3.10 and is often conceived as an intertwining of the causal effect of the exposure with causal effects attributable to other factors. A potential confounding factor is a variable that is thought to affect the outcome independently of the treatment. If this factor has a different distribution in the groups, this imbalance "explains" the lack of exchangeability between the study groups. One way to view confounding is as a way to describe in concrete terms the fact that different individuals have been selected into the two groups. In Chapter 7, the relationship between confounding and selection bias will be discussed in detail. For now, we simply observe that, among methodologists, the dividing line between confounding and selection bias has been far from clear. For example, consider a study to determine whether military service causes marital breakdown. A random sample of soldiers is compared with a random sample of civilians. If military service is voluntary, as it is currently in the United States, bias can clearly result. But is this bias attributable to the selection of the sample (selection bias) or simply to preexisting differences between the soldiers and civilians (confounding)?

The tripartite categorization of biases (selection, information, and confounding) was becoming well established by the early 1980s and was the central organizing principle for a popular epidemiologic-methods text (Kleinbaum et al., 1982). This text described these three types of bias and attempted to elaborate in algebraic terms their essential natures. To illustrate the three major categories of bias, the authors discussed several specific types of bias, but with no attempt at being comprehensive. Other popular epidemiology texts have generally followed a similar approach.

More recent discussions of bias have continued to utilize the tripartite taxonomy, while recognizing that overlap among the three categories exists. Choi and Noseworthy (1992) provide an overview of the major biases in epidemiologic research. Their discussion relates different types of bias to the specific research designs (cross-sectional, prospective cohort, retrospective cohort, case–control) that can be affected. Their article describes seven types of selection bias and six types of information bias, along with confounding bias, for which no specific subcategories are defined. For each type of bias, likely directions and methods for prevention in different study designs are addressed.

Vineis and McMichael (1998) review the types of bias that are most pertinent to the emerging area of molecular epidemiological studies. These are studies that collect genetic and other data obtained by using molecular-biological laboratory methods. These authors divide biases according to the conventional tripartite system. They highlight four specific types of selection bias and four types of information bias that have previously been identified in the literature and are especially relevant to the molecular epidemiological context. In addition, they consider issues related to *intraindividual variability* in measuring biomarkers. Two general categories of measurement error are identified: intraindividual *variation over time* and *biological sampling variation*. An example of the former is the level of a hormone, which can display a diurnal variation pattern. An example of the latter is the variability in the number of DNA adducts (damaged pieces of DNA that indicate exposure to carcinogens) across different samples from the same individual.

Another compendium of biases was assembled by Bernard Choi as a section of an encyclopedia of epidemiologic methods (Choi, 2000). Choi lists over 70 distinct types of bias. A very brief description of each type is presented. While useful as a handy reference, this catalog highlights the problem that biases can be described at various levels of generality, leading to inconsistent terminology. For example, at the most granular level we find the term *interviewer bias*, which refers to situations in which an interviewer might gather data selectively. This problem is particularly common in case–control studies; the interviewer might probe more deeply for recollection of possible exposure from a case than from a control subject. Choi categorizes interviewer bias as a subcategory of *observer bias*, which is a subcategory of *data collection bias*, which is a subcategory of *information bias*. It is conceivable that others may find it useful to identify even more specific subtypes of interviewer bias. Furthermore, interviewer bias is closely related to other types of bias. For example, *exposure suspicion bias* occurs when "knowledge of the subject's disease status may influence both the intensity and outcome of a search for exposure to the putative cause."

Delgado-Rodriguez and Llorca (2004) have provided a list of approximately 50 biases, organized both in terms of the tripartite division and also the relevant stage in the research process. Moreover, these authors suggest which types of research designs are likely to be most affected by each of these biases. This information draws heavily on the previous sources mentioned in this section and reflects the same problems of overlapping definitions and different levels of generality.

Three recent survey articles on bias are organized primarily around the tripartite scheme. These are oriented toward the most salient issues in particular areas of research: surgery (Paradis, 2008), pharmacology (Gerhard, 2008), and occupational medicine (Pearce et al., 2007). The authors describe problems of bias in general and focus on a number of specific types of bias chosen from among the plethora of possibilities that have been described in the literature.

Practicing biomedical researchers possess a great deal of practical knowledge pertaining to types of bias. The literature on biases in the biomedical context has attempted to summarize and catalog this valuable information. However, even very extensive lists of specific biases are not necessarily comprehensive and can be confusing for researchers because of the varying terminology employed by different authors. Discussions of bias have lacked an organizing principle or conceptual framework to pull together what is known about bias. The challenge laid down by David Sackett 30 years ago remains largely unmet. Neither the theoretical exposition of bias nor empirical research on the prevalence and implications of these biases has advanced markedly.

4.3 BIAS IN SOCIAL SCIENCE RESEARCH

Social scientists generally face the same problems pertaining to bias as biomedical researchers. However, the issues that are most salient in the social sciences tend to differ somewhat from those in the biomedical sciences. First, the outcomes of interest are usually measured on a numerical scale, rather than as a simple dichotomy (occurrence or nonoccurrence of a particular event). Second, the outcomes are usually related to human qualities or behaviors that can vary over time and may be difficult to measure precisely. Third, the focus of interest is often dynamic, as described in Section 3.6. That is, a comparative study seeks to determine whether a treatment can change the state of some characteristic of interest. Finally, the special problems raised by case–control studies are of very little concern to social scientists; case–control methodology is virtually nonexistent in the social sciences.

In some social science research, especially in experimental psychology, investigators can impose tightly controlled laboratory conditions. At the other extreme, field research in disciplines such as psychology, social work, sociology, anthropology, and political science often entails qualitative and naturalistic observation. Our concern in this book is with studies that attempt to compare two or more well-defined alternative conditions using quantitative methods to derive causal relationships. In a social science context, randomized experiments have occasionally been used, but observational studies are the norm. In particular, quasi-experiments have played a central role in applied social science research. Consequently, various research design elements to strengthen causal inference when randomization is not feasible have been more fully developed in the social sciences.

The most widely accepted paradigm for quasi-experimental research derives from the pioneering work of Donald Campbell and his colleagues (Campbell, 1957; Campbell and Stanley, 1963; Cook and Campbell, 1979; Shadish et al., 2002). In the social sciences, the complexity and variability of human behavior has required the development of sophisticated theories of measurement (Lord and Novick, 1968; Cronbach, 1970). Studying change (or growth) over time has demanded research designs that involve repeated measurement of the outcome variable. Furthermore, complex structures of social organization sometimes create a need for multilevel models and analyses (e.g., Bryk and Raudenbush, 1992; Gelman and Hill, 2007).

In the context of comparative studies, bias is usually discussed as a *threat to validity*. We have discussed internal validity, external validity, and construct validity in Chapter 3. We explained that the concept of construct validity extends beyond the bounds of our considerations here. A fourth type of validity is *statistical conclusion validity*, which pertains to inferential errors that result from flawed statistical analyses (Cook and Campbell, 1979; Shadish et al., 2002). Our definition of bias as a difference between the empirical effect and the causal effect ignores statistical variation attributable to finite sampling. This definition therefore precludes most threats to statistical conclusion validity.

The major threats to validity that have been recognized by social researchers are listed next. Many of these threats implicitly assume that two or more measurements using the same instrument, or "test," are obtained over the course of an intervention, including a *pretest* (baseline measurement) and a *posttest* (outcome measurement):

- History
- Maturation
- Testing
- Instrumentation
- Statistical regression
- Selection
- Mortality

History refers to the possible occurrence of an exogenous event or influence during the period under study. Such an event could affect the study groups in a similar manner or differentially. Consider our hypothetical study of SAT tutoring. Imagine that the high-school mathematics curriculum happened to be covering some of the same topics being emphasized in the tutoring program. If this outside supplementation happened to affect those in the two types of programs differently, then internal validity would be jeopardized. This situation is sometimes said to be a *selection–history interaction*. If the impact of the outside history is basically the same across groups, then internal validity may be preserved, but external validity of the study might still be threatened.

Maturation is similar to history, but it refers to internal changes that affect individual growth and development but are unrelated to the treatments. For example, the students being tutored might be developing certain skills during the normal course of their high-school studies. If the normal learning rates in the treatment groups are different, the observed difference between the study groups at posttest may be in part attributable to this *selection–maturation interaction*.

Testing is related to a potential artifact that can occur when the same measurement process is repeatedly used to assess an outcome. Familiarity with the test can lead to a falsely inflated score on later observations of the outcome. Conversely, fatigue or boredom can sometimes degrade performance in repeated testing. So the observed change during the study might not reflect the real change in the underlying construct being tested. This threat can affect internal or external validity.

Instrumentation refers to distortions that derive from the structure of the measuring instrument. One way this can occur is if the instrument itself changes in some way over time. For example, suppose that in measuring blood pressure, the investigators change from a manual to an automated technique. Another common source of instrumentation bias is the existence of a restricted range of values that can be recorded. For example, the instrument might register a maximum value, such as 800 on the SAT. If we measure a change from a pretest, the maximum amount of change will therefore be determined by the individual's starting point. A student who starts at 750 can potentially improve 50 points, whereas one who starts at 500 can improve up to 300 points. This imposition of an artificial upper limit is termed a *ceiling effect*. As a threat to internal validity, the differential growth potential across groups is termed a *selection–instrumentation interaction*.

Statistical regression, or *regression toward the mean*, is a rather complex idea that often causes confusion. As a simple example, suppose that in our tutoring study it is decided for some reason to assign the strongest students to the New Program. All students are required to take a mathematics test (pretest) and only those with scores above the mean for the whole population are admitted to the New Program. Those below this cutoff receive the Old Program. The outcome measure will be the gain between the pretest and a posttest. Assume further that the two programs are in fact equally effective. Then what would we expect to observe after the program when a posttest is administered?

There are many possibilities, depending on the nature of the pretest, posttest, and growth rates in the two groups. One possibility is that the Old Program average gain will be much greater than that for the New Program. This anomaly could occur because the pretest is an imperfect measure of the student's true mathematical ability. So the students with relatively low initial scores are really better than they appear to be. Conversely, those with relatively high scores are worse than they appear. As a result, the observed scores on the posttest for those in both groups will tend to "regress" toward the overall mean of the posttest scores for the entire student population. The apparently stronger students in the New Program will on average "improve" by a smaller amount than their apparently weaker counterparts in the Old Program.

Selection as a threat to validity is essentially the same concept as selection bias. The manner in which subjects are selected from a target population is often nonrepresentative, leading to problems of external validity. The process by which individuals are allocated to the study groups can lead to lack of internal validity. Interestingly, the classification of various specific types of selection bias has not received much attention in the social sciences. Perhaps the potential for selection bias is so endemic in the social sciences that attempting to list all specific types would be a hopeless task. On the other hand, statistical models for selection bias have been more extensively developed in the social sciences, particularly in economics. This development has been facilitated by the fact that the outcomes are usually numerical rather than categorical, enabling use of the general linear model for statistical analyses. In addition, the frequent availability of a baseline measurement (pretest) is helpful, because the selection forces affecting the pretest may be similar to those that affect the posttest. Having one or more pretest scores may allow us to identify and possibly adjust for the selection bias.

Mortality does not usually refer literally to the demise of study subjects. Rather, the term refers to the fact that subjects can drop out of a study at various stages for a number of reasons. Threats to validity can result from differences between those who drop out of a study and those who remain. In epidemiology, this threat to validity is called withdrawal bias, loss to follow-up, or attrition. Mortality can lead to bias even in a randomized experiment, because randomization ensures between-group exchangeability only at the study's outset. It is possible that attrition during the study can effectively compromise the exchangeability.

Besides the seven major threats to internal or external validity listed previously, there are threats to construct validity that may sometimes lead to bias. Construct validity in this context is primarily related to the degree of agreement between the treatment or outcome as measured and the actual construct intended by the researchers (Cook and Campbell, 1979, 64). For example, marital stress can be operationalized in various ways. Theorists might disagree about the best way to define this concept. However, even if a construct is well defined in theory, it may not be implemented in accordance with the investigator's expectations. This kind of "measurement error" does not necessarily lead to bias in the sense of a lack of exchangeability between study groups. However, it *may* generate bias indirectly by reducing the reliability of the treatment or outcome. The main construct validity issues that could lead to measurement error for either the treatment or the outcome are

- Inadequate preoperational explication
- Mono-operation bias
- Mono-method bias
- Hypothesis-guessing
- Evaluation apprehension
- Experimenter expectancies
- Levels of constructs

Inadequate preoperational explication refers essentially to lack of precision in defining the outcome operationally. *Mono-operation bias* pertains to the idea that many constructs can be operationalized in alternative closely related ways. Therefore, using several such measures can allow the investigator to "triangulate" on the true value of the treatment or outcome. Using a single measure is more likely to generate measurement errors. *Mono-method bias* is similar to mono-operational bias, but with the focus on the measuring instrument rather than the operational definition of the construct.

Hypothesis-guessing refers to distortions of responses attributable to speculation by study participants about the investigator's intent. *Evaluation apprehension* can lead to distortions resulting from the subjects' anxiety about being studied. *Experimenter expectancies* are influences on observed outcomes that result from the beliefs of those implementing a program, rather than from the program per se. Bias related to *levels of constructs* occurs when the wrong level of a treatment is implemented. For example, a tutoring program might be effective if provided daily for two hours, but ineffective if implemented for one hour on Mondays only.

We have listed seven potential threats to internal and/or external validity, and seven threats to construct validity. These threats to internal or external validity can be loosely mapped onto the tripartite system of bias classification that predominates in biomedical research. What is usually meant by *Selection* and *Mortality* tend to fall in the selection bias category. However, some of what social scientists call selection bias would be considered as confounding by many epidemiologists. *Testing* and *Instrumentation* are particular types of information bias that are prominent in social science research, especially in psychology and educational research. *History* and *Maturation* involve changing conditions that occur during the course of the study. To the extent that these changes are inherent in characteristics of study subjects that exist at baseline, these issues can be regarded as forms of confounding or selection. To the extent that they emerge only later, they can be described as intermediate causal factors. Such factors raise difficult conceptual issues that will be addressed in Chapter 8. *Statistical regression* is a concept that has engendered a great deal of confusion. In the context of comparative studies, a "regression artifact" is often regarded as a consequence of an attempt to control for confounding when the covariate is measured with error. This complex topic will be considered in Chapter 9.

Finally, the seven threats to construct validity overlap substantially with what epidemiologists might regard as potential sources of information bias. However, as mentioned previously, there are aspects of construct validity that cannot be framed simply in terms of measurement error. These issues pertain primarily to our ability to specify in clear operational terms precisely what the exposure and outcome really mean. Although far from irrelevant in the biomedical arena, such considerations are somewhat less problematic, because the constructs typically pertain more to physical rather than behavioral characteristics.

4.4 SOURCES OF BIAS: A PROPOSED TAXONOMY

So far in this chapter we have presented an overview of previous attempts to categorize and catalog types of bias. In epidemiology, methodologists have identified a wide range of specific types of bias. At a high level, these biases have been organized loosely into three broad categories: selection bias, confounding, and information bias. Scattered throughout the literature are many examples of more specific biases and some helpful discussions of how to detect or avoid them.

It seems clear now that the proliferation of terms to describe various specific biases that can arise in practice has limited utility. Resulting directories include definitions that are at different levels of generality and can overlap. Furthermore, what may be called by a particular name (e.g., diagnostic suspicion bias) may have different consequences in different study designs. In the social sciences, classification of bias has followed a different course. Rather than a dynamic proliferation of many specific types of bias, we find a very stable taxonomy of threats to validity. The basic principles and concepts were laid out about 50 years ago (Campbell, 1957; Campbell and Stanley, 1963) and have been subject to only minor reformulation over the years (Cook and Campbell, 1979; Shadish et al., 2002). The basic canons of research methodology introduced by Donald Campbell and his colleagues continue to be studied by virtually every graduate student in the social sciences. This framework still has great practical value as a general approach to the design and analysis of comparative studies. However, attempts to link these guidelines to a formal theory of causation have so far been somewhat limited.

The recent interest in counterfactual ideas in the biomedical and social sciences is a very promising development. This book is dedicated in large part to the proposition that a deeper and more practically relevant understanding of bias based on causal concepts such as exchangeability and potential outcomes is possible. After developing a theory of bias more extensively throughout the next five chapters, we will discuss in Chapter 10 the main potential *sources of bias* that tend to arise in comparative studies. We will define a source of bias formally as an aspect of a comparative study that can cause the empirical effect to deviate from the causal effect in the target population.

As discussed in previous chapters, a study would ideally collect information about the set of potential outcomes (i.e., the response pattern) for each individual in the target population. That way, any measure of causal effect could be calculated for any individual or subgroup of the target population. In reality, a comparative study is only capable of garnering data about the *actual* outcomes for a *sample* of subjects under a certain set of partially known *circumstances*. Therefore, the causal inferences that are feasible, even in an ideal study, are quite limited. Generally, they are restricted to a few conventional aggregate statistical measures [e.g., average causal effect (ACE), risk ratio, difference of means] summarizing a population-level causal effect. Most sources of bias are related to one of five primary aspects of a comparative study's design or implementation:

- Sampling of the subjects
- Assignment to exposure states
- Adherence to nominal exposure states
- Exposure ascertainment
- Outcome measurement

Sampling refers to the process, or mechanism, by which individuals are selected for participation in the study. Nonrandom sampling of subjects from the target population has the potential to distort the empirical effect by creating a study population with a different mix of individuals. Specifically, the distribution of response patterns in the study population might differ from that in the target population. The aggregate measure of causal effect can thus differ from the corresponding effect in the target population of interest. Thus, nonrandom sampling is a potential threat to external validity.

Assignment refers to the mechanism that determines the exposure state (treatment modality) for each subject. This mechanism can be a natural process, a system of social forces, or a deliberate strategy. The assignment mechanism is critically important for internal validity, because it governs the composition of the study groups. In theory, random assignment guarantees that the study groups are exchangeable in the sense defined in Chapter 2. That is, apart from random variation, the distribution of response patterns will be the same for both groups. Nonrandom assignment, on the other hand, allows the likely possibility that these distributions differ in a way that could engender bias.

We have defined a causal effect for an individual as a function of the response pattern, assuming a *fixed set of background circumstances*. Formally, we defined the relevant background circumstances as the causal context. In Section 3.7 we explained how causal mediation occurs if the exposure (risk factor, treatment modality) of interest affects not only the outcome, but also one or more of the causal factors within the causal context. In that situation, the standard causal effect estimated in the study may not be interpretable as a direct effect of the exposure. *Lack of adherence* to the nominal exposure state is a particularly common and important form of causal mediation, and it can take several forms. For example, in the context of an RCT, unwillingness or inability to comply (noncompliance) can result in a distortion of the treatment.

Exposure misclassification occurs when the true exposure status of some individuals in the study is not accurately known. In practice, this happens almost exclusively in retrospective studies. It is often a major concern in case–control studies, because exposure is often assessed long after occurrence. When a subject in one study group is misclassified, the analysis treats her data as if she were in the other study group. If misclassification is effectively random, then it should have similar effects in both study groups. In that case, the form of bias is somewhat predictable but still potentially problematic. If misclassification is nonrandom and systematically related to exposure status, the resulting havoc to the statistical analysis is complex and subtle. With even a small proportion of subjects misclassified, the resulting bias can be substantial and almost impossible to correct with confidence.

Outcome measures can be inaccurate for a variety of reasons. Such measurement error can be essentially random, or it might be systematically related to the individual's exposure state. In either case, it is theoretically possible for bias to result. The nature of this bias depends on the process of measurement used to obtain the outcome data. In common practice, certain simplifying assumptions are made, often implicitly. Causal theory can be helpful in understanding these assumptions and their implications.

The five general sources of bias listed here overlap with the more traditional division of bias types into selection bias, confounding, and information bias. The full rationale for our alternative system of classification will be clarified in the course of the following five chapters. For now, we merely suggest that the usual categories entail a great deal of ambiguity.

GUIDEPOST 4

This chapter has surveyed the various ways that types of bias have been traditionally categorized. We began with a brief overview of research design, because the form of bias can depend on the particular design. A major distinction is between randomized experiments and observational studies. We explained the special problems that pertain to case–control studies, and that can potentially compound the biases that often affect an observational study.

We then reviewed the terms in which biases have traditionally been framed in the biomedical and social sciences. The chapter concluded with a classification scheme for sources of bias that is motivated by the causal framework to be elaborated in the next five chapters. Throughout these five chapters, we discuss aspects of bias from two complementary perspectives. First, we present the traditional statistical models for dealing with bias. Second, we derive causal models to address the same issues.

In Chapter 5, we will discuss the different aspects of what is commonly termed selection bias. Discussions of selection bias are often confusing, primarily because the term has been used to describe at least three phenomena that from a causal perspective are quite distinct. In particular, the chapter aims to clarify the differences between sampling bias and assignment bias, and to explain the nature of these biases in various types of studies.

CHAPTER 5

Selection Bias

Selection bias has long been recognized as a potential threat to comparative studies. In the typhoid vaccine study over a century ago, the basic problems inherent in testing a new treatment on volunteers were well known. Yet, today there is still much confusion about the definition and implications of selection bias. Broadly speaking, selection bias can result from the manner in which individuals are chosen for inclusion in the study. However, there are at least three different aspects of a selection process that can result in bias. Discussions of selection bias often focus on one or another of these aspects. What appear to be general statements about selection bias may only be referring to a particular subcategory of selection bias. In this chapter, we attempt to unravel and clarify the various interpretations of selection bias. Our analysis will describe the issues from both the traditional statistical viewpoint and the emerging causal (counterfactual) perspective.

5.1 SELECTION PROCESSES AND BIAS

We define a *selection process* as the mechanism that determines which particular individuals from a target population become members of each particular study group (exposed and unexposed). In a widely referenced dictionary of epidemiology, Last (2001) defines selection bias as:

> Error due to systematic differences in characteristics between those who take part in a study and those who do not....Selection bias invalidates conclusions and generalizations that might otherwise be drawn from such studies. It is a common and commonly overlooked problem. (Last, 2001, 166)

This definition of selection bias is quite general—and somewhat vague. We will identify and analyze three different phenomena that have been referred to as selection bias in various contexts.

First, the study population might not be exchangeable with the target population of interest. This problem would impair the study's external validity. We have called the distortion brought about by nonrepresentative sampling from the target population *sampling bias*. Second, the study groups might not be exchangeable with each other. This problem would impair the study's internal validity. We have termed the resulting distortion *assignment bias*. Third, the data that are available for analysis may exclude individuals with particular values of the outcome variable. We will call this type of distortion *retention bias*, because it depends on which individuals are retained in the analytic sample.

Retention bias can arise in retrospective studies in which the outcomes for all individuals have occurred but have not all been reported to the investigators. The study subjects who ultimately remain available for statistical analysis may not be exchangeable with the entire study population. This problem can potentially affect both the internal and external validity. One common situation that can give rise to retention bias occurs when the data are obtained from administrative records. For example, sociologist Richard Berk offers the example of a study in which the outcome variable is a measure of the severity of incidents of wife battery (Berk, 1983). He points out that if the data are obtained from police arrest reports, less serious cases will tend to be underrepresented. In his example, the causal effect of the number of prior incidents of abuse could be distorted by this retention bias.

In this chapter, we focus primarily on the implications of the sampling and assignment mechanisms. However, the subject of retention bias will be discussed briefly in the context of our model for a numerical outcome in Section 5.7. It is worth noting that retention bias typically involves selection based on a threshold value above (or below) which the data are unavailable. Therefore, the issue has been considered only for numerical outcomes. However, a case–control study can be regarded as involving a special type of "retention" by virtue of the undersampling of controls relative to the cases. Thus, the case–control design deliberately incorporates a form of potential selection bias. As we explained in Section 4.1, use of the odds ratio rather than the risk ratio provides a way to adjust for this known form of retention bias.

Many methodologists do not regard sampling bias as an aspect of selection bias, because they consider external validity as a matter of secondary importance to be dealt with after internal validity has been established.

> It is important to keep in mind the distinction between *selecting subjects for a study* and *selection bias*. Virtually every study conducted in human populations selects study subjects from a larger population. The nature of this selection potentially affects the *generalizability* or *external validity* of the study but does not necessarily affect the validity of the comparisons made within the study or the study's *internal validity*. (Gordis, 2004, 224–225)

Should the distortion that can result from selecting subjects for a study be considered a form of bias?

The answer to this question hinges largely on our conception of the target population. If the target population represents a well-specified group of human beings for which the study findings are expected to be relevant, our main objective might be to estimate the effect *for this specific population*. For example, studying suicidality in a randomized control trial (RCT) of an antidepressant medication in clinically depressed adolescents is meant to influence treatment in the actual population expected to receive the drug. If the study results fail to generalize to this specific target population, then the term "bias" seems appropriate. On the other hand, ideas about potential generalization may be much more speculative. An investigator might hypothesize that an antidepressant's effect on depressed adolescents would extend to a population of psychotic adults.

Generally speaking, the issue of generalizability has received more consideration by social scientists than by epidemiologists (Shadish et al., 2002). In the biomedical arena, the effect of a treatment or risk factor is often expected to be fairly uniform across individuals, and the potentially affected population may be reasonably well defined. Social policies or programs, however, can depend on a wide variety of individual characteristics and on aspects of the social context. Cook and Campbell (1979) point out that:

> Generalizing *to* well explicated target populations should be distinguished from generalizing *across* populations. Each is germane to external validity: the former is crucial for ascertaining whether any research goals that specified populations have been met, and the latter is crucial for ascertaining which different populations (or sub-populations) have been affected by a treatment, i.e. for assessing how far one can generalize. (Cook and Campbell, 1979, 71)

In a particular context it may not be obvious whether the issue of generalization is one of "generalizing to" or "generalizing across." Therefore, achieving external validity might be viewed by some researchers as avoiding selection bias, and by others as extrapolating results to a qualitatively different population. For didactic purposes, we envision a real target population of immediate relevance from which sampling occurs, while recognizing that this will not always be the case. We therefore regard the difference between a causal effect in the study population and the causal effect in the target population as a type of selection bias.

From this perspective, the ultimate object of scientific interest is the causal effect in some well-defined target population. A selection process specifies for each individual in the target population two probabilities: the probability of becoming an exposed study subject and the probability of becoming an unexposed subject. Consequently, we can think of the selection process as being composed of two components: a subprocess for selecting the group of exposed subjects and a subprocess for selecting the unexposed group. For example, suppose the target population consists of 100,000 households of a certain city. We plan to "expose" 1000 households to a new program intended to improve their earning potential. These exposed households will be compared with a group of 3000 households who do not experience the intervention. The selection process determines the two selection probabilities for each individual household. Ideally, the selection process will result in two study groups that are exchangeable in the sense of having identical distributions of response patterns.

Now let us consider a particular individual household (Household A). In terms of causal analysis, Household A is completely described by its response pattern. For example, suppose that the potential outcomes for Household A are $70,000 under the treatment and $50,000 under the control condition. Furthermore, assume that, for the subset of all households with the same response pattern ($70,000, $50,000), the probability of being selected into the experimental program is 0.01 and into the control group is 0.03. More generally, let S_{jE} denote the probability for individuals with response pattern j of being included as exposed, and let S_{jU} be the probability of being included as unexposed. In theory, an individual's probability could also depend on other characteristics besides the response pattern. However, from the viewpoint of causal inference, any such factor is relevant only to the extent that it affects the response pattern. Thus, we regard all individuals with the same response pattern as essentially indistinguishable.

Bias can result when certain types of response patterns are preferentially selected into one or both study groups. Whether such "preferences" are possible is predicated on how the selection process can *distinguish* among individuals in terms of their response patterns. To analyze various types of selection processes, it is useful to recast the selection process in a different mathematical form. Let us define S_j as the probability that an individual with response pattern j is *sampled* into the study population, and A_j as the conditional probability of being *assigned* to the exposed group:

$$S_j = S_{jE} + S_{jU} \tag{5.1}$$

$$A_j = \frac{S_{jE}}{S_{jE} + S_{jU}} \tag{5.2}$$

In some situations, the *sampling mechanism* that determines which individuals are drawn from the target population into the study occurs first. Then, the *assignment mechanism* is governed by an explicit decision-making process. In all randomized experiments and most quasi-experiments, the selection process unfolds in this manner. In many observational studies, however, the exposed and unexposed subjects are selected simultaneously, and with little if any explicit consideration of the target population. For example, the hazardous effects of potentially toxic chemicals are often studied in occupational settings. A group of exposed workers is compared with a group of similar workers in the same facility, but ones who are not exposed. The study population consists of a mixture of two subcategories of employees in a particular work environment. The nature of this study population and its relationship to a broader target population may not be clear.

There are four main types of selection processes that arise in comparative studies. First, the selection process could be completely "ignorant" of relevant individual characteristics. That is, individuals are *indistinguishable* with regard to their causal response patterns. From the standpoint of causal inference, ignorance (on the part of the selection process) is truly bliss (for the researcher). Both the sampling and assignment probabilities do not depend on the response pattern. The only way that such ignorance can be assured is by sampling and assignment in a completely random manner. In our example, each household would have a 0.04 probability of being included in the sample based on a random sampling mechanism. Subsequently, a second random assignment mechanism would generate a 0.25 conditional probability of being exposed to the intervention.

A second possibility is that the selection process might distinguish among different response patterns in sampling, but not in assignment. As a result, S_j would depend on the value of j, but A_j would not. In our example, Household A might have a probability 0.04 of being sampled and a conditional probability of 0.25 of being placed in the intervention, given that it is sampled. However, Household B with response pattern ($45,000, $40,000) might have corresponding probabilities 0.08 and 0.25. Because the sampling probabilities would vary, the distribution of response patterns in the study would deviate from the distribution in the target population. In particular, the proportion of households with response pattern ($45,000, $40,000) would be overrepresented relative to those with response pattern ($70,000, $50,000). More generally, the study population and target population would not be exchangeable. However, the exposed and unexposed groups would remain exchangeable. For every response pattern in the study population, the probability of exposure is constant (0.25 in our example). Therefore, apart from the effects of random variation, each study group constitutes a microcosm of the study population.

Formally, we can express this situation using Bayes' theorem:

$$P(j|E) = \frac{P(E|j)P(j)}{\sum_j P(E|j)P(j)} = \frac{A_j P(j)}{\sum_j A_j P(j)} \qquad (5.3)$$

where $P(j|E)$ represents the conditional probability that a member of the exposed group will have response-pattern j. Essentially, this is the proportion of exposed subjects of response-type j; $P(E|j)$ is the probability of being exposed conditional on having response pattern j; $P(j)$ is the overall proportion of individuals with response pattern j *in the study population*. The summation in the denominator is the overall, or average, assignment probability in the study population. We can denote this average assignment probability as

$$\mu_A = \sum_j A_j P(j)$$

A similar expression holds for the unexposed group:

$$P(j|U) = \frac{P(U|j)P(j)}{\sum_j P(U|j)P(j)} = \frac{(1-A_j)P(j)}{\sum_j (1-A_j)P(j)} \tag{5.4}$$

But under our assumptions, $A_j = A$ for all values of j and thus Equation 5.3 reduces to $P(j|E) = P(j)$ and Equation 5.4 reduces to $P(j|U) = P(j)$. This implies that the distributions of response patterns in the two comparison groups are identical. However, they are identical to the distribution in the *study population*, which can differ from that in the target population.

A third situation occurs when the assignment mechanism can depend on the response patterns. In this case, the values of A_j can vary for the different response patterns. For example, suppose that the study directors deliberately manage to assign certain very low-income households to the experimental program. Then the value of A_j for Household B could be 0.50, while for Household A this assignment probability could be only 0.25. As a result, the ratio of households like B to households like A in the experimental group will be higher than the corresponding ratio in the control group. Clearly, the two study groups will in general not be exchangeable. The distributions of response patterns in the two groups are determined by Equations 5.3 and 5.4 and do not reduce to a simpler form. Lack of exchangeability in this situation creates the possibility of bias. Actual bias will depend on the effect measure and the nature of the nonexchangeability.

In principle, it is possible that the sampling mechanism will be effectively random even though the assignment is not. However, it is far more common to have nonrandom sampling when the assignment is nonrandom. Even if sampling were random, the lack of internal validity would usually be deemed the critical problem to be addressed. Unless this problem can be overcome, external validity is not considered important, or even meaningful. After all, how can we conceptualize the generalizability of an "effect" that is not truly a causal effect?

The distinction between internal and external validity appears to be well defined. However, in observational studies, the two concepts are not completely separable. For example, in the studies of supported work it was common to regard the volunteers as the target population of interest. The aim of some studies was to determine the effect of supported work programs for those who would actually volunteer for the program. This "effect of treatment on the treated" would ideally be estimated by performing a randomized experiment that included a representative sample of the target population (volunteers). But most studies actually compared the experience of volunteers against that of a control group drawn nonrandomly from the general population. Attempts were then made to correct for the bias by various methods of statistical adjustment. For now, suppose that we decided naïvely to make a simple comparison between the groups. What would be the nature of the resulting bias?

The actual study population would be a mixture of the exposed group of volunteers (representative of the target population) and the unexposed group, whatever it might represent. Suppose first that the volunteers and controls were exchangeable. Then we would clearly have internal validity. However, we would also have external validity, because both the study groups, and thus the whole study population, would be exchangeable with the target population. Now suppose the volunteers and controls were not exchangeable. Then there would be neither internal nor external validity. More generally, the "closer" the distributions of the response patterns in each study group are to the distribution in the target population, the smaller will be *both* the internal and the external validity. Thus, the heuristic idea that a sharp distinction between internal and external validity can be drawn is somewhat misleading.

Most statistical adjustment methods attempt to eliminate internal bias but ignore external bias. This approach is reasonable if the individual causal effects are believed to be quite uniform. In that case, any measure of effect will not depend on which particular individuals have been sampled into the study population. However, when these effects can vary, the causal effect in the study population can sometimes differ substantially from the true causal effect of interest.

Finally, we mentioned previously that there is a type of selection bias that can occur when the data for individuals with certain outcome values tend to be eliminated from the study. Of course, such selection based on the outcome can occur only if the actual outcomes have already occurred. A distinction is sometimes made between two possible ways that the outcome can affect selection. First, an individual's outcome value might explicitly determine whether she is retained in the study. This situation has been termed *explicit selection* (Gulliksen, 1950; Lord and Novick, 1968; Goldberger, 1981) or *direct selection* (Berk, 1983). Alternatively, retention could be determined by another variable that is correlated with the outcome. Then the outcome *effectively* determines the selection probability for each individual. This situation is termed *incidental selection* or *indirect selection*.

Suppose that for some reason we decided not to include any households with an income above $60,000 at the end of the experiment. Then Household A would be retained in the control group if assigned to it, because its observed value would be $50,000. However, if assigned to the intervention group, the resulting income of $70,000 would lead to exclusion from the study. This would represent an example of explicit selection based on the outcome. Alternatively, suppose we excluded households whose total income tax payments exceeded $15,000. Then, the incidental impact of this selection mechanism would be to exclude many higher-income households. This would be an example of incidental selection. Whether the criterion for retention entails explicit or incidental selection, the result can be to alter the distributions of response patterns in the two groups.

The remainder of this chapter discusses the theory of selection bias and possible ways to assess its existence and magnitude. Most of this chapter presents these ideas in the context of a dichotomous outcome. We begin with the traditional model often used by epidemiologists to describe selection bias in this case. The main insights and limitations of this approach are explicated. We then move on to present a causal perspective on bias. Each of the four types of selection processes outlined above is analyzed using a simple causal model for a dichotomous outcome (see Section 3.4). Finally, we develop a similar approach to conceptualizing selection bias when the outcome is numerical and the empirical effect is the difference between the group means. We show that the basic mechanisms that can produce selection bias are similar for dichotomous and numerical outcomes, but the mathematical expressions of these mechanisms are quite different.

5.2   TRADITIONAL SELECTION MODEL: DICHOTOMOUS OUTCOME

We begin by discussing the traditional perspective on selection bias when the outcome is a dichotomous variable. Table 4.1 displayed the data from a study in the cross-tabular form familiar to epidemiologists and is reproduced here, with slightly different notation, as Table 5.1. Note that we have suppressed the dependence of the exposed and unexposed proportions (q_{WE} and q_{ZU}) on the sample. From the traditional viewpoint, there is no need to make this explicit, because the counterfactual idea that the exposure states for the two groups could have been switched is not part of this traditional conceptual framework.

Assume that each cell of Table 5.1 contains the number of individuals in some *source population* who would fall into a particular combination of exposure status (E or U) and outcome (D or H). This source population could be either the study population or some other target population of interest. More generally, we could assume that the outcome has more than two possible values and may be numerical instead of categorical. However, the main issues can be shown more clearly by using the basic model for illustrative purposes.

In Table 5.1, q_E represents the probability of being a *case* (i.e., experiencing an event D) if an individual is exposed, and q_U represents the probability of

Table 5.1   Standard 2 × 2 Table

	Disease	Healthy
Exposed	$q_E N_E$	$(1 - q_E)N_E$
Unexposed	$q_U N_U$	$(1 - q_U)N_U$

being a case if unexposed; N_E and N_U represent the total number of exposed and unexposed individuals in the target population. In Section 4.1, we explained how the risk ratio can be approximated by the odds ratio when the values of q_E and q_U are small. Since this is usually the case in epidemiological research, the odds ratio is often adopted as a working measure of effect. However, many epidemiologists view the odds ratio as merely a convenient approximation to the risk ratio and not as the ultimate parameter of interest (Greenland, 1987b).

From the traditional perspective, the problem of selection bias pertains to the distortion in estimating the effect measure that can result from nonrepresentative sampling from the source population (Kleinbaum et al., 1981). Let OR_{true} be the "true" odds ratio of interest. The nature of this true effect is usually not defined explicitly, but it can be conceptualized as the value of the effect that would be obtained if the entire source population were available for analysis. Selection bias occurs when the process of sampling from this source population results in a distortion of the true effect.

Selection bias is defined as the difference between the expected value of the observed effect (essentially what we have termed the empirical effect) and the true effect. In terms of odds ratios, selection bias can be defined as the difference between (or ratio of) OR_{emp} and OR_{true} that is caused by the selection process. In our notation we express OR_{true} as

$$OR_{true} = \frac{q_E(1-q_U)}{q_U(1-q_E)} \qquad (5.5)$$

To express OR_{emp} we must first define the selection probabilities for various categories of individuals. Let S_{ED} be the probability that an exposed case in the source population is included in the study; let S_{EH} be the corresponding probability that an exposed control is selected; S_{UD} and S_{UH} are similarly defined for unexposed cases and controls. Then the resulting study population is shown in Table 5.2.

Let us define a quantity ψ as the cross-product of the selection probabilities:

$$\psi = \frac{S_{ED}S_{UH}}{S_{EH}S_{UD}} \qquad (5.6)$$

Now based on Table 5.2 we can derive OR_{emp} as

Table 5.2 Standard 2 × 2 Table under Selection

	Disease	Healthy
Exposed	$q_E S_{ED} N_E$	$(1-q_E)S_{EH}N_E$
Unexposed	$q_U S_{UD} N_U$	$(1-q_U)S_{UH}N_U$

$$\mathrm{OR_{emp}} = \frac{q_E(1-q_U)}{q_U(1-q_E)}\,\psi \tag{5.7}$$

Therefore, ψ is a measure of the selection bias and $\mathrm{OR_{emp}}$ is unbiased if and only if $\psi = 1$.

In principle, this expression can be used to help in estimating the potential bias in actual studies (e.g., Kleinbaum et al., 1981). However, the usefulness of this approach is limited:

> In practice, estimation of the selection or loss probabilities (or even of their ratios) is quite difficult, since it usually requires either information from another study…or knowledge of such selection probabilities for related studies. For these reasons, the investigator must realize that selection bias may exist even though it cannot be precisely determined or adjusted for. (Kleinbaum et al., 1982, 214)

The main problem with this traditional model is that the nature of the "true" source population is unclear. Specifically, there is no way to separate conceptually the possible reasons for the association between exposure and outcome in the source population. To what extent does the association represent a causal effect and to what extent is it the result of other factors? Therefore, the traditional model expresses the relationship between the empirical effect in one entire population and the corresponding empirical effect in the particular subpopulation selected for the study. However, the empirical effect in any actual source population may already differ from the causal effect of interest.

In a sense, we are not really interested in the *actual* version of the source population, but rather in a hypothetical version that would have occurred if exposure had been assigned at random. The relative risk in such a randomized target population would be the true object of interest, in other words the *causal effect*. But the traditional model has no way to represent explicitly this causal effect. Therefore, the "bias" represented in this model is *relative* to a possibly biased, or even undefined, effect in the source population. As such, it does not satisfy our definition of bias as a difference between an empirical effect and the corresponding causal effect.

5.3   CAUSAL SELECTION MODEL: DICHOTOMOUS OUTCOME

We now introduce a model for selection processes that is based on the counterfactual perspective. Once again we will restrict consideration to the simple case of a dichotomous exposure and outcome. The basic counterfactual model for this situation was described in Section 3.4 (see Table 3.2). As mentioned in Section 3.4, the causal effects can be written as

$$\mathrm{RD_{caus}} = P_2 - P_3 \tag{5.8}$$

Table 5.3   Standard 2 × 2 Table: Causal Model

	Disease	Healthy
Exposed	$(P_1 + P_2)N_E$	$(P_3 + P_4)N_E$
Unexposed	$(P_1 + P_3)N_U$	$(P_2 + P_4)N_U$

$$\text{RR}_{\text{caus}} = \frac{P_1 + P_2}{P_1 + P_3} \tag{5.9}$$

Unlike the traditional model, which can explicitly represent only empirical quantities, the counterfactual model is framed in terms of causal parameters (P_1, P_2, P_3, P_4). These four parameters specify the entire distribution of response patterns in the target population. This distribution determines the causal effect, such as the causal risk ratio or odds ratio, but does not fully determine the empirical effect. The empirical risk ratio depends in addition on the selection process and possibly as well on other sources of bias.

If the selection process is such that the study groups are both representative of the target population, then the standard 2 × 2 table of data can be represented as shown in Table 5.3. We will now consider how this table is affected by various types of selection processes. In Section 5.1, we noted that the selection process could be characterized by specifying for each response pattern j a sampling probability S_j and an assignment probability A_j. In Section 3.4 we explained how for a dichotomous outcome there are only four response patterns. The full selection process can thus be expressed by four selection probabilities and four assignment probabilities. For example, the selection probability for *doomed* individuals (those for whom the outcome event always occurs) can be denoted S_1 and the assignment probability A_1.

To illustrate the counterfactual model, let us consider first the case of an ideal study, one that employs randomization in both sampling and assignment. Because of random sampling, every member of the target population has the same probability S of being selected into the study. Because of random assignment, every selected subject has the same probability A of being exposed. Consider now the standard 2 × 2 table and focus on the exposed-case (ED) cell. To end up in this cell, a person must be *doomed* or *causal*, must be admitted into the study, and must be assigned to the exposed group. Therefore, this cell will contain $AS(P_1 + P_2)N$ individuals. (Technically, this expression represents the *expected value* of the number in the cell, but we are ignoring finite sampling considerations.)

Similar logic applies to the other three cells of the 2 × 2 table, resulting in the observed data given in Table 5.4. Consequently, the values of q_{WE} and q_{ZU} observed in the study can be expressed as

$$q_{WE} = \frac{AS(P_1 + P_2)N}{ASN} = P_1 + P_2 \tag{5.10}$$

Table 5.4   Standard 2 × 2 Table: Ideal Study

	Disease	Healthy
Exposed	$AS(P_1 + P_2)N$	$AS(P_3 + P_4)N$
Unexposed	$(1 - A)S(P_1 + P_3)N$	$(1 - A)S(P_2 + P_4)N$

$$q_{ZU} = \frac{(1-A)S(P_1 + P_3)N}{(1-A)SN} = P_1 + P_3 \qquad (5.11)$$

The difference between q_{WE} and q_{ZU} is the empirical risk difference RD_{emp} and the ratio is the empirical risk ratio RR_{emp}. From Equations 5.10 and 5.11, these can be expressed as

$$RD_{emp} = P_2 - P_3 = RD_{caus} \qquad (5.12)$$

$$RR_{emp} = \frac{P_1 + P_2}{P_1 + P_3} = RR_{caus} \qquad (5.13)$$

Thus, in an ideal study, the empirical effects would be identical to the causal effects, as we would expect. Moreover, from Table 5.4 it can easily be seen that any other measure of causal effect will also be correctly measured. This follows from the fact that the distributions of response patterns in the two study groups are identical.

5.4   RANDOMIZED EXPERIMENTS

An ideal study requires both random sampling and random assignment. However, in most randomized experiments, the subjects cannot be sampled randomly from the target population. As a practical matter, it is almost never feasible to implement random sampling, even when the target population is very well specified. Furthermore, the study eligibility criteria typically restrict the sample in certain ways, such as excluding individuals for whom the treatments may be harmful. The restricted sample that results may not be representative of (i.e., exchangeable with) the target population.

In terms of the traditional model, we can express the results of a randomized experiment using Table 5.2. However, because of the random assignment, the conditional probability of being exposed, given that a person is a study participant, is independent of any individual attribute. In particular, it must be independent of whether D will occur, so that

$$\frac{S_{ED}}{S_{ED} + S_{UD}} = \frac{S_{EH}}{S_{EH} + S_{UH}} \qquad (5.14)$$

Table 5.5   Possible Response Patterns under Selection

Response Pattern	Exposed	Unexposed	Proportion
1: Doomed	D	D	$S_1 P_1 / \mu_S$
2: Causal	D	H	$S_2 P_2 / \mu_S$
3: Preventive	H	D	$S_3 P_3 / \mu_S$
4: Immune	H	H	$S_4 P_4 / \mu_S$

From Equation 5.14 it follows that $\psi = 1$, so there is no selection bias according to this model. That is, nonrepresentative sampling does not appear to induce bias as long as there is random assignment. In effect, the traditional model only considers selection bias to occur if there is a threat to internal validity. This limitation of the traditional model was discussed in Section 3.4. Satisfying the condition $\psi = 1$ means that the empirical effect in the study is equal to the empirical effect in the source population, but does not address the issue of bias as a difference between an empirical effect and a corresponding causal effect.

Now consider the counterfactual representation of a randomized experiment. In terms of the model presented in Section 5.3, a common assignment probability A for all subjects can still be assumed. However, without random sampling from a target population, the assumption of a common sampling probability S may not be valid. How can the variation in the sampling probabilities distort the empirical effect? We have previously defined S_j to be the probability that someone with response pattern j is selected for the study. Furthermore, let $\mu_S = \sum_1^4 S_j P_j$ the average sampling probability across the whole target population. Then the proportions of subjects in the study population no longer equal their corresponding proportions in the target population. For any response pattern j the proportion in the study population is Q_j, where

$$Q_j = \frac{S_j P_j}{\mu_S} \tag{5.15}$$

The situation just described is summarized in Table 5.5. The total number of subjects in the study population is $N\mu_S$ and the number of individuals in the ED cell of the standard 2×2 table is given by

$$N_{ED} = A(Q_1 + Q_2)N\mu_S = A(S_1 P_1 + S_2 P_2)N \tag{5.16}$$

Applying similar logic we can obtain the entire 2×2 cross-table. From Table 5.6, we obtain expressions for the event rates in the exposed and unexposed populations as

$$q_{WE} = Q_1 + Q_2 = \frac{S_1 P_1 + S_2 P_2}{\mu_S} \tag{5.17}$$

Table 5.6   Standard 2 × 2 Table: Randomized Study

	Disease	Healthy
Exposed	$A(S_1P_1 + S_2P_2)N$	$A(S_3P_3 + S_4P_4)N$
Unexposed	$(1 - A)(S_1P_1 + S_3P_3)N$	$(1 - A)(S_2P_2 + S_4P_4)N$

$$q_{ZU} = Q_1 + Q_3 = \frac{S_1P_1 + S_3P_3}{\mu_S} \tag{5.18}$$

The empirical risk difference and risk ratio are simply

$$\text{RD}_{\text{emp}} = Q_2 - Q_3 = \frac{S_2P_2 - S_3P_3}{\mu_S} \tag{5.19}$$

$$\text{RR}_{\text{emp}} = \frac{Q_1 + Q_2}{Q_1 + Q_3} = \frac{S_1P_1 + S_2P_2}{S_1P_1 + S_3P_3} \tag{5.20}$$

From Equations 5.19 and 5.20 we can see how the empirical effect depends on the sampling mechanism.

Note first that Equations 5.17 and 5.18 show how the rates of events observed in each group depend on the sampling probabilities. For example, both rates will decrease as the value of S_1 is reduced. This seems logical because *doomed* individuals contribute events regardless of exposure status. The value of q_{WE} also depends on S_2 because *causal* individuals have events when exposed. The value of q_{ZU} depends on S_3 because *preventive* individuals have events when unexposed. Both rates also depend on S_4 indirectly, through μ_S. Indeed, in many practical situations the proportion of *immunes* P_4 is much larger than the proportions for any of the three other response patterns. As a result, the value of μ_S may be determined primarily by the value of S_4.

From Equations 5.8 and 5.19 we see that RD_{emp} will tend to be greater than RD_{caus} to the extent that $S_2 > \mu_S$ and/or $S_3 < \mu_S$. Conversely, bias will be nega-tive when $S_3 > \mu_S$ and when $S_2 < \mu_S$. In words, when sampling favors *causal* individuals, the empirical effect tends to be positively biased; when sampling favors *preventives*, the bias tends to be negative. However, the sampling prob-abilities for *doomed* and *immune* also come into play through their effects on μ_S. For example, consider the situation summarized in Table 5.7. Here the sampling mechanism favors *causal* individuals and disfavors *preventive* indi-viduals. As a result, RD_{emp} equals 0.128 rather than the causal effect of 0.030. But suppose that we also happened to oversample *immunes*, to create the situ-ation shown in Table 5.8. Now the various sampling effects "cancel out" to fortuitously produce the correct value of 0.030 for RD_{caus}.

For the risk ratio, the situation is somewhat different. Comparing Equations 5.9 and 5.20 we see that the sampling probability for the *doomed* plays a more

Table 5.7 Nonrandom Sampling Example 1

Response Pattern	p_j	S_j	$S_j p_j / \mu_S$
1: Doomed	0.050	0.010	0.0459
2: Causal	0.050	0.030	0.1376
3: Preventive	0.020	0.005	0.0092
4: Immune	0.880	0.010	0.8073

$RD_{caus} = .030 \quad RD_{emp} = .128$
$RR_{caus} = 1.43 \quad RR_{emp} = 3.33$

Table 5.8 Nonrandom Sampling Example 2

Response Pattern	p_j	S_j	$S_j p_j / \mu_S$
1: Doomed	0.050	0.010	0.0108
2: Causal	0.050	0.030	0.0325
3: Preventive	0.020	0.005	0.0022
4: Immune	0.880	0.050	0.9544

$RD_{caus} = .030 \quad RD_{emp} = .030$
$RR_{caus} = 1.43 \quad RR_{emp} = 3.33$

Table 5.9 Nonrandom Sampling Example 3

Response Pattern	p_j	S_j	$S_j p_j / \mu_S$
1: Doomed	0.050	0.100	0.0988
2: Causal	0.050	0.030	0.0296
3: Preventive	0.020	0.005	0.0020
4: Immune	0.880	0.050	0.8696

$RD_{caus} = 0.030 \quad RD_{emp} = 0.028$
$RR_{caus} = 1.43 \quad RR_{emp} = 1.27$

direct role. The factor $S_1 P_1$ appears in both the numerator and denominator of the expression for RR_{emp}. Therefore, increasing the value of S_1 moves the value of RR_{emp} closer to the null value of 1.0. Furthermore, the probability S_4 for the *immunes* is completely irrelevant to the risk ratio, although it affects the absolute rates of events observed in the two study groups. For the situations shown in Tables 5.7 and 5.8, the risk ratios are both equal to 3.33, although the event rates differ because of the different proportions of *immunes*. From Table 5.7, the risk in the exposed group is 18.4% and in the unexposed group is 5.5%; from Table 5.8, the corresponding risks are 4.3% and 1.3%.

Now suppose that S_1 is increased from 0.01 to 0.10. The results are displayed as Table 5.9. Note that the empirical risk ratio decreases to 1.27, which is smaller than the causal effect of 1.43. Thus, the increase in the proportion of *doomed* has attenuated the empirical effect.

In an actual study, we will not know the values of the sampling probabilities, or even their relative magnitudes. So, we will not be able to calculate the bias. However, there may be information that can at least provide qualitative insight. Suppose we know that the effect is monotone, with $P_3 = 0$ in the target population. Then Equation 5.20 reveals how the magnitude of the empirical risk ratio RR_{emp} would depend on the values of S_1 and S_2. For example, we might also know that the study attempted to "enrich" the sample by overincluding those susceptible to experiencing the event (*doomed*). Because S_1 would be relatively large, we would expect the risk ratio estimated in the study population to be smaller than the causal risk ratio applicable to the entire target population.

In the more general case when $P_3 > 0$ a variety of possible scenarios can be envisioned. Some of these are considered in Chapter 10, where we discuss the potential sources of bias. In some situations of practical importance, it is even possible for the empirical effect in the study population to be "reversed" by nonrandom sampling (Weisberg et al., 2009). As a simple example, suppose that an exposure is beneficial "on balance" although harmful for some, so that $P_3 > P_2$. The *preventives* might appear to be at higher risk than *causals* prior to the study, and therefore excluded more frequently. Then Equation 5.19 indicates that $Q_3 < Q_2$ whenever $S_2/S_3 > P_3/P_2$. So, sampling of *causals* at high enough rates relative to *preventives* can outweigh the preponderance of *preventives* in the target population.

5.5   OBSERVATIONAL COHORT STUDIES

In an observational study, there is neither random sampling from the target population nor random assignment to determine exposure status. Therefore, it becomes far more speculative to interpret the empirical effect as a measure of causal effect. The critical issue is whether the assignment mechanism induces lack of exchangeability between the study groups. The traditional model for an observational study is simply the general model presented as Table 5.2. The condition for absence of selection bias is that $\psi = 1$, which is often impossible to verify.

The full counterfactual model for an observational study is presented in Table 5.10. The expressions in each cell show how both sampling and assignment probabilities generate the number of observations in each of the four cells. The event rates can now be expressed as

$$q_{WE} = \frac{A_1 S_1 P_1 + A_2 S_2 P_2}{\sum_1^4 A_j S_j P_j} \tag{5.21}$$

$$q_{ZU} = \frac{(1 - A_1) S_1 P_1 + (1 - A_3) S_3 P_3}{\sum_1^4 (1 - A_j) S_j P_j} \tag{5.22}$$

Table 5.10 Standard 2 × 2 Table: Observational Study

	Disease	Healthy
Exposed	$(A_1S_1P_1 + A_2S_2P_2)N$	$(A_3S_3P_3 + A_4S_4P_4)N$
Unexposed	$[(1 - A_1)S_1P_1 + (1 - A_2)S_3P_3)]N$	$[(1 - A_2)S_2P_2 + (1 - A_4)S_4P_4)]N$

Table 5.11 Standard 2 × 2 Table: Observational Study

	Disease	Healthy
Exposed	$(A_1Q_1 + A_2Q_2)\mu_s N$	$(A_3Q_3 + A_4Q_4)\mu_s N$
Unexposed	$[(1 - A_1)Q_1 + (1 - A_3)Q_3)]\mu_s N$	$[(1 - A_2)Q_2 + (1 - A_4)Q_4)]\mu_s N$

Therefore, the empirical risk difference and risk ratio are

$$\text{RD}_{\text{emp}} = \frac{A_1S_1P_1 + A_2S_2P_2}{\sum_1^4 A_jS_jP_j} - \frac{(1-A_1)S_1P_1 + (1-A_3)S_3P_3}{\sum_1^4 A_jS_jP_j} \tag{5.23}$$

$$\text{RR}_{\text{emp}} = \frac{A_1S_1P_1 + A_2S_2P_2}{(1-A_1)S_1P_1 + (1-A_3)S_3P_3} \frac{\sum_1^4 (1-A_j)S_jP_j}{\sum_1^4 A_jS_jP_j} \tag{5.24}$$

These expressions seem too complex to offer much insight. But suppose that we rewrite Equations 5.21 and 5.22 in terms of the response-pattern distribution *in the study population*. Recall that the proportion of individuals with response pattern j in the study population is $Q_j = S_jP_j/\mu_s$. Furthermore, let us define the average assignment probability for individuals in the study population as $\mu_A = \sum_1^4 A_jQ_j$ Then Table 5.10 simplifies to Table 5.11. Furthermore, Equations 5.21 and 5.22 become

$$q_{WE} = \frac{A_1Q_1 + A_2Q_2}{\mu_A} \tag{5.25}$$

$$q_{ZU} = \frac{(1-A_1)Q_1 + (1-A_3)Q_3}{1-\mu_A} \tag{5.26}$$

We can rewrite Equations 5.23 and 5.24 as

$$\text{RD}_{\text{emp}} = \frac{A_1Q_1 + A_2Q_2}{\mu_A} - \frac{(1-A_1)Q_1 + (1-A_3)Q_3}{1-\mu_A} \tag{5.27}$$

$$\text{RR}_{\text{emp}} = \frac{A_1Q_1 + A_2Q_2}{(1-A_1)Q_1 + (1-A_3)Q_3} \frac{(1-\mu_A)}{\mu_A} \tag{5.28}$$

The implications of these relationships can be explored by setting up a spreadsheet that embodies Equations 5.25–5.28 and plugging in different values for the various parameters. However, we can also derive some general insight directly from the equations. Suppose that the absolute event rates are low, so that the proportion of *immune* subjects Q_4 is close to 1.0. What would happen if we were to move some subjects from the unexposed to the exposed group?

From Equation 5.25 we see that if we transfer a *doomed* or *causal* individual, the numerator will increase slightly. The denominator will also increase, but by a much smaller amount in proportion to its value, because of the assumption that Q_4 is large. As a result, the outcome event rate in the exposed group will increase slightly. Similarly, Equation 5.26 implies that the event rate in the unexposed group will *decrease* slightly if we move a *doomed* or *preventive* individual from the unexposed to the exposed group. As a result, the risk difference and risk ratio will tend to be biased upward by an overrepresentation of *doomed, causal,* or *preventive* individuals in the exposed group. Conversely, these measures will be biased downward by an underrepresentation of any of these.

At first impression, this result might seem paradoxical. If increasing the propensity for *causals* to become exposed increases the empirical effect, how can increasing this propensity for the *preventives* have the same effect? Let us consider the risk ratio. If more *causals* are assigned to the exposed group, then the numerator of RR_{emp} increases. If more *preventives* are exposed, the denominator decreases. The net result is an increased value of RR_{emp}, but the reasons are different. For the *doomed* individuals, both of these mechanisms apply. Increasing the value of A_1 tends to increase the numerator and decrease the denominator. Thus, there is a "double impact" on the empirical effect when relatively more *doomed* individuals are exposed. In this sense, the allocation of *doomed* subjects is a more powerful determinant of the empirical effect than the allocation of either *causals* or *preventives*.

Equations 5.27 and 5.28 can sometimes provide qualitative insight by focusing attention on exactly how bias might be induced in a given situation. For example, in a particular context we might ask whether there is any reason to suspect that *doomed* individuals are especially likely to become exposed. Are they more likely to volunteer for treatment or to seek medical care than, say, an *immune* person? Such questions can sometimes help in properly interpreting a study's findings.

It is important to remember that the previous discussion is predicated on the assumption of low absolute event rates (i.e., $Q_4 \cong 1.0$). If the absolute rates are increased, the results of shifting different response types can vary, as can be verified by trying different values for the parameters. For example, suppose that the proportion of *doomed* is similar to the proportion of *immune*. Then it is possible that moving a *causal* individual from unexposed to exposed would increase the risk ratio, but moving a *preventive* would decrease it.

Finally, we consider a special case of particular interest; assume that the event is rare and that the exposed and unexposed groups are of equal size, meaning $\mu_A = 0.5$. This situation is common for many RCTs and quasi-experiments. In that case Equations 5.25–5.28 become

$$q_{WE} = 2[A_1 Q_1 + A_2 Q_2] \tag{5.29}$$

$$q_{ZU} = 2[(1 - A_1)Q_1 + (1 - A_3)Q_3] \tag{5.30}$$

$$RD_{emp} = 2[(2A_1 - 1)Q_1 + A_2 Q_2 - (1 - A_3)Q_3] \tag{5.31}$$

$$RR_{emp} = \frac{A_1 Q_1 + A_2 Q_2}{(1 - A_1)Q_1 + (1 - A_3)Q_3} \tag{5.32}$$

Suppose that there are no *preventives*, so that $Q_3 = 0$. Then the causal effect must be positive, because $q_{WE} > q_{ZU}$ according to Equations 5.17 and 5.18. However, from Equations 5.29 and 5.30 we can see that for small enough values of A_1 the direction of the effect can be reversed. For example, suppose that $Q_1 = Q_2 = 0.5$; then $RR_{caus} = 2.0$. If the value of A_1 decreases, while A_4 increases slightly to keep the overall assignment probability at $\mu_A = 0.5$, then the value of RR_{emp} will decrease. If A_1 becomes sufficiently reduced, then this risk ratio will fall below 1.0. In general, if a low enough proportion of *doomed* subjects are exposed, the impact of the resulting selection bias can nullify, or even reverse, the causal effect.

5.6 TRADITIONAL SELECTION MODEL: NUMERICAL OUTCOME

There exists an extensive literature on statistical models to study causal effects when the outcome is on a numerical scale. Much of the methodological development has been generated by econometricians and other quantitative social scientists, who are typically concerned with causal relationships among continuous variables. In biomedical research, numerical outcomes are not uncommon, although event data are more prominent because occurrence of a disease or adverse event is often of primary interest. When an outcome is numerical, analyses usually entail the *general linear model* (GLM) and variables are often assumed to have *normal* distributions unless this assumption is clearly implausible. *Linear regression* models are used to estimate the causal effect of an independent (or exogenous) variable on a dependent (or endogenous) outcome variable. Furthermore, the effect of interest is usually assumed to be a constant. The *structural equation modeling* (SEM) approach can extend the linear modeling approach to more complex situations involving causal interrelationships among several variables. The rather restrictive mathematical assumptions and models within this framework allow the derivation of valuable insights, but also limit the generality of this knowledge.

Selection bias in the context of comparative studies has been analyzed extensively within this standard paradigm. To understand the traditional approach, it is helpful to begin with the simple situation in which the exposure has no effect. The classical linear model to express this basic situation can be written as follows:

$$Y_{iE} = \mu + e_{iE} \tag{5.33}$$

$$Y_{iU} = \mu + e_{iU} \tag{5.34}$$

where Y_{iE} is the outcome value for individual i in the exposed group and Y_{iU} is the outcome for individual i in the unexposed group. (Note that we will use the more conventional Y instead of V to represent the outcome variable within the context of traditional statistical modeling.) The quantity μ is the mean value of the outcome in both study populations. The terms e_{iE} and e_{iU} are random "errors" that represent the deviations of individual values from μ. The mean value of these error terms is zero across the entire study population, but not necessarily across either of the study groups. Let μ_{eE} and μ_{eU} represent the mean values of the errors in the exposed and unexposed groups, respectively. The difference of group means can be expressed as

$$\mu_{YE} - \mu_{YU} = \mu_{eE} - \mu_{eU} \tag{5.35}$$

If the assignment of subjects to study groups is random, we will have $\mu_{eE} = \mu_{eU} = 0$. Thus, the usual difference of means would correctly estimate a zero effect. Otherwise, the estimate could be biased.

Now suppose the treatment has a constant additive effect α. The standard way to express this situation is as follows:

$$Y_i = \mu + \alpha F + e_i \tag{5.36}$$

The variable Y_i represents the outcome for individual i; and F has the value 1 if individual i is exposed, and 0 if i is not exposed. Note that if $\mu_{eE} = \mu_{eU} = 0$, Equation 5.36 is a classical linear regression model. Thus, α is simply the coefficient of F in this model and can be estimated straightforwardly by least-squares regression analysis. Therefore, in a randomized experiment or effective equivalent, we would have internal validity.

What about external validity? Within this framework, the study will have external validity if the value of α in the study population has the same value as in the target population. In general, assessing whether this condition holds is based on intuition, sometimes augmented by data on characteristics of the study and target populations. However, there is no explicit representation of the sampling process in the statistical model.

In an observational study, the group means may well be unequal ($\mu_{eE} \neq \mu_{eU}$). From a classical linear modeling perspective, the resulting problem is the correlation between the variables F and e. Therefore, straightforward least-squares regression will yield a biased estimate of the coefficient α in this model. Within the GLM framework, it is usual to conceptualize the difference between the values of μ_{eE} and μ_{eU} as attributable to differences in the mean values of certain covariates. For simplicity, suppose there is just one such covariate X. Then Equation 5.36 can be written

$$Y_i = \mu + \alpha F + \beta X_i + e_i \qquad (5.37)$$

If the variable(s) included in the model "explain" the differences between μ_{eE} and μ_{eU}, then the errors will be uncorrelated with F. Therefore, least-squares regression will produce an unbiased estimate of α. This statistical approach is sometimes known as the *analysis of covariance* (ANCOVA). The practical problem, of course, is to find a set of covariates for which the variables F and e become uncorrelated. ANCOVA is one of several approaches for dealing with selection bias. Most of these strategies entail more sophisticated mathematical models and various assumptions about the selection process.

Finally, we consider the issue of explicit selection. For concreteness, we will discuss the well-known problem of *truncated data*. Table 5.12 represents a hypothetical set of data for the SAT tutoring study. For simplicity, assume that students have been randomly assigned to the two study groups. The mean outcome score for students in the Old Program is 591 and the mean for the New Program is 641. Suppose, however, that a malicious gnome sneaks into the tutoring center and somehow manages to remove any posttest with a score above 700. Then the scores of three students in the New Program group will be missing, but only one of the Old Program scores will be excluded. The results are displayed in Table 5.13. As a result, the mean value in the New Program group will be reduced by 44 points, but the mean value in the Old Program group by only 18 points. The observed effect is reduced to 24 points from the 50-point effect in the complete data.

Table 5.12   SAT Tutoring Results

New Program Group	Old Program Group
800	750
720	670
710	660
680	630
660	610
640	590
590	540
570	520
560	510
480	430
New-group mean:	*Old-group mean:*
641	*591*

Table 5.13   Truncated SAT Tutoring Results

New Program Group	Old Program Group
800	750
720	670
710	660
680	630
660	610
640	590
590	540
570	520
560	510
480	430
New-group mean:	*Old-group mean:*
597	*573*

From the traditional perspective, each score in the Old Program group results from a "random error" around the "true" Old Program mean. Each score in the New Program group results from a random error around the true New Program mean. In the full data set, the error distributions for the two groups are identical. However, in the truncated data, relatively more large positive errors are excluded in the New Program group than in the Old Program group. As a result, the observed group means move closer together. The amount of "attenuation" of the estimated effect depends on the probability distribution of the errors. Certain methods to correct for explicit selection bias are based on particular assumptions about the errors, for example that the errors are normally distributed.

5.7   CAUSAL SELECTION MODEL: NUMERICAL OUTCOME

In the traditional statistical model, the outcome values are regarded as random variables. The mean value of a population is the main object of scientific interest, and the difference between study group means is the primary measure of effect. Individual outcome values are viewed as essentially indistinguishable random observations from a hypothetical infinite population. Even if some recognition is given to the idea that effects can vary across individuals, the average effect (ACE) is nearly always the focus of attention.

As explained in the previous section, statistical methods based primarily on the GLM are often used to adjust for potential bias. When the assumptions underlying these techniques hold, bias can be completely eliminated. However, when residual bias remains, the traditional model offers little guidance to assess its magnitude and direction. From a causal perspective, assessment of potential bias requires speculation about the nature of deviations from exchangeability. Gaining intuition (or relevant data) about this critical issue is very difficult without explicit reference to potential outcomes.

The counterfactual perspective conceptualizes the individual potential outcomes V_{iE} and V_{iU} not as random deviations around a mean, but as the primary reality of interest. Understanding as much as possible about the response patterns for individuals, or at least classes of them, is the ultimate objective. The practical problem, of course, is that we can only estimate certain aggregate causal effects, such as the ACE. These aggregate effects can be of limited utility as applied to specific individuals. For example, we might know that a certain drug works much better for individuals who are at higher baseline risk. Then the overall average effect might be highly misleading for a person who is at very high risk. In general, our ideal objective is to find the best estimate of the effect for each individual, taking all of the available information about him or her into account.

As in the case of a dichotomous outcome, we conceptualize the target population as a distribution of response patterns. However, instead of only four possible response patterns, we imagine a large (but finite) number of distinct patterns. (We could also consider the case in which the outcome is a continuous variable. The essential logic would be identical, but more advanced mathematical methods would be necessary.) In the following discussion, it will be convenient to modify our notation and terminology somewhat. Let J be the number of distinct response patterns and let P_j be the proportion of individuals with the jth possible response pattern ($j = 1,2,3,...,J$). We will denote the jth response pattern as (V_{jE}, V_{jU}). Here V_{jE} and V_{jU} represent the (common) potential outcomes for all individuals with response pattern j in the target population. We will say that a particular individual is of Type j if she has response pattern j.

For any Type-j person, we define the individual causal effect (ICE) to be the arithmetic difference between her potential outcome values under the exposed and unexposed conditions. The ICE for a Type-j individual can be expressed as

$$\Delta_j = V_{jE} - V_{jU} \tag{5.38}$$

Furthermore, let Δ represent the random variable that assumes the value Δ_j for all Type-j individuals in the target population. Then the ACE for the target population is the average of the ICEs. This target ACE can be expressed as

$$\text{ACE} = \sum_j P_j \Delta_j = \sum_j P_j V_{jE} - \sum_j P_j V_{jU} = \mu_{VE} - \mu_{VU} \tag{5.39}$$

In an ideal study (random sampling and assignment) μ_{WE} and μ_{ZU} in the study population would equal (apart from sampling variation) μ_{VE} and μ_{VU}, respectively. Therefore, the empirical effect $\mu_{WE} - \mu_{ZU}$ would be an unbiased estimate of the ACE.

Randomized Experiment

In Section 5.4, we described the selection process for a randomized experiment when the outcome is dichotomous. We defined sampling probabilities corresponding to each of the four response patterns. We can extend this idea by defining a sampling probability for each possible response pattern in the target population. For each response pattern j in the target population, let S_j be the associated sampling probability. These S_j values define a variable S over the target population. Let μ_S be the mean of S in the target population:

$$\mu_S = \sum_j S_j P_j \tag{5.40}$$

Furthermore, the proportion of Type-j individuals in the study population is $Q_j = S_j P_j / \mu_S$. So, the ACE in the study population can be represented as

$$\text{ACE}_{\text{study}} = \sum_j \Delta_j Q_j = \frac{\sum_j \Delta_j S_j P_j}{\mu_S} \tag{5.41}$$

We can reexpress Equation 5.41 in a form that provides additional insight:

$$\text{ACE}_{\text{study}} = \text{ACE}_{\text{target}} + \frac{\text{Cov}_t(S, \Delta)}{\mu_S} \tag{5.42}$$

The proof of this result is presented in the Appendix to this chapter. The subscript t on the covariance in Equation 5.42 indicates that the distributions of S and Δ are over the target population. Intuitively, Equation 5.42 means that the ACE is biased if and only if there exists a statistical association between the sampling probability and the ICE. For example, if response patterns with large positive values of Δ tend to have relatively high sampling probabilities, the bias will be positive.

The covariance is a measure of statistical association that is a scaled version of the correlation coefficient. Technically, the covariance equals the correlation times the product of the standard deviations for Δ and S. Therefore, if either Δ is a constant or S is a constant, there can be no bias. In particular, there will be no bias under random sampling, because S would be constant.

Let Bias_S be the bias attributable to (nonrandom) sampling from the target population. Then from Equation 5.42 we have

$$\text{Bias}_S = \frac{\text{Cov}_t(S, \Delta)}{\mu_S} \tag{5.43}$$

Note that Bias_S represents the difference between the ACE in the study population and the ACE in the target population. Sampling that favors individuals who have relatively high (or low) ICEs produces an estimate that lacks external validity. However, the empirical effect in an RCT is an unbiased estimate of the causal effect in the study population. So, a randomized study still has internal validity. In an observational study, there is typically bias attributable to nonrandom assignment in addition to Bias$_S$.

Observational Study

Suppose we wish to estimate the ACE based on an observational study. We define A_j to be the assignment probability for a Type-j individual, conditional on being in the study. We show in the Appendix that, for the exposed group, the expected value μ_{WE} can be expressed as

$$\mu_{WE} = \mu_{VE} + \frac{\text{Cov}_s(A, V_E)}{\mu_A} \tag{5.44}$$

Here μ_A is the mean value of A *in the study population*; $\text{Cov}_s(A,V_E)$ is the covariance between A and V_E in the study population.

We see from Equation 5.44 that μ_{WE} is an overestimate of μ_{VE} if the assignment variable A is correlated positively with the potential-outcome values under exposure. It is an underestimate if this correlation is negative. Similarly, the expected value μ_{ZU} in the unexposed group is

$$\mu_{ZU} = \mu_{VU} + \frac{\text{Cov}_s(1 - A, V_U)}{1 - \mu_A} \tag{5.45}$$

To gain additional insight, let us define M_j to be the *mean individual response* for a Type-j individual:

$$M_j = \frac{V_{jE} + V_{jU}}{2} \tag{5.46}$$

We show in the Appendix that Equation 5.46 can be expressed as

$$\mu_{WE} = \mu_{VE} + \frac{\text{Cov}_s(A, M) + (1/2)\text{Cov}_s(A, \Delta)}{\mu_A} \tag{5.47}$$

According to Equation 5.47, the overestimate (or underestimate) of μ_{WE} has two components. One component depends on the covariance between A and the mean response M. If assignment to the exposed group favors subjects with relatively high mean individual responses, the exposed group mean will be inflated. The second component depends on the correlation between A and the ICE. If assignment to exposure favors subjects with relatively large ICEs, the exposed group mean will be inflated.

A similar expression can be derived for the unexposed group:

$$\mu_{ZU} = \mu_{VU} + \frac{\text{Cov}_s(1 - A, M) + (1/2)\text{Cov}_s(1 - A, \Delta)}{1 - \mu_A} \tag{5.48}$$

The net assignment bias for the difference between μ_{WE} and μ_{ZU} can be obtained from Equations 5.47 and 5.48 as Bias_A:

$$\text{Bias}_A = \left[\frac{\text{Cov}_s(A, V_E)}{\mu_A} - \frac{\text{Cov}_s(1-A, V_U)}{1-\mu_A} \right] \qquad (5.49)$$

In the Appendix, we demonstrate that after some algebraic manipulation using Equations 5.47 and 5.48 we can represent Equation 5.49 as

$$\text{Bias}_A = \frac{1}{\mu_A(1-\mu_A)} \left[\text{Cov}_s(A, M) + \frac{1}{2}(1-2\mu_A)\text{Cov}_s(A, \Delta) \right] \qquad (5.50)$$

This expression simplifies further in three situations of practical interest:

- The individual effect is uniform.
- The exposure is infrequent.
- The study groups are of equal size.

If the effect is uniform, then the value of Δ is a constant, so Equation 5.50 reduces to

$$\text{Bias}_A = \frac{1}{\mu_A(1-\mu_A)}[\text{Cov}_s(A, M)] \qquad (5.51)$$

The assignment bias depends only on the relationship between A and M. If the effect is variable, then the value of the average assignment probability μ_A plays a role. If exposure is relatively rare, as occurs in many epidemiologic studies, μ_A will be small, so Equation 5.55 reduces approximately to

$$\text{Bias}_A = \frac{1}{\mu_A} \left[\text{Cov}_s(A, M) + \frac{1}{2}\text{Cov}_s(A, \Delta) \right] \qquad (5.52)$$

In this case, the covariance between A and Δ can have a significant impact on the magnitude of assignment bias. On the other hand, if the study groups are equal in size ($A = 0.5$), then Equation 5.50 would reduce to

$$\text{Bias}_A = 4\text{Cov}_s(A, M) \qquad (5.53)$$

According to Equation 5.53, with study groups the same size, the bias of the ACE estimate does not depend on the allocation of the ICEs in the study population! The distributions of Δ in the two study groups are relevant only to the extent that the group sizes are unequal. In other words, whether the more "causal" individuals or the more "preventive" individuals tend to become exposed has no bearing on the bias of the ACE. To understand how such a counterintuitive result arises, let us consider a simple illustration.

Suppose first that we have two exchangeable study groups of unequal size; the exposed group contains 49 individuals, and the unexposed group contains 201. Assume that the average outcome value is 60 in each of the groups, so the true causal effect is zero. That is, the average value of Δ in each group must be zero. Now imagine that there is an individual in the unexposed group with a response pattern of (80, 40). Suppose that, prior to the study, we decide to move this person from the unexposed to the exposed group. What happens to the bias? In this case, the average outcome value in the exposed group would now increase from 60 to 60.4. The average in the unexposed group would also increase, but from 60 to 60.1. Therefore, a bias of 0.3 would result.

Let us alter this scenario slightly by changing the group sizes. Suppose that the exposed group contained 99 subjects, and the unexposed contained 101. After the transaction, both groups have 100 subjects. Now, however, the average outcome in each group becomes 60.2, and no bias results. Note that the bias equals zero only because the resulting group sizes are equal. Furthermore, the absence of bias does not mean that the hypothetical transfer would have no impact at all. Rather, the means in both groups would change, but by the same amount (0.2 in our example). The fact reflected in Equation 5.53 that the bias depends only on the distributions of M in the two groups, but not on the distributions of Δ, is specific to the choice of the ACE as the measure of effect.

Finally, we note that Bias_A is the assignment bias relative to the causal effect $\text{ACE}_{\text{study}}$ in the study population. However, $\text{ACE}_{\text{study}}$ can still suffer from sampling bias, namely the Bias_S specified in Equation 5.43. Therefore, with respect to the causal effect $\text{ACE}_{\text{target}}$ for the target population, the total bias is given by

$$\text{Total Bias} = \text{Bias}_S + \text{Bias}_A \qquad (5.54)$$

where Bias_S can be interpreted as the "external bias" component of the study and Bias_A as the "internal bias" component.

Selection on the Basis of the Outcome Value

Finally, we consider selection that depends on the value of the outcome. We mentioned in Section 5.1 that such selection is called explicit if retention in the sample is determined directly by the outcome, but incidental if the *probability* of being retained is a function of the outcome. We begin by assuming that the study would be ideal (random sampling and assignment) in the absence of the explicit or incidental selection. For an individual with response pattern j we define R_{jE} to be the probability of being *retained* in the study if she is exposed. Similarly, R_{jC} is the probability of being retained in the study if she is not exposed

In the Appendix, we show that

$$\mu_{WER} = \mu_{WE} + \frac{\text{Cov}_E(R_E, W_E)}{\mu_{RE}} \tag{5.55}$$

$$\mu_{ZUR} = \mu_{ZU} + \frac{\text{Cov}_U(R_U, Z_U)}{\mu_{RU}} \tag{5.56}$$

Here μ_{RE} is the mean value of R_E and $\text{Cov}_E(R_E, W_E)$ the covariance between R_E and W_E *in the exposed group*. Similarly, μ_{RU} is the mean value of R_U and $\text{Cov}_U(R_U, Z_U)$ the covariance between R_U and V_U *in the unexposed group*. Furthermore, μ_{WER} is the mean outcome in the exposed group after explicit (or incidental) selection, and μ_{ZUR} is the corresponding mean outcome in the unexposed group. From Equations 5.55 and 5.56, it follows that

$$\mu_{WER} - \mu_{ZUR} = \mu_{WE} - \mu_{ZU} + \frac{\text{Cov}_E(R_E, W_E)}{\mu_{RE}} - \frac{\text{Cov}_U(R_U, Z_U)}{\mu_{RC}} \tag{5.57}$$

From Equation 5.57 we see that the mean outcome in the exposed group is increased if subjects with relatively high outcome values are more likely to be retained. Similarly, the mean outcome in the unexposed group will increase if subjects with relatively high outcome values tend to be retained. Conversely, these means will both decrease if individuals with relatively low values are retained. The net bias in the empirical effect is the difference between the changes in the two groups. We designate this component of bias as retention Bias$_R$:

$$\text{Bias}_R = \frac{\text{Cov}_E(R_E, W_E)}{\mu_{RE}} - \frac{\text{Cov}_U(R_U, Z_U)}{\mu_{RU}} \tag{5.58}$$

The direction and magnitude of Bias$_R$ depend on the nature of the "retention processes" in each of the study groups.

As an illustration, consider again the situation in which there is truncation of observations with values above some cutoff. Assume that the study groups are exchangeable and the ICE is a positive constant α. In the exposed group, the value of R_{jE} would be 1 for values of W_{jE} up to the cutoff, and 0 for values above this cutoff. A similar situation would prevail for the unexposed group. In this case, the retention probabilities would be either 0 or 1, and the higher value (1) would be associated with higher values of the outcome. As a result, the covariance in each group would be negative. The values of the two covariances in Equation 5.58 would depend on the distributions of W_E and Z_U. Because we are assuming a constant positive effect, there would be more subjects who exceed the cutoff ($R_{jE} = 0$) in the exposed group than in the unexposed group. Consequently, the covariance in the exposed group would be lower (more negative) than the covariance in the unexposed group, resulting in a negative bias.

It follows from Equation 5.57 that Bias_R is the additional bias, above sampling and assignment bias, that may affect the value of the empirical effect $\mu_{WE}-\mu_{ZU}$. Therefore, we have the following general result:

$$\text{Total Bias} = \text{Bias}_S + \text{Bias}_A + \text{Bias}_R \qquad (5.59)$$

This equation summarizes the three levels of distortion that can affect a comparative study. The fact that the total bias is a simple sum of three components derives from the special mathematical properties of the ACE as a measure of effect. For another measure, such as the difference of medians, this simple additive breakdown would not be possible. Nonetheless, we can still regard the total selection bias conceptually as being composed of three components whose "sum" constitutes the total bias.

GUIDEPOST 5

This chapter has developed a conceptual and mathematical framework for various issues that have traditionally fallen under the heading of selection bias. We have identified three conceptually separate levels of bias that pertain to the selection of subjects for inclusion: nonrandom sampling, nonrandom assignment, and explicit selection. Simple mathematical models were developed to clarify the nature of the bias resulting from these sources, when the outcome is dichotomous and when it is numerical. These models can provide general insight and serve as a foundation for methods to avoid or adjust for bias.

Our discussion in this chapter has focused on the underlying structure of selection processes. As a practical matter, the assignment mechanism is typically the primary concern, except in those special situations in which randomization is feasible. We emphasized exchangeability as the key property that ensures unbiased estimation, and we showed that the bias depends on the extent to which certain types of individuals are preferentially assigned to the exposed group. In the simple case of a dichotomous outcome, the magnitude of bias is determined primarily by the assignment probability for *doomed* individuals and secondarily by the assignment probability for *causal* and *preventive* individuals.

Lack of exchangeability is one way to answer the question of when a valid comparison is impossible. However, traditionally, most methodologists have framed the issue as one of differences between the study groups in terms of particular factors. These covariates, or confounding factors, are imagined to be responsible for that part of the observed outcome difference that is not attributable to the exposure. However, the way in which these confounders operate to create this bias, and can be analyzed to correct for the bias, has been a vexed issue. In Chapter 6, we will explain the various ways in which confounding has been interpreted. In Chapter 7, we will relate these traditional interpretations of confounding to the causal perspective adopted in this book. We will argue that confounding is not a separate *type* of bias distinct from selection bias (i.e., nonrandom assignment). Rather, confounding describes the same phenomenon, but with more explicit attention to factors that explain why the study groups are not exchangeable.

APPENDIX

This Appendix provides a derivation of the main results presented in Section 5.7 in the form of a theorem and corollary.

Theorem. The total bias in estimation of the average causal effect in a comparative study can be expressed as the sum of three components:

$$\text{Total Bias} = \text{Bias}_S + \text{Bias}_A + \text{Bias}_R$$

The three components are defined as follows:

$$\text{Bias}_S = \frac{\text{Cov}_t(S, \Delta)}{\mu_S}$$

$$\text{Bias}_A = \left[\frac{\text{Cov}_s(A, V_E)}{\mu_A} - \frac{\text{Cov}_s(1-A, V_U)}{1-\mu_A} \right]$$

$$\text{Bias}_R = \left[\frac{\text{Cov}_E(R_E, W_E)}{\mu_{RE}} - \frac{\text{Cov}_U(R_U, Z_U)}{\mu_{RU}} \right]$$

Corollary. Bias_A can be expressed in the form

$$\text{Bias}_A = \frac{1}{\mu_A(1-\mu_A)} \left[\text{Cov}_s(A, M) + \frac{1}{2}(1-2\mu_A)\text{Cov}_s(A, \Delta) \right]$$

Proof of Theorem. In the study population, the proportion of Type-*j* individuals is given by Q_j:

$$Q_j = \frac{P_j S_j}{\sum_j P_j S_j} = \frac{P_j S_j}{\mu_S} \tag{A.5.1}$$

It follows that

$$\text{ACE}_{\text{study}} = \sum_j Q_j \Delta_j = \frac{1}{\mu_S} \sum_j P_j S_j \Delta_j = \frac{E(S\Delta)}{\mu_S} \tag{A.5.2}$$

From the definition of covariance

$$E(S\Delta) = \text{Cov}_t(S, \Delta) + \mu_S \text{ACE}_{\text{target}} \qquad (\text{A.5.3})$$

Following from Equations A.5.2 and A.5.3 is Eq. A.5.4:

$$\text{ACE}_{\text{study}} = \text{ACE}_{\text{target}} + \frac{\text{Cov}_t(S, \Delta)}{\mu_S} \qquad (\text{A.5.4})$$

We define Q_{jE} to be the proportion of Type-j subjects in the exposed group and define Q_{jU} to be the proportion of Type-j subjects among the unexposed:

$$Q_{jE} = \frac{Q_j A_j}{\sum_j Q_j A_j} = \frac{Q_j A_j}{\mu_A} \qquad (\text{A.5.5})$$

Then the mean value μ_{WE} in the exposed group is given by

$$\mu_{WE} = \sum_j Q_{jE} V_{jE} = \frac{1}{\mu_A} \sum_j Q_j A_j V_{jE} = \frac{E(AV_E)}{\mu_A} \qquad (\text{A.5.6})$$

Here the expectation $E(AV_E)$ is over the distribution of AV_E in the study population. From the definition of the covariance and Eq. A.5.6, we then obtain

$$\mu_{WE} = \mu_{VE} + \frac{\text{Cov}_s(A, V_E)}{\mu_A} \qquad (\text{A.5.7})$$

For the unexposed group the probability of assignment is $1 - A_j$ for a Type-j individual. Therefore, we have

$$Q_{jU} = \frac{Q_j(1 - A_j)}{\sum_j Q_j(1 - A_j)} = \frac{Q_j(1 - A_j)}{1 - \mu_A} \qquad (\text{A.5.8})$$

Therefore, the mean value μ_{ZU} in the unexposed group is given by

$$\mu_{ZU} = \sum_j Q_{jU} V_{jU} = \frac{1}{(1 - \mu_A)} \sum_j Q_j(1 - A_j) V_{jU} \qquad (\text{A.5.9})$$
$$= \frac{E[(1 - A)V_U]}{1 - \mu_A}$$

It follows that

$$\mu_{ZU} = \mu_{VU} + \frac{\text{Cov}_s(1 - A, V_U)}{1 - \mu_A} \qquad (\text{A.5.10})$$

From Equations A.5.7 and A.5.10 we obtain Eq. A.5.11:

$$\mu_{WE} - \mu_{ZU} = \mathrm{ACE}_{\mathrm{study}} + \frac{\mathrm{Cov}_s(A, V_E)}{\mu_A} - \frac{\mathrm{Cov}_s(1 - A, V_U)}{1 - \mu_A} \quad (A.5.11)$$

Now we define Q_{jER} and Q_{jUR} to be the proportions of Type-j subjects among the exposed and unexposed, respectively, *after selection based on the outcome values*:

$$Q_{jER} = \frac{Q_{jE} R_{jE}}{\sum_j Q_{jE} R_{jE}} = \frac{Q_{jE} R_{jE}}{\mu_{RE}} \quad (A.5.12)$$

$$Q_{jUR} = \frac{Q_{jU} R_{jU}}{\sum_j Q_{jU} R_{jU}} = \frac{Q_{jU} R_{jU}}{\mu_{RU}} \quad (A.5.13)$$

Then the mean value μ_{WER} of the outcomes for those *remaining* in the treatment group is given by

$$\mu_{WER} = \sum_j Q_{jER} W_{jE} = \frac{1}{\mu_{RE}} \sum_j Q_{jE} R_{jE} W_{jE} = \frac{E(R_E W_E)}{\mu_{RE}} \quad (A.5.14)$$

Here the expectation $E(R_E W_E)$ is over the distribution of $R_E W_E$ in the exposed group. It follows that

$$\mu_{WER} = \mu_{WE} + \frac{\mathrm{Cov}_E(R_E, W_E)}{\mu_{RE}} \quad (A.5.15)$$

Similarly, for the mean μ_{ZUR} in the unexposed group, we obtain

$$\mu_{ZUR} = \mu_{ZU} + \frac{\mathrm{Cov}_U(R_U, Z_U)}{\mu_{RU}} \quad (A.5.16)$$

It follows from Equations A.5.15 and A.5.16 that

$$\mu_{WER} - \mu_{ZUR} = \mu_{WE} - \mu_{ZU} + \frac{\mathrm{Cov}_E(R_E, W_E)}{\mu_{RE}} - \frac{\mathrm{Cov}_U(R_U, Z_U)}{\mu_{RU}} \quad (A.5.17)$$

Proof of the Theorem follows directly from Equations A.5.4, A.5.11, and A.5.17. $\square$

Proof of Corollary. We note that

$$V_{jE} = M_j + \frac{\Delta_j}{2} \quad (A.5.18)$$

$$V_{jU} = M_j - \frac{\Delta_j}{2} \qquad \text{(A.5.19)}$$

Substituting these expressions for V_{jE} and V_{jU} into Eq. A.5.11 yields

$$\text{Bias}_A = \frac{\text{Cov}(A, M + \Delta/2)}{\mu_A} - \frac{\text{Cov}(1 - A, M - \Delta/2)}{1 - \mu_A} \qquad \text{(A.5.20)}$$

Applying the basic probability rules that govern covariances then yields

$$\text{Bias}_A = \frac{\text{Cov}(A, M)}{\mu_A} + \frac{\text{Cov}(A, \Delta)}{2\mu_A} + \frac{\text{Cov}(A, M)}{1 - \mu_A} - \frac{\text{Cov}(A, \Delta)}{2(1 - \mu_A)} \qquad \text{(A.5.21)}$$

The Corollary is obtained by placing each of the four terms over a common denominator and consolidating terms. $\quad\square$

CHAPTER 6

Confounding: An Enigma?

The concept of statistical confounding is understood intuitively by most practicing epidemiologists and social scientists. However, attempts to define confounding explicitly have encountered numerous difficulties. In Chapter 5, we explained how selection bias can occur when the study groups are formed in a manner that does not entail randomization. Viewed as a problem of selection, this noncomparability of the groups leads naturally to consideration of the assignment mechanism. In the current chapter, the differences between study groups are seen in a somewhat different light. Rather than focusing on the assignment mechanism, we ask *how* the individuals differ across these groups.

Consider the observational studies that appeared to show a benefit of hormone replacement therapy (HRT) on coronary heart disease (CHD). The women who opted for HRT were thought by many to be different in certain important respects from those who did not. Generally, the HRT-users appeared somewhat healthier and more "health-conscious" than the nonusers. Consequently, the apparent reduction of CHD risk may have been caused by one or more preexisting health-related factors rather than to HRT. That is, the effects of HRT and these other factors could have been confounded.

In many situations like the HRT studies, biases can be ascribed *either* to selection or to confounding. The choice is to some extent a matter of preference and convention. Indeed, many methodologists, especially in the social sciences, nearly always describe bias resulting from nonequivalent control groups as a form of selection bias. Others, primarily in the biomedical sciences, usually refer to confounding. Some epidemiologists also distinguish *between* confounding and bias. They argue that, although confounding distorts a causal effect, it is not inherently a flaw in the research design (e.g., Gordis, 2004; Szklo and Nieto, 2007). Furthermore, confounding can be said to differ from selection bias because the confounding factors exist in the world and are not produced by the research design. From this perspective, methodological error (and hence bias) occurs only if the researchers fail to adjust properly for the (preexisting) confounding. Throughout this book, we regard any discrepancy between the empirical effect and the causal effect as a type of bias. Whether such a discrepancy is the "fault" of the investigators does not seem to be an essential aspect of "bias" and may be very difficult to determine.

6.1 WHAT IS THE REAL PROBLEM?

Various proposed formulations seem incomplete or ambiguous when subjected to rigorous critical examination. Consequently, there is much confusion about the true nature of this important problem. The essence of confounding has been framed in a variety of different ways. Most discussions of confounding bias highlight one or more of the following aspects of comparative studies:

1. Distortion of the observed statistical association by *extraneous causes*.
2. Inadequate *statistical control* for confounding variables.
3. Imperfect *comparability* between the study groups.
4. Nonrandom *assignment* of individuals to the study groups.
5. Incorrect *specification* of a structural model.

The intuition behind the term "confounding" is related to all of these ideas. Each can be regarded as a different window on what is essentially the same reality.

> Any conflict is apparent rather than real, since the word "confounding" may be defined any way a user wishes. Some definitions will be useful, and some will be sterile in the usual research context. (Grayson, 1987, 546)

In this chapter, we do not endorse one or another of the definitions of confounding, or attempt to offer a single "correct" definition of this term. The aim is rather to explain the ideas underlying these various attempts to define confounding. In the following chapter, we present a perspective on confounding grounded in the causal ideas and models developed earlier in this book. Our hope is to approach the underlying issue in a way that reconciles apparently conflicting points of view.

6.2 CONFOUNDING AND EXTRANEOUS CAUSES

Perhaps the oldest and most widespread view of confounding is based on the idea of extraneous causal influences. This "classical" perspective assumes that the observed relationship between exposure and outcome is the sum of various causal forces acting in concert. The goal of research is to isolate the effect of one particular factor from the effects of other extraneous factors that may be operating concurrently. Unfortunately, traditional statistical theory does not explicitly define what is meant by a causal factor. Most discussions of confounding seem to assume an intuitive understanding of causation that can suffice for practical purposes.

Suppose that a classically trained epidemiologist or social scientist is trying to make sense of data from an observational study. For example, the observational studies of HRT found a strong and consistent benefit (RR < 1.0) with respect to CHD. Let us assume that any sources of selection bias or information bias can be safely ignored. Furthermore, assume that the observed effect achieves a *p*-value of 0.0001. Clearly, it is important to know whether this "real" effect was "caused" by HRT or by something else. On the one hand, a correlation between HRT and CHD does not necessarily reflect causation. On the other hand, an epidemiologist might believe that the data do in fact justify a causal interpretation. One way to argue for this position would be to consider alternative possible explanations for the observed effect. If all plausible alternatives can be ruled out, we can assert that the observed effect must indeed be causal.

This approach assumes that an observed effect would have to be causal *but for* some other explanatory factors that generate a "spurious correlation" between the exposure and the outcome. After all, *something* must have caused the effect. If not the exposure under study, then what other factors might be responsible? If we can identify and rule out or adjust for the effects of all such possible causes, then the "remainder" must be the true causal effect of the exposure.

An extraneous cause may or may not actually engender bias in a given study. If it does lead to bias, then we call the extraneous cause a *confounding factor*, a *confounding variable*, or simply a *confounder*. Methodologists have struggled for decades with how to explain precisely what conditions are necessary to declare that a particular variable is in fact a confounder. These efforts have produced a working definition that (with some variations) is widely accepted as a rough guide for researchers but remains somewhat ambiguous under careful scrutiny. A typical formulation of the classical definition of a confounding factor is given in a recent textbook (Szklo and Nieto, 2007, 154):

> *The confounding variable is causally associated with the outcome*
> *and*
> *noncausally or causally associated with the exposure*
> *but*
> *is not an intermediate variable in the causal pathway between exposure and outcome.*

This statement has three components. First, a confounding factor must be a cause of the outcome. Second, this variable must have a statistical association, though not necessarily causal, with the exposure. Third, the causal influence exerted by the confounder on the outcome must not be itself caused by the exposure.

Causal Association between the Confounder and the Outcome

The traditional perspective offers no clear theoretical definition of causality. In practice, many data analysts rely in part on whether a statistical association (e.g., odds ratio or correlation) between the confounder and the outcome exists. This association might be found in the study at hand, or possibly from an external source of data. However, such an association may not be causal; it can be affected by selection bias or information bias, or itself be confounded by other unknown factors. Thus, there is no way conceptually, let alone empirically, to determine whether an association between a potential confounder and the outcome is causal in nature.

Furthermore, reliance on a purely statistical association between a potential confounder and the outcome to suggest a causal connection entails another complication. Imagine a situation in which a potential confounding variable X has no causal effect on the outcome Y, but the risk factor F and X are highly correlated. The causal effect of F could induce a statistical association between X and Y. In the absence of outside knowledge to the contrary, X would appear to be a confounding factor. To avoid such situations, epidemiologists sometimes add the further qualification that a statistical relationship between X and Y should exist *conditional on the exposure state*.

Another conceptual issue pertains to the role of a possible *proxy*, or *marker*, for a "true" confounder. For example, suppose that a model of child development on some cognitive dimension of interest involves several psychological and social factors that are correlated with age. Then the child's age may be a useful "confounder" in a study to compare alternative programs that aim to enhance the developmental process. But the pure passage of time that defines a person's chronological age is often a marker for various underlying maturational processes that may be difficult to observe or even specify. Therefore, age per se may not exert a causal effect, but may still be considered a confounding variable for purposes of statistical adjustment.

Furthermore, it is possible that the "true" causal variable is measured with error. In that case, the value of the measured variable does not accurately indicate the corresponding value of the actual causal factor. Because the nature of causation implied by the classical definition of confounding is not explicit, even for a perfectly measured variable, what criterion could we possibly employ to judge whether an imperfectly measured covariate exerts a causal effect on the outcome?

Association between the Confounder and Exposure

Suppose that a purported confounder displays no statistical association with the outcome of interest. For example, in an observational study of HRT, it is possible that HRT users and nonusers might have very similar distributions of "health consciousness." It may seem obvious that even if health consciousness can strongly affect CHD, the study's results would not necessarily be biased in any way. So, an association between the potential confounding factor and the outcome would appear to be a necessary condition for confounding to occur. However, this statement requires some qualification.

Suppose there are several variables that can lay claim to being confounders. Indeed, there is virtually no limit to the number of such variables, or to the complexity of intercorrelations among them. This entire set of variables may comprise a single multivariate confounder, which would have to satisfy the condition of having different distributions in the two study groups. However, this condition need not necessarily apply to any one of the constituent variables. Whether an individual variable can be considered a possible confounder would depend on its conditional distribution. After conditioning on all of the other potential confounders, is there still a difference in the variable's distribution across the groups? If so, it might be a confounding factor. So, the property of being a confounder depends on the other confounders that are included in the multivariate confounder.

For concreteness, consider the death penalty study described in Chapter 1. Recall that the outcome for any defendant was whether or not a death sentence was imposed. The "exposure" in this situation was whether the victim was white. The crude relative risk associated with the victim's being white was 8.3. An important legal issue was the extent to which this large effect was confounded by various alternative causes, especially those that corresponded to legitimate aggravating or mitigating factors. In particular, were the circumstances of the white-victim and black-victim murders different enough to explain away the observed effect?

One of the factors considered was that the victim was a stranger to the defendant. Suppose (strictly hypothetically) that white murder victims were more likely to be strangers than black victims. Further, suppose that death sentences were more likely to be rendered when the victim was a stranger. We can even assume that this association was causal in nature. Then stranger-victim would appear to be a confounding variable. Stranger-victim satisfies the standard criteria and therefore distorts the "true" white-victim effect.

Another possible confounding factor was whether or not the murder also involved rape. Suppose this factor strongly increased the chances of receiving a death sentence, and that the statistical relationship between stranger-victim and death sentencing was entirely spurious, generated by the fact that the murdered rape victims were usually strangers to their assailants. Then after stratifying the cases into two strata, raped-victim cases and other cases, we could analyze the data separately within each stratum. In this hypothetical scenario, the statistical relationship would disappear. Is stranger-victim truly a confounder? The answer depends on which other variables are included in our conceptual–statistical model.

Intermediate Variables between the Exposure and Outcome

Suppose that a variable Z has a different distribution in the exposed and unexposed groups. Further, assume that Z exerts a causal influence (however defined) on the outcome. Suppose, however, that Z is itself caused by F. Here Z is an *intermediate variable* on the causal pathway between exposure and outcome. Therefore, the effect of Z can be regarded as part of the causal effect exerted by F on Y.

As an example, suppose hypothetically that HRT use could cause a decrease in the level of blood pressure for a certain category of women. Elevated blood pressure, or hypertension (HTN), is a known risk factor for CHD. By decreasing blood pressure levels, HRT might reduce the rate of HTN. Should we consider an intermediate variable like HTN in this example to be a potential confounding variable? Ordinarily not, because we are primarily concerned about the total effect of exposure, rather than the specific pathways through which the effect operates.

Suppose, however, that the intermediate variable Z is only partially determined by F. For example, there may be a difference in average blood pressure between the users of HRT and nonusers that is *not* caused by HRT use. In such a situation, Z would normally be regarded as a confounding variable. The problem in practice is that there is no way to separate the observed association between Z and F into a component "caused" by F and a component that is not.

6.3 CONFOUNDING AND STATISTICAL CONTROL

Several times throughout this book, we have alluded to the notion of *statistical control* for a confounder (or set of confounders). If X has been identified as a confounding variable, it may be possible to compensate, or control, for its effect in order to obtain an unbiased measure of effect. To accomplish this feat, we must appropriately *adjust* the empirical effect to *account for* the impact of X. Technical details of such adjustment can be ignored for the moment.

Statisticians tend to regard confounding as a problem of controlling for *covariates*. Confounding exists if adjustment for the necessary covariates does not occur, or is not adequate to remove the bias. The locus of the problem shifts from defining the nature of the variables that can *generate the problem* (of confounding) to identifying the variables that *need to be controlled* (to avoid confounding). From this angle, confounding arises because the set of variables actually controlled (if any) is incomplete. Imagine that the true effect is like a painted image that has been obscured by decades of accumulated grime. The data analyst's job is to restore the pristine image by removing the distorting buildup, layer by layer. *Residual bias* is what remains if this cleaning process (statistical adjustment) fails to remove some layers (i.e., adjust for all relevant confounders).

Traditionally, this pragmatic approach has been very appealing to statisticians, because it concentrates primarily on fixing the problem. Of course, the challenge of selecting appropriate variables for statistical adjustment is ever present, but causality tends to recede into the background. Explicit questions about causality are relegated to a philosophical realm that most practicing statisticians prefer to avoid. In fact, discussions of confounding by statisticians often dispense with causal notions entirely (e.g., Kleinbaum et al., 1982; Grayson, 1987; Hauck et al., 1991). A confounding variable is not viewed as necessarily *causing* the outcome value, but as any variable whose control would reduce bias.

In effect, this perspective assumes that a relevant covariate is one that is capable of *predicting* the outcome value. However, this predictive ability is generally assessed in the abstract. For example, a frequently used covariate in psychology is a pretest score. Performance on a pretest does not usually play a direct role in determining the score on the posttest. However, the pre-score is often highly predictive of the posttest score. From the classical viewpoint, the pre-score reflects some latent characteristic such as "level of the skill being tested." This latent variable may be considered to be the "real" extraneous causal variable in the sense set forth in Section 6.2. However, many statisticians might simply appeal to the notion that differences between groups on the average pre-score would be predictive of post-score differences in the absence of an intervention. Therefore, it may be considered intuitively obvious that controlling for the pretest will reduce bias. Metaphorically, adjusting for the pretest can help "peel away" some (or ideally all) of the layers of confounding "grime" obscuring the true effect.

To understand how statistical control works in practice, consider a basic situation: estimating the risk ratio from a standard 2×2 table in the presence of a single dichotomous confounder. Suppose that the exposure is hormone replacement therapy (HRT) and the outcome is a CHD event. Let age be the covariate that is considered necessary to control in order to avoid confounding bias. Furthermore, suppose that the postmenopausal women in the study are categorized crudely into two categories: relatively older women and relatively younger women. Ignoring age, the hypothetical results of such a study are presented in Table 6.1. The observed risk ratio is 0.74, suggesting a protective effect of HRT. Suppose, however, that we *stratify* our analysis by performing two separate analyses, one for the older women and one for the younger. By *conditioning* on age in this way, the covariate has been *held constant*. Assuming

Table 6.1 Hypothetical HRT Study Results

	CHD	No CHD	
HRT	245	3005	7.5%
No HRT	690	6060	10.2%
		RR = 0.74	

Table 6.2   Stratified Hypothetical HRT Study Results: Version 1, Older Women

	CHD	No CHD	
HRT	35	215	14.0%
No HRT	570	4180	12.0%
		RR = 1.17	

Table 6.3   Stratified Hypothetical HRT Study Results: Version 1, Younger Women

	CHD	No CHD	
HRT	210	2790	7.0%
No HRT	120	1880	6.0%
		RR = 1.17	

that age is the only relevant confounder, the estimated effect must be unbiased within each of the strata. The study data conditional on the value of age is called the *stratum-specific*, or *age-specific*, data. In this example, Table 6.2 shows the age-specific results for the stratum of older women. Similarly, Table 6.3 displays the age-specific data for the younger women.

In this example, the age-specific risk ratio has the value 1.17 for both strata. Under our assumptions, 1.17 is thus the unbiased (true) risk ratio. A statistician might view the crude risk ratio of 0.74 as a statistical artifact that resulted from failure to control for age. She might also point out that the reversal of the observed effect after conditioning on age is an example of the celebrated *Simpson's paradox* (Simpson, 1951). The idea that conditional associations between variables can differ from the corresponding unconditional associations has been known since the dawn of modern statistical theory (Yule, 1903). Simpson's article focused attention on the fact that the direction of the association can even be reversed, as in our example. In one sense there is nothing remarkable in this reversal per se to the student of basic probability theory; it is a mathematical fact of statistical life. However, this mathematical phenomenon becomes interesting when the conditional effects are interpreted as causal. Then it might seem surprising that the overall (unadjusted) effect can be negative when the causal effect is positive.

In our hypothetical HRT study, control for age by stratification has "uncovered" the presumably causal RR of 1.17. Suppose, however, that the two stratum-specific risk ratios had turned out to be 0.74 instead of 1.17. In that case, we could simply collapse the two strata into one by pooling the data, without altering the risk ratio. From the perspective of confounding as a lack of statistical control, there would be no confounding bias. The adjustment for age was not really necessary, as the empirical effect did not change as a result of controlling for age.

Table 6.4   Stratified Hypothetical HRT Study Results: Version 2, Older Women

	CHD	No CHD	
HRT	145	505	22.3%
No HRT	640	3710	14.7%
	RR = 1.52		

Table 6.5   Stratified Hypothetical HRT Study Results: Version 2, Younger Women

	CHD	No CHD	
HRT	100	2500	3.8%
No HRT	50	2350	2.1%
	RR = 1.85		

Now we consider a more complex (and realistic) possibility illustrated by Tables 6.4 and 6.5. Assume that the 2×2 table is exactly the same (see Table 6.1) as in the previous example. Suppose, however, that both of the age-specific risk ratios differ from the overall risk ratio and are not equal to each other. For example, the 2×2 table conditional on being older might look like Table 6.4. The corresponding risk ratio is 1.52. The data for the younger women, on the other hand, could be as shown in Table 6.5. The risk ratio for the younger women is 1.85.

Of course, it is possible that the difference between the stratum-specific risk ratios is the result of random variation. In that case, the apparent disparity between the effects in the two strata would be ignored. Let us assume, however, that we are dealing with the true values of these RRs. Then we may be interested in an overall, or average, causal effect that applies to the whole study population. In principle, we would like to simulate the empirical effect (e.g., risk ratio) that would have been obtained if the distribution of the confounding factor in each treatment group had been identical to the distribution in a target population. Such an estimate is said to be *adjusted* for the differences between the distributions of the covariate(s) in the study groups. If the adjusted empirical effect happens to be equal to the unadjusted effect, then the data are considered to be *collapsible*. The special case described previously in which the stratum-specific effects and the adjusted effect are all equal has been termed *strict collapsibility* (Greenland, Robins, and Pearl, 1999).

Consider first the situation when the outcome is numerical and the average causal effect (ACE) is the parameter of interest. Furthermore, suppose that there is a single covariate X. There are several techniques that can be employed to accomplish statistical adjustment. One common method involves subdividing the study population into a number of categories, or strata, based on the value of the covariate. Each stratum includes individuals who are relatively homogeneous with respect to their values of X. A widely used rule of thumb is to divide the range of X values into five categories, each of which contains approximately the same number of subjects (Cochran, 1968b). The ACE is then estimated as a simple average of the estimated effects for the individual strata. This approach is called *subclassification* or *stratification*. It is not difficult to see that the overall ACE for the study population is equal to the average of the stratum-specific effects.

For a dichotomous outcome, the same general approach can be applied. However, the interpretation of the results is less straightforward. Consider the simple example discussed earlier. The estimated risk ratio for older women is 1.52 and for younger women is 1.85. Because the two strata are of equal size, we can average them to obtain an overall estimate of 1.69. However, this risk ratio cannot be interpreted straightforwardly as an average of individual causal effects (Greenland, 1987b). Indeed, it is not obvious exactly what causal parameter is being estimated.

In this example, the age-adjusted risk ratio is 1.69, compared with the crude RR of 0.74. So, the strata in this example are not collapsible. Had the age-adjusted risk ratio been 0.74, we would have said that the strata were collapsible, at least with respect to the risk ratio. It is evident that collapsibility and confounding are intimately related concepts. Therefore, it has been tempting to define confounding in terms of noncollapsibility. Indeed, as a practical matter, such a definition is in most circumstances not very misleading (Greenland, Robins, and Pearl, 1999; Pearl, 2000, 193–194). However, collapsibility cannot serve as the *defining* element of confounding.

First, collapsibility can depend on the particular effect measure chosen. Possible alternative measures include the risk difference and the odds ratio, with many others theoretically of interest (Grayson, 1987). For example, it is possible for the strata to be collapsible for the risk ratio, but not for the odds ratio or the risk difference (Greenland, Robins, and Pearl, 1999). If a criterion for confounding is to have any general conceptual value, it should refer to a phenomenon that cannot vary according to the particular effect measure, which after all is somewhat arbitrary.

Second, collapsibility can depend on the particular methodology chosen to perform the statistical adjustment. In our example, the technique employed was *stratification*. Other possible analytic strategies include *direct standardization*, *indirect standardization*, *matching*, and techniques based on *least-squares regression*, *logistic regression modeling*, or *log-linear modeling* (for an introduction to these methods, see Anderson et al., 1980). It is possible that collapsibility could be obtained when using stratification, as in our example, but not when using logistic regression.

Perhaps the major problem with the collapsibility-based definition of confounding is conceptual. Consider, once again, our hypothetical HRT study summarized by Tables 6.4 and 6.5. We assumed that the focus was on an overall effect applicable to the study population. However, for some purposes, scientific interest might concentrate on the stratum-specific effects. For example, the study's practical implications (e.g., treatment recommendations) may be different for relatively older women than for younger ones. Why calculate an average effect across the entire population that may not be applicable to either of the identifiable subgroups? Interest in collapsibility is motivated in part by the denial of effect modification as a likely and/or important possibility. From a statistical viewpoint, especially when finite sampling is considered, effect modification is a major complicating factor. However, the assumption that there exists a well-defined "true" causal effect independent of the covariate(s) cannot always be justified. This issue will be considered in depth from a counterfactual perspective in the next chapter.

Setting aside this conceptual issue, collapsibility seems most useful as a criterion for ruling out certain potential confounders from consideration. Suppose that statistical adjustment produces an estimate that is altered very little from the unadjusted value. Then it would seem that the potential confounder is not in fact causing a substantial bias, whether or not it is "truly" a confounder in some abstract sense. For example, in the Yale Hemorrhagic Stroke Project (HSP) case–control study, several potential confounders were considered for inclusion in the adjustment model. After inclusion of a core set based on biological considerations, other covariates were tried. Only one covariate (educational level) produced a change of at least 10% in the estimated odds ratio and was consequently added to the adjustment model (Kernan et al., 2000).

So far, our discussion has been overly simplistic, because we have concentrated on controlling for a single variable, such as age. In practice, there may be other variables in addition to age that should arguably be controlled as well. In our example, HRT users might be healthier in certain respects than nonusers. When multiple potential confounding variables are under consideration, the impact of adjusting for any individual variable can be complex. It is quite possible that adjustment for a particular variable might not change the estimated effect, but that adjusting for this variable in combination with another could have a dramatic impact. Conversely, adjustment for a variable might appear to change the empirical effect, but adding another covariate could "cancel out" this change.

> The dilemma posed by statistical adjustments is that no matter what variables we include in the analysis … , there may be an omitted variable … that together with the included variables constitutes a confounding variable. Moreover, it is not enough to demonstrate that all plausible confounding variables excluded have similar distributions across groups. As with our example, such a variable may still be important in combination with others. So the analyst must be fairly certain that no variable has been left out which mediates the effect of those variables included. (Anderson et al., 1980, 49)

Thus, a strategy of testing individual potential confounders for lack of collapsibility may not be adequate.

6.4 CONFOUNDING AND COMPARABILITY

The classical idea of extraneous causes and the more operational notion of inadequate statistical control have proven extremely useful. However, over time it became evident that these two intellectual pillars could not provide sufficient support for a comprehensive theory of confounding. In epidemiology, the dissatisfaction with traditional definitions of confounding led to the idea that confounding is not fundamentally related to particular confounding variables. Rather, its essence inheres in the extent to which different study groups are *equivalent*, or *comparable*.

The basic idea that emerged around 1980 was that the ideal standard of comparison for the exposed group "should be an adequate proxy for the performance of those receiving the treatment—the treatment group—if they had not received the treatment" (Anderson et al., 1980, 2). An influential article titled "Confounding: Essence and Detection" argued for a perspective that would define bias as the "lack of comparability" between the treatment group data and the "hypothetical outcomes that would have materialized had the exposure had no effect" (Miettinen and Cook, 1981).

These ideas inspired biostatisticians Wickramaratne and Holford (1987) to attempt a more formal analysis of the "adequacy of the control group." Their mathematical model was based on the notion of a hypothetical outcome distribution that would have been observed in the exposed group without treatment. Based on this model, they derived conditions for absence of confounding and showed them to be equivalent to the more heuristic conditions arrived at inductively by Miettinen and Cook.

Wickramaratne and Holford stayed within the conventional statistical framework: populations were the primary focus, and causal effects were defined only in terms of parameters of the outcome distributions. At virtually the same time their ideas were taking shape, a more revolutionary approach to comparability was being developed by Greenland, Robins, and their colleagues (Greenland and Robins, 1986; Robins and Morgenstern, 1987). Greenland and Robins noted that to their knowledge "no one has presented a theory of epidemiological confounding based on a model for individual effects." Their basic approach was similar to that developed a decade earlier by statistician Donald Rubin, but was tailored to epidemiological problems rather than those in the social sciences. In particular, they focused on dichotomous outcomes rather than the numerical outcomes featured in Rubin's "potential-outcome" modeling.

Greenland and Robins saw the main issue involved in confounding as one of parameter *identifiability* in the statistical sense. Loosely speaking, a statistical model is said to be identifiable if the estimates of its parameters converge to unique values as sample sizes increase. Lack of identifiability typically arises because the model contains too many parameters relative to the available data. As a result, the observed data are consistent with more than one underlying model. For example, suppose we are estimating a linear regression model with a single independent variable. Imagine that the regression coefficient β is actually the sum of two components $\beta_1 + \beta_2$ Then without some additional information we could estimate $\beta_1 + \beta_2$, but could not identify (distinguish) the two individual parameters. Now, suppose that we could somehow obtain a second data set in which the value of β_1 was known to be zero. Then β_2 could be estimated from this supplementary data set, and $\beta_1 + \beta_2$ estimated from the original data, allowing us to identify both parameters.

Greenland and Robins applied essentially this concept to the data arising from a comparative study. Suppose we are trying to estimate the risk ratio or risk difference from two study groups, an exposed group labeled W and an unexposed group labeled Z. One way to estimate the effect of exposure would be to compare the rate of events in Group W against the rate in Group Z. For response Type j, let P_{jW} be the proportion of subjects in Group W and let P_{jZ} be the corresponding proportion in Group Z. In the exposed Group W, the proportion who experience the outcome event can be expressed as $P_{1W} + P_{2W}$. The counterfactual proportion that would occur if Group W were unexposed would be $P_{1W} + P_{3W}$. The actual proportion of individuals with events in Group Z would be $P_{1Z} + P_{3Z}$.

Greenland and Robins observed that equality of $P_{1W} + P_{3W}$ and $P_{1Z} + P_{3Z}$ would be sufficient to identify any causal effect (e.g., the RR or RD) *for the population represented by Group W*. They regarded this condition as one of *partial exchangeability* (Greenland and Robins, 1986). Complete exchangeability would mean that we could also reverse the process. That is, we could identify the causal effect *for the population represented by Group Z* by comparing the observed rate in Group Z with the observed rate in Group W. This approach would succeed if $P_{1W} + P_{2W} = P_{1Z} + P_{2Z}$. Thus, complete exchangeability requires two conditions:

$$P_{1W} + P_{2W} = P_{1Z} + P_{2Z} \qquad (6.1)$$

$$P_{1W} + P_{3W} = P_{1Z} + P_{3Z} \qquad (6.2)$$

Resting confounding on the foundation of exchangeability cast a clarifying light on several issues. Perhaps most iconoclastic was the assertion that confounding is not fundamentally related to specific covariates or confounders! Rather, it can be viewed as pertaining essentially to the different compositions of the groups being compared; confounders are simply convenient dimensions that describe how the groups differ. Statistical control for suspected extraneous causal factors can be viewed as an attempt to attain exchangeability *conditionally*. Collapsibility is regarded not as an end in itself, but rather as a usual (though not invariable) consequence of confounding in this more basic sense of comparability.

Greenland and Robins were interested in the conditions that were necessary and sufficient for identifying causal effects. In this book, we have advocated a stronger (i.e., more restrictive) definition of exchangeability: identical distributions of response patterns. This stronger form of exchangeability captures the essence of what intuitively is meant for study groups to be completely "equivalent in all relevant respects." It also describes the critical property that is uniquely guaranteed by random assignment. However, such exchangeability is excessively stringent from the viewpoint of being theoretically necessary to achieve comparability between study groups (Greenland, Robins, and Pearl, 1999).

Consider the case of dichotomous outcomes analyzed by Greenland and Robins. Their two conditions for complete exchangeability are stated in Equations 6.1 and 6.2. These conditions are not sufficient to guarantee that the response-pattern distributions in the two groups are identical. For example, suppose that $P_{1Z} = P_{2Z} = P_{3Z} = 0.1$ and $P_{4Z} = 0.7$ and that $P_{1W} = 0.2, P_{2W} = P_{3W} = 0$, and $P_{4W} = 0.8$. In this case, Equations 6.1 and 6.2 would be satisfied, but the stronger definition of exchangeability would not. Note that Equations 6.1 and 6.2 imply that the *marginal distributions* of potential outcomes in the two groups (in this case the proportions of individuals with events under each exposure state) are identical. Exchangeability in our strongest sense means that the *joint distributions* of the potential outcomes are identical.

6.5 CONFOUNDING AND THE ASSIGNMENT MECHANISM

A second powerful idea that emerged in the 1980s was the importance of the assignment mechanism as a key to understanding and dealing with confounding bias. This insight and its implications grew out of research on statistical methods for analyzing observational studies conducted in the 1970s by statistician Donald Rubin (Rubin, 1974, 1977, 1978). Rubin had been a graduate student under William G. Cochran at Harvard, and he has acknowledged being greatly influenced by Cochran's pioneering work in the design and analysis of observational studies (Rubin, 1984).

Three important themes related to confounding emerged in Rubin's early work. First, causal effects can be defined at an individual level in terms of potential outcomes. Second, confounding can fruitfully be regarded as a result of nonrandom assignment of subjects to treatment groups. Third, the goal of statistical control for confounding should be to simulate the results that would have been observed if randomization had been employed. Rubin and Paul Rosenbaum published a landmark article in 1983 that laid out the key practical implications of this line of thinking (Rosenbaum and Rubin, 1983a).

Many of the applications that Rosenbaum and Rubin were working on at the time pertained to evaluation of educational and other social programs. Estimating an average effect was the primary objective of such evaluation studies. Consequently, they focused primarily on developing methodology for unbiased estimation of the average effect. Our discussion here merely scratches the surface of their more rigorous and extensive discussion in several important publications (Rosenbaum, 1984, 2002; Rosenbaum and Rubin, 1983a).

Rubin had previously termed an assignment mechanism *ignorable* given a particular covariate (or set of covariates) X if the marginal distributions of potential outcomes in both study groups conditional on X were identical

(Rubin, 1978). In our notation, assignment is ignorable, given the value of X, when the distributions of W_E and Z_E are identical and the distributions of W_U and Z_U are identical. The assignment mechanism is called ignorable because after conditioning on X we can obtain an unbiased estimate of any causal effect that depends only on the marginal distributions of potential outcomes. In terms of the discussion of comparability in the previous section, ignorability of the assignment mechanism is equivalent to complete exchangeability, conditional on X.

In their 1983 article, Rosenbaum and Rubin introduced the concept of *strong ignorability*. Essentially, the assignment mechanism is strongly ignorable conditional on X if the *joint distribution* of potential outcomes is independent of treatment assignment. Strong ignorability is therefore equivalent to identical distributions of response patterns, if we fix the value of X. That is, strong ignorability corresponds to exchangeability in the restrictive sense used throughout this book. In the Rubin–Rosenbaum framework, a comparison is considered to be "unconfounded" when the assignment mechanism is strongly ignorable, conditional on all the covariates controlled in the analysis. In effect, confounding is seen as the *residual bias* that remains if the measured covariates fail to remove all nonrandomness inherent in the assignment mechanism. By equating confounding with residual exchangeability (in the restrictive sense), Rubin and Rosenbaum were able to craft an entirely new approach to statistical control for confounding. Their novel methodology was based on the idea of the *propensity score*.

The propensity score is defined as the probability of exposure, conditional on the individual's value of X. Strong ignorability means that the assignment mechanism is *effectively random* given the value of X. Rosenbaum and Rubin demonstrated that, under strong ignorability, conditioning on the propensity score would be theoretically equivalent to controlling for X. Moreover, this result holds true even if X is a set comprising many variables. Furthermore, they proved that it is feasible to estimate the propensity scores from observable data and to use the estimated scores in place of the true propensity scores. Under certain quite general conditions, the resulting adjusted effect estimate will converge toward the causal effect as the sample size increases.

Rubin and Rosenbaum have suggested that adjusting on the basis of an estimated propensity score offers several advantages over more traditional statistical methods, such as analysis of covariance (ANCOVA) and logistic regression (e.g., Rubin, 1997). Perhaps most important, using propensity scores makes techniques based on matching and stratification much more feasible. By controlling for the single propensity-score variable rather than perhaps dozens of suspected confounders, the need for complex statistical adjustment procedures can be avoided. For example, suppose we have n confounders that are all dichotomous. Their joint distribution could be represented as a contingency table with 2^n cells. With 10 such confounders there would be 1024 cells. Conditioning on a fixed value for the n covariates (i.e., a particular cell of the contingency table) might result in a very small sample size in each individual cell. Traditional statistical adjustment methods employ complex mathematical models to circumvent this "sparse data" problem. The propensity-score approach, on the other hand, effectively reduces the vector of covariates to a single dimension (variable) and thereby greatly reduces the sparse data issues that arise when there are many potential confounders.

According to its supporters, the propensity-score methodology is relatively straightforward to understand and implement, is not dependent on unverifiable modeling assumptions, and is, to some degree, capable of providing warnings that the available data may not support a valid causal estimate. On the other hand, skeptics have countered that these theoretical advantages may be somewhat illusory in practice. Estimating the propensity score on the basis of available data may sometimes prove quite difficult. The results obtained from a propensity-score analysis may be rather sensitive to the method of estimation used, casting doubt that the correct effect has been identified.

Various methods for testing the validity of the estimated propensity score have been proposed and their merits debated. In Chapter 1, we described briefly the ongoing controversies about the applicability of the propensity-score approach in the context of the National Supported Work (NSW) program. The interchanges between proponents and skeptics focus largely on whether propensity scores are better at correcting for confounding bias than other statistical approaches. Somewhat lost in these fairly technical discussions are the contributions of propensity-score theory to conceptual understanding of confounding. Regarding the nonrandom assignment process as the source of confounding bias has important implications. First, this perspective leads naturally to the concept of strong ignorability (exchangeability) as a theoretical definition of nonconfounding. Second, the idea that confounding does not depend on the particular measure of effect becomes transparent, because ignorability pertains to response-pattern *distributions* rather than to specific parameters. Third, the strong-ignorability concept of confounding pertains clearly to the comparability of study groups, rather than to the existence of particular confounders.

Individual covariates are relevant only as predictors of treatment assignment. From the standpoint of bias correction, whether a covariate is "truly" a cause of the outcome or merely a proxy does not matter. In fact, the inadvertent inclusion of variables that facilitate better prediction of assignment, but are causally unrelated to the outcome, are harmful only in the sense of inefficient estimation (e.g., wider confidence intervals). It is arguably better to err on the side of possible overinclusion of such factors than to risk omitting an important causal factor (e.g., Rubin and Thomas, 1996; Rubin, 1997).

6.6 CONFOUNDING AND MODEL SPECIFICATION

The ideas of exchangeability and ignorability highlight lack of comparability between study groups as the essential aspect of confounding. Specific covariates that tend to distinguish these groups are regarded as reflections of nonrandom assignment, rather than as independent causes of the outcome. The comparability perspective tended to divert attention away from individual causal factors and their interrelationships. Ironically, at virtually the same time (mid-1980s) that these ideas were being introduced, an alternative paradigm for causality emerged that drove counterfactual thinking in the opposite direction. Judea Pearl, along with several colleagues, was in the process of formalizing a methodology for unraveling the causal interconnections among a network of several variables. Confounding was interpreted as a special case of a more general problem amenable to analysis using the mathematical and graphical "language" of Bayesian causal networks.

Recall that we have defined the causal effect of a treatment or exposure as a change in the individual's outcome, assuming that *all other circumstances are held fixed*. The causal network approach tries to encompass a much more general problem. A causal effect is imagined to be embedded in a larger network of interrelated causal influences that can be explicitly represented. Pearl and his colleagues analyzed how, and under what conditions, it could be possible to *identify* this effect. Pearl essentially defined a causal effect of variable V_1 on V_2 as a change in the distribution of outcomes for V_2 resulting from an *intervention* in which we modify the value of V_1, otherwise leaving all else the same. However, the concept of "leaving all else the same" is interpreted within the framework of the specified causal network. Heuristically, we can imagine that while "setting" the value of V_1, we allow all other causal relationships in the system to continue as before, except that causal influences on V_1 are "pruned" from the causal network (Pearl, 2000, 70).

In the context of a comparative study, this approach to causality is formally equivalent to the potential outcomes framework. However, an approach based explicitly on potential outcomes would have been mathematically intractable for analyzing the causal relationships among several variables. To overcome this obstacle, a graphical "shorthand" was developed to convey the causal logic in any system of interrelated variables. This graphical approach to causality is extremely powerful, but at the price of simplification in three critical respects.

First, these methods are "nonparametric" in the sense of specifying only whether or not a causal relationship exists between two variables; the precise mathematical form of this effect is not made explicit in the graph. To include such quantitative information, Pearl and like-minded colleagues appeal to the mathematical framework of *structural equation modeling* (SEM); SEM allows the causal effects to be quantified, but usually at the cost of restricting the mathematical form in certain convenient ways (e.g., additivity, linearity).

The second simplification is that the entire causal network (or set of structural equations) can be specified. That is, not only can we observe all the relevant variables, but we can also diagram the flow of causal influences among them. This requirement is considerably more demanding than simply specifying relevant covariates. Third, a causal network assumes that the causal influences can be summarized as probabilistic relationships among *variables*. In effect, the potential outcomes are stochastic, as discussed in Section 3.3. This idea implicitly regards individuals as indistinguishable with respect to the causal effect of a particular variable. Thus, the idea of an interactive effect that can depend on an individual's causal factors is suppressed in this framework.

Currently, there is active discussion among methodologists about the extent to which the graphical and SEM framework for causal modeling is of practical utility. Indisputable, however, is the enormous impact this approach has exerted on thinking about causality. Within the limitations we have described, this approach offers valuable tools for clarifying the exact conditions under which the effect of a particular variable can be identified. These conditions provide an operational definition of confounding within the context of a given causal network and provide *in principle* a way both to test for confounding and to perform necessary statistical adjustment (Pearl, 2000, 184–185). If the structure of a full causal model can be correctly specified, the presence or absence of confounding can be determined definitively. Of course, this "if" is a very big one.

For actually estimating causal effects, Pearl favors the SEM paradigm. However, he is highly critical of attempts in recent decades by econometricians and other quantitative social scientists to reduce structural modeling to a mathematical–statistical exercise. In his view, SEM was originally developed by geneticists (Wright, 1921) and economists (e.g., Haavelmo, 1943; Koopmans and Reiersol, 1950) in order that "qualitative cause–effect information could be combined with statistical data to provide quantitative cause–effect assessment relationships among variables of interest" (Pearl, 2000, 135). He posits a number of reasons why this attitude has fallen out of fashion, and he highlights the widespread confusion among SEM practitioners owing to the divorce of SEM from its causal roots.

In the field of econometrics, recent infiltration of counterfactual thinking has been substantial. Economists James Heckman and Charles Manski have been in the forefront of these developments. Heckman's work with Richard Robb in the 1980s drew upon the econometric paradigm of *switching models* as a way to introduce counterfactual concepts into the standard regression framework (Heckman and Robb, 1985). A number of econometricians have built on this foundation to extend the modeling of causal effects in various directions, as summarized in a recent monograph (Morgan and Winship, 2007).

Manski has approached the problem of causal identification from a different direction, focusing on the trade-off between the strength of assumptions and the credibility of resulting inferences (e.g., Manski, 2007). With strong enough assumptions, such as the mathematical form of the statistical model and the sufficiency of variables in the model, a causal effect becomes identifiable. However, the validity of the underlying assumptions may be harder to justify. This perspective has generated new methods that produce *bounds* on a causal effect instead of a point estimate. The narrowness of these bounds is a function of the particular assumptions made. The usual definition of identifiability can be seen as a limiting case when the data and assumptions are adequate to sustain a point estimate. Note that the bounds are not confidence intervals, which quantify the uncertainty attributable to random variability. Rather, these bounds reflect systematic uncertainty resulting from limited data and/or weak (i.e., less restrictive) assumptions.

GUIDEPOST 6

For several decades, methodologists have struggled to explain clearly the meaning of confounding. This chapter has surveyed five intellectual currents that have run through the discourse on this confusing issue. Each of these themes captures an important aspect of confounding but fails to articulate fully the connotations of this somewhat enigmatic term.

The notion of confounding as distortion of the observed effect by extraneous causes begs the questions of what exactly is a "cause" and how do several causes jointly determine the outcome. Confounding as inadequate statistical control can perhaps tell us when a particular covariate is unnecessary, based on the criterion of collapsibility, but cannot explain what it means for adjustment to account for all "relevant" confounders. The idea of confounding as lack of comparability has been formulated in terms of exchangeability between the study groups. However, exchangeability fails to consider explicitly the intuition that confounding has something to do with particular factors that differ between the groups. Likewise, confounding as nonrandom assignment of subjects to the study groups, which essentially equates confounding and selection bias, fails to address explictly the interpretation of "confounding factors." Finally, confounding as incorrect model specification achieves precision in the explication of confounding by adopting simplifying conceptual and mathematical assumptions.

In the next chapter, a conception of confounding will be attempted that is grounded in counterfactual causal theory. This proposed framework aims to illuminate the essence of this fundamental methodological problem. Each of the five aspects of confounding will be placed in a broader context and related to the counterfactual framework espoused in this book. We will then discuss the possible approaches for dealing with the problem of confounding in light of the insights provided by our causal modeling perspective.

CHAPTER 7

Confounding: Essence, Correction, and Detection

The classical tradition in epidemiology has viewed confounding as a systematic distortion of a true effect by extraneous causal factors. The observed effect is imagined as a sum of the exposure effect and the effects of these confounders. A confounder is defined as a variable that is associated with exposure, is "causally" related to the outcome, and is not an intermediate variable between exposure and the outcome. The collapsibility perspective traditionally favored by many statisticians, on the other hand, does not address the issue of causality directly. Rather, confounding is said to be present if adjustment for additional relevant covariates would cause the estimated effect to change. Failure to observe such change, after adjusting for additional variables, is taken as an indication that confounding is absent. Identifying potential confounders is based on an intuitive determination of the causal factors that may be at work.

In 1981, an article with the title "Confounding: Essence and Detection" challenged the adequacy of the extraneous-variable and collapsibility formulations of confounding (Miettinen and Cook, 1981). Miettinen and Cook proposed that confounding is best understood as a matter of comparability between study groups. We will consider the essence and detection of confounding from a comparability perspective in light of the counterfactual framework. In addition, we have explicitly added consideration of *correction* for confounding. In virtually every observational study, some sort of statistical adjustment is employed. Detection then concerns the problem of *residual confounding* that remains after such statistical control. Conversely, suspicion of residual confounding calls for additional corrective measures, which can take a variety of forms. Correction and detection are therefore two sides of the same coin; both are fundamentally concerned with the implications of conditioning on covariates.

Bias and Causation: Models and Judgment for Valid Comparisons, By Herbert I. Weisberg
Copyright © 2010 John Wiley & Sons, Inc.

Unfortunately, detection of residual confounding is generally accorded much less attention than correction. Rather than attempting to assess the nature and extent of possible confounding bias, researchers hope that bias can be removed by sophisticated statistical analyses. However, no purely statistical criterion is capable of providing absolute assurance of success. In the end, whether any meaningful bias remains will be a judgment call; there will always be a residue of uncertainty that needs to be addressed frankly:

> Forthright acknowledgment of ambiguity should be the norm, but it is distressingly rare. The scientific community rewards those who produce strong novel findings. The public, impatient for solutions to its pressing concerns, rewards those who offer simple analyses leading to unequivocal policy recommendations. These incentives make it tempting for researchers to maintain assumptions far stronger than they can persuasively defend, in order to draw strong conclusions. (Manski, 2007, 7)

Ultimately, our ability to correct for bias often hinges on being able to discern whether residual confounding remains. Obtaining evidence relevant to this critical question may call for the sort of creative sleuthing that distinguishes the best scientific research:

> Efforts to detect hidden bias involve collecting data that have a reasonable prospect of revealing a hidden bias if it is present. If treatments are not assigned to subjects as if at random ... bias might leave visible traces in data we have or could obtain. Efforts to detect hidden bias use additional information beyond the treated and control groups. ... Detecting hidden bias entails checking that treatment effects appear where they should, and not elsewhere. (Rosenbaum, 2002, 206)

Appraising the adequacy of statistical adjustments is not simply a matter of mathematical technique. A researcher (or critical reviewer) may need to place the study in a larger context within which potential sources of hidden bias (i.e., residual confounding) can be identified and tested. Do the results hold water in the light of all the challenges we can muster based on background knowledge, external data, logic, etc.?

Most methodologists focus primarily on technique rather than on this broader assessment of validity. Much effort has gone into developing mathematically complex variants of methods such as matching, stratification, and regression. In contrast, rather little research has been devoted to the technically less sophisticated but ultimately more challenging conundrum of determining what residual bias may remain after statistical adjustment. This question is heavily dependent on context and much less amenable to theoretical analysis and purely technical solutions.

7.1 ESSENCE: THE NATURE OF CONFOUNDING

We have seen that, at one level, assignment bias and confounding represent alternative ways to describe the same phenomenon. However, viewing confounding merely as an issue of nonrandom assignment ignores the role of causal factors. It is reasonable to assert that confounding corresponds to different distributions of response patterns in the study groups. However, this definition does not attempt to explain how these differences arise and why statistical control for covariates may be useful. To address these questions, we will bring to bear the model for individual causation developed in Section 2.2.

A Causal Model for Confounding

From the counterfactual perspective, the causal effect *for a particular individual* is some function of his response pattern, such as the arithmetic difference between his potential outcomes, assuming that we *hold fixed everything else besides exposure status*. We have said that a risk factor *F* exerts a causal effect on an individual if and only if his potential outcomes under the exposed and unexposed conditions differ. We have emphasized that this individual causal effect is conditional on holding "everything else" constant.

In our causal model, each individual is portrayed as a pie chart whose sectors represent the possible causal *factors* that together determine his outcome. This model is quite general in the sense that no assumptions are made about the specific mechanisms by which this effect is exerted. Rather, the model embodies the simple and realistic idea that a confluence of relevant circumstances somehow determines whether the outcome event takes place, or more generally determines the value of the outcome variable. In the context of a comparative study, one of these causal factors is the risk factor under investigation. All other factors *in combination* determine the potential outcomes and therefore the response pattern. The pattern of values for factors other than the risk factor represents "everything else" that is imagined to be held fixed. We have referred to this fixed pattern for each individual as the *causal context* within which exposure takes place.

In a comparative study, we have two study groups of individuals. Each member of a study group possesses a particular causal context. For illustrative purposes, we have assumed previously that there were five possible components, including the risk factor *F*, and that each was dichotomous. We also assumed the outcome was dichotomous. Occurrence of the outcome event *D* was indicated by shading the pie chart. Adopting these same conventions, we can represent a comparative study as in Figs. 7.1 and 7.2. Figure 7.1 portrays an exposed study group with eight subjects. Four of the subjects experience the event, and four do not. Therefore, the observed risk is 0.50.

Figure 7.2 represents the unexposed group to which the group in Fig. 7.1 is being compared. The two study groups are exchangeable, because the two distributions of response patterns are identical. However, at the level of underlying causal factors, there is much variation among the individuals. For example, the two *doomed* individuals in the exposed group have different causal contexts. For one of these subjects, Factors 1 and 2 are present; for the other, Factors 2 and 3 are present. From the standpoint of causal modeling, these

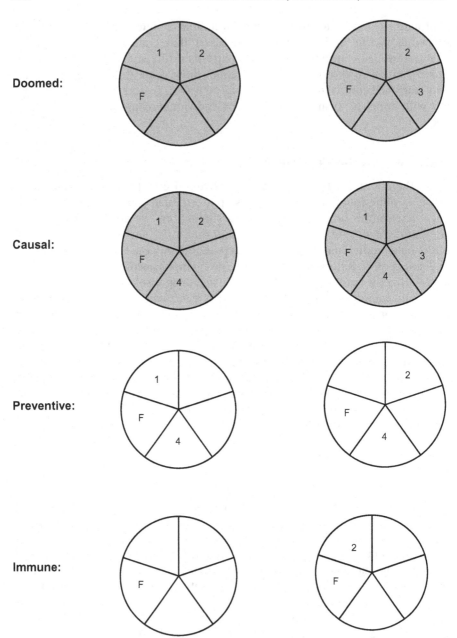

Figure 7.1. An illustration of an *exposed* group in which half of the subjects experience the outcome event. The underlying causal model that determines the outcomes is shown in Table 7.1. The unexposed group in the same study is shown in Fig. 7.2. The study groups are exchangeable, because the distributions of response patterns are the same.

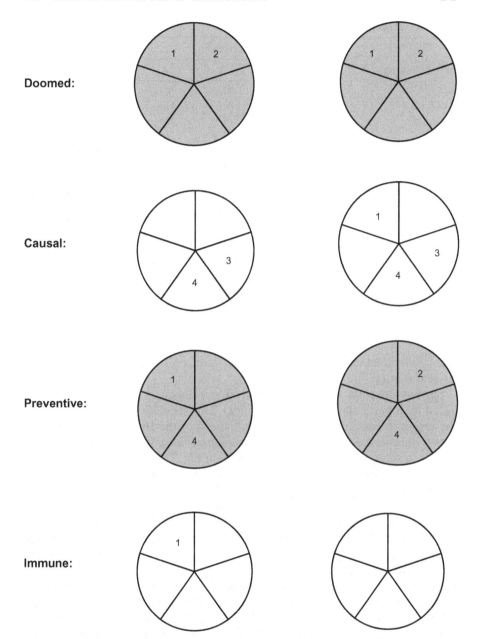

Figure 7.2. An illustration of an *unexposed* group in which half of the subjects experience the outcome event. The underlying causal model that determines the outcomes is shown in Table 7.1. The exposed group in the same study is shown in Fig. 7.1. The study groups are exchangeable, because the distributions of response patterns are the same.

Table 7.1   Hypothetical Causal Model

Causal Context	Causal Factors				Response Pattern
	1	2	3	4	
1	No	No	No	No	Immune
2	Yes	No	No	No	Immune
3	No	Yes	No	No	Immune
4	No	No	Yes	No	Immune
5	No	No	No	Yes	Immune
6	Yes	No	No	Yes	Preventive
7	No	Yes	No	Yes	Preventive
8	Yes	Yes	No	Yes	Causal
9	Yes	No	Yes	Yes	Causal
10	No	Yes	Yes	Yes	Causal
11	Yes	Yes	Yes	Yes	Causal
12	No	No	Yes	Yes	Causal
13	Yes	Yes	No	No	Doomed
14	Yes	No	Yes	No	Doomed
15	No	Yes	Yes	No	Doomed
16	Yes	Yes	Yes	No	Doomed

latent individual differences are not relevant; only the resulting response patterns matter.

The outcomes for various individuals in our illustrative example were generated by a particular hypothetical model. This model is shown in Table 7.1. The response pattern that corresponds to each of the 16 possible causal contexts is indicated. For example, the exposed *preventive* individual with Factors 1 and 4 present and Factors 2 and 3 absent in Fig. 7.1 is unshaded (no event), whereas the unexposed subject with the same causal context is shaded (event). Figures 7.1 and 7.2 illustrate a situation in which the study groups are exchangeable. Now we modify the assignment mechanism to create the alternative scenario shown in Figs. 7.3 and 7.4. In this situation, the groups are no longer exchangeable. There is a tilt toward exposure for *doomed* and *causal* individuals, and toward nonexposure for *preventive* and *immune*. The assignment probabilities for *doomed* and *causal* subjects are now 0.75, whereas the assignment probabilities for *preventive* and *immune* are 0.25. As a result, the empirical risk ratio becomes 1.5 instead of the true causal effect of 1.0.

The traditional perspective regards confounding as attributable to certain *variables*. These confounding variables are conceived as representing (or at least being correlated with) characteristics that directly influence the outcome. But the precise causal nature of these confounders is unclear. The causal model described in Section 2.2 provides a useful way to conceptualize the underlying sources of nonexchangeability between study groups. The distribution of response patterns in each group clearly depends on the causal factors that pertain to the members of each group. In this sense, the factors represent

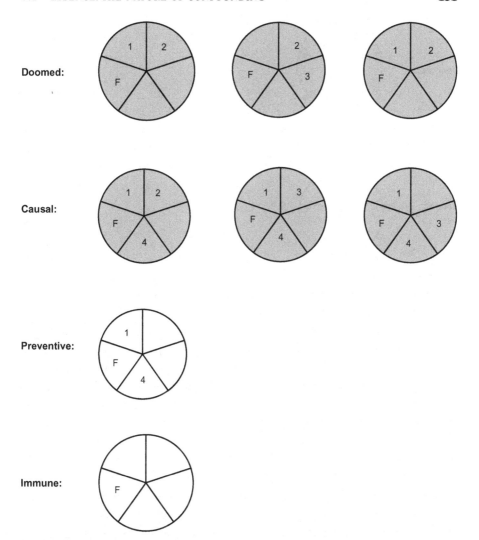

Figure 7.3. An illustration of an *exposed* group in which three-quarters of the subjects experience the outcome event. The underlying causal model that determines the outcomes is shown in Table 7.1. The unexposed group in the same study is shown in Fig. 7.4. The study groups are not exchangeable, because the distributions of response patterns are not the same.

confounders. However, the net effect of a particular factor on the outcome is not simply a function of of that factor's value in isolation. Rather, this effect depends on the causal configuration as a whole, including the risk factor in conjunction with the causal context.

The causal context possesses the three properties traditionally associated with confounding variables. First, the distributions of these contexts may differ between study groups. Second, the causal context exerts a "causal influence"

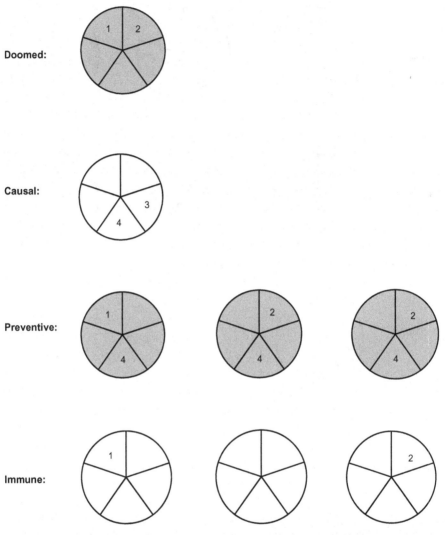

Figure 7.4. An illustration of an *unexposed* group in which half of the subjects experience the outcome event. The underlying causal model that determines the outcomes is shown in Table 7.1. The exposed group in the same study is shown in Fig. 7.3. The study groups are not exchangeable, because the distributions of response patterns are not the same.

in the sense of determining the response pattern for each individual. Third, because the causal context is fixed, it cannot be an intermediate variable affected by the risk factor. Thus, within this framework, the causal context in its entirety can be considered the confounder. From this perspective, treating a factor as a variable in the usual way would be a gross simplification. Only in very special circumstances could the causal effect of such a factor be abstracted from the causal context.

From a traditional statistical viewpoint, this logic has a "through the looking glass" quality. In ordinary probability theory, the distinction between a variable and what we are calling a factor does not exist. Most statisticians would perhaps interpret the idea of a causal context as meaning that the dependence of the outcome on the independent variables is *highly interactive*. Such highly interactive relationships are thought to be somewhat exotic, and usually not very plausible. Moreover, even if subtle patterns of interaction exist, it is believed that their effects must be small relative to the effects of a few main relationships. Unfortunately, there is no good reason for the belief that the workings of nature respect our penchant for simplicity.

Traditional Definitions of Confounding Revisited

In Section 6.1, we discussed five ways in which the problem of confounding has been framed:

1. Distortion of the statistical association by *extraneous causes*
2. Inadequate *statistical control* for confounding variables
3. Imperfect *comparability* between the study groups
4. Nonrandom *assignment* of individuals to the study groups
5. Incorrect *specification* of a structural model

We will now analyze each of these aspects of confounding from a causal perspective.

Consider a typical confounding variable such as age, gender, or educational level. In what sense do any of these represent an *extraneous cause* of the outcome? In terms of our causal model, age might be a factor that, in conjunction with a risk factor F and the remainder of the causal context, is thought to determine the outcome value. That would imply that the variable "age" is causally related to the outcome in some manner. However, as explained previously, the exact impact of age might vary across individuals, depending on their causal contexts. To the extent that such variability occurs, we would not expect to find a direct, or separate, causal relationship between age and the outcome.

Rather, the causal "effect" of this factor operates through its ability to modify the response patterns for various individuals in both study groups. The actual confounding bias that results can be considered the net effect of these causal modifications. Therefore, the causal factors that modify the response patterns (causal modifiers) can be considered the true "confounders." However, the complex, and indirect, nature of their effects on the empirical effect cannot (usually) be reduced to a simple relationship between an observable variable (e.g., age) and the outcome.

Many methodologists appear to believe that extraneous causal factors have some sort of objective existence. However, the causal context in which a risk factor operates is an abstraction from reality. Any causal model, no matter how insightful or elaborate, is always approximate and provisional. Its utility as an intellectual device depends on the construct validity of the causal factors and on the theoretical framework within which these constructs have meaning. This underlying theoretical framework reflects what some have termed a *causal mechanism*:

> A causal mechanism is a scientific theory describing aspects of the various biological, chemical, physical, or social processes by which the treatment produces its effects. For example, the causal mechanism by which cigarette smoking causes lung cancer includes a description of the inhalation of carcinogens contained in cigarette smoke. To say that a causal mechanism is a scientific theory is to say it is a tentative, typically incomplete description that is subject to refutation on the basis of evidence. (Rosenbaum, 1984, 42)

There are no extraneous causes that exist "out there" as objective realities waiting to be discovered. Any causal mechanism is a mental construct that serves as a working model of relevant aspects of the real world. The classical definition of confounding tends to downplay this subjective and metaphysical aspect of causality.

The idea of confounding as the result of inadequate *statistical control* regards failure to condition on an adequate set of covariates as the root of the problem. Confounding is believed to derive from the differing distributions of a confounder X in the two study groups. It is imagined that controlling for X would *restore* the true causal effect by undoing, or compensating for, the distorting effect of X. This "restoration" is typically accomplished by resolving the total study population into subgroups (strata) within which the value of X is constant. Within any stratum the distortion caused by the imbalance in X is presumed to disappear.

As explained in Section 6.3, this set of ideas is also sometimes called the *collapsibility* definition of confounding. The basic notion is that conditioning on X should yield estimated within-stratum effects that are all unbiased. The within-stratum estimated effects can then be combined in some fashion to produce an overall estimate. The strata are weakly collapsible if the adjusted effect obtained in this way does not differ from the unadjusted effect. They are strongly collapsible if the within-stratum effects all individually equal the unconditional effect. In that case, adjustment for the covariate was apparently unnecessary (i.e., no confounding existed) because adjustment failed to alter the estimated effect. However, if the stratum-specific effects are found to be homogeneous but different from the unadjusted effect, confounding must have been present.

This reasoning has led some analysts to test a possible confounder by adjusting for it (possibly after previous adjustment for other confounding variables) and observing whether the estimated effect is altered. If there is very little change of the observed effect (<10% is an arbitrary criterion sometimes adopted), then control of that variable is considered unnecessary. If,

Table 7.2 Two Causally Homogeneous Strata (Conditional Exchangeability Assumed)

	$X = 1$		$X = 0$	
	Exposed	Unexposed	Exposed	Unexposed
Doomed	$N_1A_1P_1$	$N_1(1 - A_1)P_1$	$N_0A_0P_1$	$N_0(1 - A_0)P_1$
Causal	$N_1A_1P_2$	$N_1(1 - A_1)P_2$	$N_0A_0P_2$	$N_0(1 - A_0)P_2$
Preventive	$N_1A_1P_3$	$N_1(1 - A_1)P_3$	$N_0A_0P_3$	$N_0(1 - A_0)P_3$
Immune	$N_1A_1P_4$	$N_1(1 - A_1)P_4$	$N_0A_0P_4$	$N_0(1 - A_0)P_4$

however, a material change occurs, the additional variable is included in the statistical adjustment model.

The logic of this collapsibility argument appears straightforward, but in fact reflects a failure to appreciate certain distinctions that flow from an explicit causal framework. Specifically, we must be aware of the difference between a causal effect and an empirical effect, and between a distribution of response patterns and a particular effect measure based on this distribution. It is certainly true that stratifying on a suitable covariate set can result in unbiased within-stratum causal (and hence also empirical) effects (i.e., conditional exchangeability). However, it is not true that the strata will be *causally homogeneous*. By causally homogeneous, we mean that each stratum contains the same distribution of response patterns. Thus, even though exchangeability exists within each stratum, the empirical effects, although unbiased, will generally not be identical across strata.

To understand why, suppose that the study groups are exchangeable within every stratum, so that all conditional empirical effects are unbiased. We will say that statistical adjustment has been *successful* when this situation prevails. Furthermore, assume that the strata are causally homogeneous. This scenario is represented in Table 7.2 in the case of a dichotomous covariate X with two possible values. Let A_1 be the assignment probability for all four response patterns when $X = 1$, and let A_0 be the assignment probability when $X = 0$. Furthermore, let N_1 and N_0 be the numbers of individuals in the two strata. Within each stratum, let P_j represent the proportion of individuals with response-type j. Throughout this chapter, we will not be concerned with the possibility of sampling bias, which can of course additionally complicate matters. In effect, we assume that the response-pattern distributions represent those prevailing in both the study population and the target population.

The (expected) data for the two strata are displayed in Table 7.2. It is clear from Table 7.2 that, within each stratum, the exposed and unexposed groups have an identical distribution of response patterns, so the study groups within each stratum are exchangeable. Suppose now that we perform an unconditional analysis based on the overall sample. The resulting dataset is shown in Table 7.3. Because the response-pattern distributions in the two study groups are identical, there is exchangeability unconditionally. That is, causal

Table 7.3 Merging of Two Causally Homogeneous Strata (Conditional Exchangeability Assumed)

	Exposed	Unexposed
Doomed	$(N_1A_1 + N_0A_0)P_1$	$[N_1(1 - A_1) + N_0(1 - A_0)]P_1$
Causal	$(N_1A_1 + N_0A_0)P_2$	$[N_1(1 - A_1) + N_0(1 - A_0)]P_2$
Preventive	$(N_1A_1 + N_0A_0)P_3$	$[N_1(1 - A_1) + N_0(1 - A_0)]P_3$
Immune	$(N_1A_1 + N_0A_0)P_4$	$[N_1(1 - A_1) + N_0(1 - A_0)]P_4$

homogeneity of the strata implies the *absence* of confounding, even without conditioning on X. Therefore, if confounding exists, successful statistical adjustment for a covariate could not result in causally homogeneous strata. Consequently, the causal effects within strata will ordinarily vary as well, unless some special circumstances prevail. Thus, statistical control does not typically restore some "true" causal effect that is uniform across the strata. Rather, it partitions the total population into strata that represent subpopulations for which different causal effects occur.

Conceptions of confounding as related to *imperfect comparability* and *lack of random assignment* should by now be quite familiar to the reader. We have explained that both concepts define selection bias as being related to the assignment of subjects to study groups. In this sense, confounding and selection bias are the same. However, the notion of confounding implicitly attributes this nonexchangeability to specific characteristics of individuals whose differing distributions across groups "explain" the lack of exchangeability. These characteristics are typically conceptualized as simple variables (blood pressure, age, educational level, etc.). We have suggested that the relevant individual characteristics can be better envisioned as causal factors, but these factors do not directly influence the outcome. Rather, the "real" confounding factor is the entire causal context.

Randomized assignment of subjects to the study groups is the cornerstone of modern experimental design. Randomization is often described as a method for assuring internal validity by equalizing, or balancing, the distributions of all potential confounding factors across comparison groups. For statisticians, it is quite natural to attribute confounding to the lack of randomization. Statistical adjustment is thus perceived as a method intended to compensate for the lack of randomization. After control for appropriate covariates, it is hoped that effect estimates can be obtained *as if randomization had occurred*. Residual confounding occurs when this "as if randomized" assumption cannot be justified, even after adjusting for all known and measured confounders.

From the causal perspective, randomization works because every causal context has the same probability of becoming exposed. Equivalently, the distribution of all causal contexts is identical across study groups. As a result, the distributions of response patterns are identical, so the groups are exchangeable. In this sense, randomization and exchangeability are closely related concepts. However, randomization guarantees not only exchangeability, but also *conditional exchangeability* when "controlling" for any covariate. By equalizing the assignment probabilities at the underlying level of causal contexts, random assignment guarantees exchangeability for *any* partitioning of the study population, however specified.

On the other hand, after conditioning on a covariate, exchangeability within strata is not necessarily maintained. In more conventional statistical parlance, if an assignment mechanism is strongly ignorable, conditional on a given set of covariates, it may become nonignorable after conditioning on an additional covariate. Thus, strong ignorability (exchangeability) is a weaker condition than random assignment. However, as a practical matter, conditional strong ignorability is sufficient to ensure that statistical adjustment is successful in the sense defined earlier. Rosenbaum and Rubin formulated the concept of strong ignorability to capture the aspect of randomization that is necessary to permit successful adjustment (Rosenbaum and Rubin, 1983a). Strong ignorability is what we are hoping to obtain by conditioning on covariates when true randomization is not feasible.

The notion of confounding as *incorrect model specification* has achieved currency in the context of causal networks and structural equation models (SEM). As discussed in Section 6.6, this school of thought, with deep roots in computer science, econometrics, and the social sciences, utilizes graphs and sophisticated mathematical models to describe causal networks. Within this framework, confounding can be regarded as an important special case of the more general phenomenon of model misspecification. The graphical–SEM approach achieves great power to clarify many subtleties pertaining to causal inference, providing that certain simplifying assumptions are made.

The graphical–SEM framework focuses on an explicit mathematical model of interrelationships among variables. Causality is a property of the deterministic or probabilistic *system* in which these variables are embedded. The metaphor underlying this viewpoint is a network of mechanical or electrical components that are physically interconnected. For example, Pearl has likened causal graphs and structural equation models to engineering diagrams that describe complex electrical circuits (Pearl, 2000, 344–345). An exposure or treatment is seen as a kind of switch (dichotomous) or dial (numerical) that can change the state of the system when operated. The rest of the system remains untouched, except that preexisting causal forces that influenced exposure status are deleted from the system (Pearl, 2000, 70).

Unless a variable is fixed by some action or constraint, the values of the variables are considered to be governed by a joint probability distribution. This distribution represents what is known about the variables, based upon current information:

We will adhere to the Bayesian interpretation of probability, according to which probabilities encode degrees of belief about events in the world and data are used to strengthen, update, or weaken those beliefs. In this formalism, degrees of belief are assigned to propositions ... and those beliefs are combined and manipulated according to the rules of probability calculus. (Pearl, 2000, 2)

The data from a comparative study consist of the observed values of the variables for the subjects in the study. The values for each individual comprise one particular realization of the system's operation. A statistical association between two variables (e.g., exposure and outcome) is a reflection of the system's causal structure as encoded in the probability distributions of the variables. In this framework, confounding is defined to exist when the observed association between exposure and outcome does not equal the causal effect (Pearl, 2000, 184).

This mechanical metaphor is most apt when there are a manageable number of causal factors that tend to act independently in well-understood ways. However, the metaphor is much less appropriate when the causal factors can operate in different ways for different individuals. In that case, the causal effects can be expected to vary in accordance with certain individual characteristics. The graphical–SEM approach works well when the causal mechanism can be accurately described in terms of a system of interconnected variables. On the other hand, it is less useful when the causal mechanism implies that the precise causal dynamics can differ across individuals.

The insights facilitated by the graphical–SEM framework have been described extensively elsewhere (e.g., Pearl, 2000; Morgan and Winship, 2007). Psychologist Steven Sloman has provided a very readable nontechnical introduction (Sloman, 2005). We have chosen to emphasize an alternative causal perspective that allows us to address the implications of individual variability more directly and comprehensively. In particular, this viewpoint is essential for a clearer understanding of a central issue that has caused much confusion: the relationship between confounding and effect modification.

Confounding and Effect Modification

Confounding and effect modification both pertain to the consequences of *conditioning on covariates*. However, these two phenomena are usually regarded as quite distinct:

> The term interaction is used in epidemiology to describe a situation in which two or more risk factors modify the effect of each other....This phenomenon is also known as *effect modification* and is distinguished from the phenomenon of confounding ... *confounding* refers to a situation in which a variable that is associated with both the exposure and the outcome of interest is responsible for the entirety or part of the statistical association between the exposure and the outcome. Interaction between a given variable (*effect modifier*) and a given exposure is a different phenomenon. (Szklo and Nieto, 2007, 183)

We have demonstrated, however, that from the counterfactual perspective this sharp distinction between confounding and effect modification is vacuous. In

Table 7.4   Two Causally Heterogeneous Strata (Conditional Exchangeability Assumed)

	$X = 1$		$X = 0$	
	Exposed	Unexposed	Exposed	Unexposed
Doomed	$N_1 A_1 P_{11}$	$N_1(1 - A_1)P_{11}$	$N_0 A_0 P_{10}$	$N_0(1 - A_0)P_{10}$
Causal	$N_1 A_1 P_{21}$	$N_1(1 - A_1)P_{21}$	$N_0 A_0 P_{20}$	$N_0(1 - A_0)P_{20}$
Preventive	$N_1 A_1 P_{31}$	$N_1(1 - A_1)P_{31}$	$N_0 A_0 P_{30}$	$N_0(1 - A_0)P_{30}$
Immune	$N_1 A_1 P_{41}$	$N_1(1 - A_1)P_{41}$	$N_0 A_0 P_{40}$	$N_0(1 - A_0)P_{40}$

Table 7.5   Merging of Two Causally Heterogeneous Strata (Conditional Exchangeability Assumed)

	Exposed	Unexposed
Doomed	$N_1 A_1 P_{11} + N_0 A_0 P_{10}$	$N_1(1 - A_1)P_{11} + N_0(1 - A_0)]P_{10}$
Causal	$N_1 A_1 P_{21} + N_0 A_0 P_{20}$	$N_1(1 - A_1)P_{21} + N_0(1 - A_0)]P_{20}$
Preventive	$N_1 A_1 P_{31} + N_0 A_0 P_{30}$	$N_1(1 - A_1)P_{31} + N_0(1 - A_0)]P_{30}$
Immune	$N_1 A_1 P_{41} + N_0 A_0 P_{40}$	$N_1(1 - A_1)P_{41} + N_0(1 - A_0)]P_{40}$

the previous subsection, we explained that successful adjustment for covariates would *necessarily* result in causal heterogeneity across strata. This underlying variability at the level of response-pattern distributions will ordinarily (but not necessarily) be echoed in corresponding variability of stratum-specific causal effects, such as the risk ratio or risk difference. Furthermore, because the study groups are exchangeable within strata under the assumption of successful adjustment, the empirical effects will equal the corresponding causal effects. So, effect modification (both causal and empirical) is an intrinsic aspect of successful control for confounding!

What about the converse? Does the absence of causal homogeneity imply confounding? Let us consider the situation portrayed in Table 7.4. Within each of the strata formed by conditioning on X, the study groups are exchangeable. However, the two strata have different distributions of response patterns. Therefore, the strata are causally heterogeneous. The data set formed by merging these strata is shown in Table 7.5. The two study groups are not exchangeable, except perhaps for particular values of the parameters. To understand this more intuitively, consider the numerical example displayed in Table 7.6. The distributions of response patterns in the study groups differ substantially. As a result, the risk ratios also differ (2.00 in the exposed group versus 1.50 in the unexposed).

Now suppose we analyze the unstratified population. Combining the two strata would result in an overall risk ratio of 2.25, as shown in Table 7.7. However, this empirical RR is not a causal effect, because the response-pattern distributions in the exposed and unexposed groups are different.

Table 7.6   Example of Two Causally Heterogeneous Strata (Conditional Exchangeability Assumed)

	X = 1		X = 0	
	Exposed	Unexposed	Exposed	Unexposed
Doomed	25	475	300	200
Causal	25	475	600	400
Preventive	0	0	300	200
Immune	200	3800	1800	1200
	RR = 2.00		RR = 1.50	
	OR = 2.25		OR = 1.71	
	RD = 0.100		RD = 0.100	

Table 7.7   Response-Pattern Distributions for Example of Merged Causally Heterogeneous Strata Shown in Table 7.6

	Exposed	Unexposed
Doomed	0.100	0.100
Causal	0.192	0.130
Preventive	0.092	0.030
Immune	0.615	0.741
	RR = 2.25	
	OR = 2.77	
	RD = 0.163	

As illustrated by this example, when adjustment for a covariate is successful, causal heterogeneity does indeed signal that confounding must have existed prior to the adjustment. So, we have shown that successful statistical adjustment results in causal heterogeneity if and only if there is confounding. Put differently, confounding is *equivalent* to the existence of causal heterogeneity after successfully conditioning on a covariate. This insight contradicts the widespread belief that confounding can be distinguished from effect modification empirically, at least with some degree of confidence:

> When there is a confounding effect, the associations seen across strata of the potentially confounding variable are of similar magnitude to each other but are all different from the crude estimate. (Szklo and Nieto, 2007, 166)

As discussed previously, the fallacy in this statement lies in the belief that conditioning on appropriate confounders can reveal the "true" causal effect that has somehow been "distorted" by the confounders. This true effect is envisioned as invariant across levels of the confounder. But, except in one

Table 7.8 Two Causally Heterogeneous Strata with Assignment Homogeneity: $A_1 = A_0 = A$ (Conditional Exchangeability Assumed)

	$X = 1$		$X = 0$	
	Exposed	Unexposed	Exposed	Unexposed
Doomed	$N_1 A P_{11}$	$N_1 (1 - A) P_{11}$	$N_0 A P_{10}$	$N_0 (1 - A) P_{10}$
Causal	$N_1 A P_{21}$	$N_1 (1 - A) P_{21}$	$N_0 A P_{20}$	$N_0 (1 - A) P_{20}$
Preventive	$N_1 A P_{31}$	$N_1 (1 - A) P_{31}$	$N_0 A P_{30}$	$N_0 (1 - A) P_{30}$
Immune	$N_1 A P_{41}$	$N_1 (1 - A) P_{41}$	$N_0 A P_{40}$	$N_0 (1 - A) P_{40}$

Table 7.9 Merging of Two Causally Heterogeneous Strata with Assignment Homogeneity: $A_1 = A_0 = A$ (Conditional Exchangeability Assumed)

	Exposed	Unexposed
Doomed	$A(N_1 P_{11} + N_0 P_{10})$	$(1 - A)(N_1 P_{11} + N_0 P_{10})$
Causal	$A(N_1 P_{21} + N_0 P_{20})$	$(1 - A)(N_1 P_{21} + N_0 P_{20})$
Preventive	$A(N_1 P_{31} + N_0 P_{30})$	$(1 - A)(N_1 P_{31} + N_0 P_{30})$
Immune	$A(N_1 P_{41} + N_0 P_{40})$	$(1 - A)(N_1 P_{41} + N_0 P_{40})$

special (albeit important) situation, equal values for the conditional (within-stratum) effects can occur if and only if confounding does not exist!

This special situation is shown in Tables 7.8 and 7.9. Here the assignment probability is the same for both strata: $A_1 = A_0 = A$. In this case, the distributions of response patterns differ for the two strata, so there is causal heterogeneity. As a result, the response-pattern distributions in the study overall differ from both of the stratum-specific distributions. However, the exposed and unexposed groups are exchangeable (i.e., there is no confounding). In general, if the assignment probability is constant across strata, causal heterogeneity does not imply confounding. This result has some very important practical implications, as will be seen in the discussion of propensity scores in Section 7.2.

The counterfactual perspective also helps to clarify the relationship between confounding and collapsibility. Recall that collapsibility in its strong form implies that the within-stratum effects are all equal to the overall effect for the study population. Causal collapsibility is thus a stronger version of causal homogeneity; not only is the response-pattern distribution identical across strata, but this same distribution is also found in the study population as a whole. As noted previously, collapsibility of estimated effects is sometimes used as a criterion for selection of covariates.

Suppose that confounding is believed to be absent, possibly after adjustment for a tentatively chosen set of covariates, but we wish to test an additional candidate covariate. After conditioning on this proposed confounder, the stratum-specific empirical effects all equal the crude effect. How can we interpret these results? One possibility is that both the unadjusted and stratum-specific effects are biased, but just happen to be equal. Alternatively, adjustment has been successful and the empirical effects are unbiased, but there really was no confounding prior to adjustment. Under the assumption of successful adjustment, observed collapsibility does provide evidence for lack of confounding.

Suppose, however, that strong collapsibility is *not* observed after adjusting for the additional covariate. Furthermore, assume that adjustment has been successful. Does lack of collapsibility imply the existence of confounding? The answer depends on whether the assignment probabilities vary across the different strata. If they do (assignment heterogeneity), we have the situation portrayed in Tables 7.3 and 7.4. There is confounding. However, if assignment homogeneity occurs, then the study groups must have been exchangeable prior to statistical adjustment (see Tables 7.8 and 7.9). In this special situation, there is no confounding, but the overall response-pattern distribution differs from each of the within-stratum distributions. As a result, the stratum-specific causal effects differ from the unadjusted causal effect. Ordinarily, we would not expect assignment homogeneity except in a randomized experiment. Therefore, as a practical matter, lack of collapsibility after successful adjustment would imply the existence of confounding. Of course, the close relationship between confounding and collapsibility depends on the assumption of successful adjustment.

We have summarized the relationships between causal homogeneity, (strong) collapsibility, and confounding when statistical control for a confounder is successful in Table 7.10. If there is causal homogeneity across strata, then there is no confounding and collapsibility occurs. This statement is true regardless of whether the assignment probabilities are uniform across strata or not. If there is causal heterogeneity, there is always noncollapsibility. However, the existence of confounding depends on whether the assignment probabilities are uniform. With varying assignment probabilities across strata, there is confounding, but with uniform assignment probabilities there is no confounding.

Unfortunately, we cannot observe the response-pattern distributions directly. So, heterogeneity and/or collapsibility can be observed only for a particular *measure* of causal effect, such as the risk ratio or risk difference.

Table 7.10   Relationships among Causal Homogeneity, Effect Collapsibility, and Confounding (Conditional Exchangeability Assumed)

	Assignment Homogeneous	Assignment Heterogeneous
Causal Homogeneity	Collapsibility No confounding	Collapsibility No confounding
Causal Heterogeneity	Noncollapsibility No confounding	Noncollapsibility Confounding

The underlying causal relationships among heterogeneity, collapsibility, and confounding that exist in terms of response-pattern distributions are not necessarily echoed in terms of particular effects. These relationships can depend on the particular mathematical properties of the effect measure (Geng et al., 2001). Miettinen and Cook rejected the collapsibility criterion in part because the existence of effect collapsibility could depend on the particular effect measure being employed. They recognized that the notion of collapsibility was not "rooted in first principles of confounding" that should be independent of the measure selected (Miettinen and Cook, 1981).

Finally, we note that this discussion has focused on *strong* collapsibility. Model-based statistical adjustment procedures generally do not explicitly produce stratum-specific effects. Rather, these techniques yield an overall effect estimate that is interpreted as a uniform effect or some sort of typical effect applicable to the entire population. Therefore, we can only observe whether or not *weak* collapsibility appears to hold. That is, does the adjusted effect yielded by the technique equal the crude effect? These model-based adjustments submerge the issue of possible causal heterogeneity. So, the results cannot reveal whether causal homogeneity would have occurred if simple stratification had been employed. It is not clear, therefore, what exactly we can learn from observing weak collapsibility.

Two Famous Paradoxes

The interrelationships among confounding, effect modification, and statistical control (conditioning) have caused a great deal of confusion. Much of this confusion derives from the lack of an explicit causal model within classical statistical theory. The nature of this confusion is exemplified by two well-known statistical anomalies that cannot be satisfactorily understood without a causal model. The more famous of the two is Simpson's paradox, which refers to the modification, and even possible reversal, of a statistical association between two variables by conditioning on a third variable. In the context of a comparative study, we might find that conditional effects are very similar within each stratum of the third factor (e.g., men and women), but opposite to the direction of the overall effect. The question then arises of which effect (adjusted or unadjusted) represents a causal effect. Usually, it is assumed that the more "refined" conditional analysis represents the "true" causal effect, reflected in the common effect within strata, whereas the unadjusted effect results from confounding.

The second anomaly arose originally in the context of epidemiological case–control studies. Joseph Berkson presented a hypothetical model for the selection of hospital controls that could, under certain circumstances, result in apparent bias (Berkson, 1946). Berkson posited a situation in which the exposure E was cholecystitis, the outcome event D was diabetes, and the controls consisted of patients who came to the same hospital for refractive errors. He assumed that there was no "true" association between cholecystitis and diabetes in the general population. Moreover, whether an individual would be hospitalized, given that she had any of the three conditions, was assumed to be independent of whether she also had any of the other two. On the surface, the use of hospital patients with refractive errors as controls appears reasonable. There seems no reason for people with refractive errors to have rates of cholecystitis that differ from those in the general population.

In his hypothetical scenario, Berkson demonstrated that the effect of exposure in the hospitalized population was in fact positive. This "paradox" came to be known as Berkson's fallacy or Berkson's bias. In Berkson's original example, the positive association arises from the fact that an individual's probability of being hospitalized varies according to how many of the three conditions she has. If she has all three, she is almost three times as likely to reach the hospital. Berkson's bias is often defined more broadly to describe any situation involving an apparent effect in a subpopulation, although there is no true effect in the source population. The term is sometimes used even more generally to encompass any situation in which the selection of individuals into a subpopulation can distort the true relationship in the source population. In this broader sense, it is equivalent to selection bias as defined in terms of the traditional statistical model of bias (see Section 5.2).

What makes each of the phenomena discussed by Simpson and Berkson potentially paradoxical is the ambiguity about causation. In the Simpson case, it seems intuitive that the within-stratum effect and the overall effect cannot *both* be causal. In particular, if a positive causal effect occurs in each subpopulation, an unconditional negative effect must be somehow biased. But in what sense is this statement true? In the Berksonian situation, there is no causal effect for the source population. So, it seems intuitive that an observed effect in a subpopulation must be biased. After all, how could a true positive or negative effect materialize "from thin air" within a subpopulation?

From a counterfactual viewpoint, both paradoxes arise from ambiguity about the nature of the underlying causal models. Tables 7.11 and 7.12 represent the most general situation pertaining to a population that is composed of two strata. These tables can be used to calculate the empirical effects under any set of assumptions about the various parameters. In fact, it is highly informative to translate these tables into spreadsheets and to insert various values of the parameters. Such an exercise can reveal how situations such as those described by Simpson and Berkson can arise from certain causal assumptions.

Table 7.11 General Causal Model for Two Strata

	Stratum 1		Stratum 0	
	Exposed	Unexposed	Exposed	Unexposed
Doomed	$N_1 A_{11} P_{11}$	$N_1(1 - A_{11})P_{11}$	$N_0 A_{10} P_{10}$	$N_0(1 - A_{10})P_{10}$
Causal	$N_1 A_{21} P_{21}$	$N_1(1 - A_{21})P_{21}$	$N_0 A_{20} P_{20}$	$N_0(1 - A_{20})P_{20}$
Preventive	$N_1 A_{31} P_{31}$	$N_1(1 - A_{31})P_{31}$	$N_0 A_{30} P_{30}$	$N_0(1 - A_{30})P_{30}$
Immune	$N_1 A_{41} P_{41}$	$N_1(1 - A_{41})P_{41}$	$N_0 A_{40} P_{40}$	$N_0(1 - A_{40})P_{40}$

In particular, this experiment can provide insight regarding conditions under which the empirical effect based on the total study population or on either of the strata can be interpreted as causal effects.

As one informative example, we have created a set of causal parameters and sample sizes that generate the data originally presented in Section 6.3 to illustrate Simpson's paradox from a traditional perspective (see Tables 6.1– 6.3). The results are displayed in Table 7.13. The risk ratio in each subpopulation is 1.17. The response-pattern distributions for the exposed and unexposed groups in the study as a whole are displayed in Table 7.14, which shows that

Table 7.12   General Causal Model for Population Comprising Two Strata

	Exposed	Unexposed
Doomed	$N_1A_{11}P_{11} + N_0A_{10}P_{10}$	$N_1(1 - A_{11})P_{11} + N_0(1 - A_{10})P_{10}$
Causal	$N_1A_{21}P_{21} + N_0A_{20}P_{20}$	$N_1(1 - A_{21})P_{21} + N_0(1 - A_{20})P_{20}$
Preventive	$N_1A_{31}P_{31} + N_0A_{30}P_{30}$	$N_1(1 - A_{31})P_{31} + N_0(1 - A_{30})P_{30}$
Immune	$N_1A_{41}P_{41} + N_0A_{40}P_{40}$	$N_1(1 - A_{41})P_{41} + N_0(1 - A_{40})P_{40}$

Table 7.13   Data for Simpson's Paradox Scenario in Section 6.3

	Older Women		Younger Women	
	HRT	No HRT	HRT	No HRT
Doomed	0.12	0.12	0.06	0.06
Causal	0.02	0.02	0.01	0.01
Preventive	0	0	0	0
Immune	0.86	0.86	0.93	0.93
Total N	250	4750	3000	2000
	RR = 1.17		RR = 1.17	
	OR = 1.19		OR = 1.18	
	RD = 0.020		RD = 0.010	

Table 7.14   Overall Response-Pattern Distributions for Simpson's Paradox Scenario Shown in Section 6.3

	HRT	No HRT
Doomed	0.065	0.102
Causal	0.011	0.017
Preventive	0	0
Immune	0.925	0.881
Total N	3250	6750
	RR = 0.74	
	OR = 0.72	
	RD = −0.027	

the overall empirical risk ratio is 0.74. So, we have a situation in which there is effect homogeneity, but not collapsibility.

Does this example violate the principle that causal homogeneity must imply collapsibility? No, because the two strata are not *causally* homogeneous, even though their risk ratios happen to be equal. In fact, other effect measures, such as the risk difference and the odds ratio, are *not* equal in the two strata. The parameters in this example were deliberately contrived to produce *effect* homogeneity specifically for the risk ratio, despite the absence of *causal* homogeneity.

This example illustrates a sense in which our naïve intuitions about causality are validated. If we have underlying causal homogeneity (i.e., identical response-pattern distributions) for the two subpopulations, this common response-pattern distribution will necessarily be replicated in the overall population (see Tables 7.2 and 7.3). Thus, the causal effects in the strata cannot differ from that in the total population. The "paradox" arises because it *is* possible to have two equal conditional *effects* that differ markedly from the overall empirical effect in the total population. However, this situation is an artifact that results from a fortuitous combination of the response-pattern distributions in the two strata and the particular effect measure employed.

Now let us consider Berkson's bias. Suppose that there is no observed association between the risk factor and outcome in a certain population. Assume further that this null association is presumed to represent a "real" absence of causal relationship in the population. How can this situation be interpreted? One possibility is that there is no causal effect for any individual. Under this "sharp null hypothesis" there are no *causal* or *preventive* individuals in the population. There can, therefore, be no *causals* or *preventives* in any subpopulation, and no causal effect in the subpopulation. However, it is possible that in the subpopulation the assignment probability for the *doomed* individuals could be higher than that for the *immune* individuals, resulting in a positive empirical effect. Thus a positive empirical effect for the hospitalized patients would be possible, but would necessarily be attributable to assignment bias. In this sense, our naïve intuitions are validated: something cannot come from nothing, simply by selecting a particular subpopulation.

An alternative interpretation is that there is a "null effect" in the population. In that case, some individuals *can* be causally affected, but the effects happen to balance out across the population ($P_2 = P_3$) Is it possible that a nonnull causal effect could occur in a subpopulation (e.g., hospitalized individuals) but not in the general population? If so, that would mean that an observed nonnull effect would not necessarily be a distortion of the "true" effect. Rather, the observed effect would estimate a *different* causal effect applicable to the hospitalized individuals.

To see how this could occur, let X be an indicator of whether the individual is included in the subpopulation. Let $X = 1$ for a person who is hospitalized, and let $X = 0$ for one who is not hospitalized. Assume that in the general population the exposed and unexposed groups are exchangeable. Furthermore, assume that exchangeability holds also within each of the two strata. From the preceding discussion, this situation could arise only if the strata were strongly collapsible; the causal effect could not deviate from the null effect in the general population. So, it would seem that a nonnull effect observed for the hospitalized patients must be the result of bias.

However, suppose we assume the rate of exposure to be unrelated to individual characteristics. Then the assignment probability would be the same for each stratum as it is in the general population ($A_1 = A_0 = A$). We would have the situation represented by Tables 7.8 and 7.9. It would be possible to have $P_{21} > P_{31}$, but $P_{20} < P_{30}$ in such a way that the two stratum-specific causal effects effectively "neutralize" each other, resulting in an overall risk ratio of 1.0. From Table 7.9, the condition necessary to balance the proportions of *causal* and *preventive* individuals in the total population is seen to be $N_1 P_{21} + N_0 P_{20} = N_1 P_{31} + N_0 P_{30}$. So, if the effect in the general population results from a balance between *causals* and *preventives*, rather than a complete absence of any effect, the nonnull effect observed for the hospitalized patients might represent a causal effect.

Effect Modification and Causal Modification

For each individual, the response pattern is determined by his values of the underlying causal factors that form the causal context. Successful control for confounding means that after conditioning on certain covariates, the study groups are exchangeable. That is, the bias produced by the different distributions of causal contexts is eliminated after conditioning on the covariates. We have shown that successful adjustment would normally produce effect modification, with varying causal effects across different strata. Moreover, because of within-stratum exchangeability when adjustment is successful, the empirical effects would equal the causal effects, and therefore would vary as well. So, there is a close relationship between confounding and effect modification. Now we will consider the nature of effect modification more generally.

Ideally, we could adjust for confounding by conditioning on the values of all causal factors that make up the causal context. For example, we could condition on all four causal factors in the pie charts we have used for illustrative purposes. Controlling in this way for the entire causal context would determine a unique response pattern shared by all the individuals in each stratum. Now suppose that there is a "relevant" subset of the factors (say, Factors 1 and 3) that actually vary across the groups and generate the nonexchangeability. Other factors (Factors 2 and 4) are irrelevant in the sense that the study groups are exchangeable within strata formed by conditioning on Factors 1 and 3. Then it would be sufficient to condition on the values of the relevant subset of causal factors (Factors 1 and 3) to correct for confounding.

Because the within-stratum exchangeability holds only if the relevant factors (e.g., Factors 1 and 3) are fixed, the values of these factors must affect the response patterns for some individuals, conditional on the values of other factors (e.g., Factors 2 and 4). In other words, the relevant factors must together comprise a causal modifier. Let X represent this (possibly multivariate) causal modifier. For simplicity, we will assume that X is a single dichotomous factor. Suppose we control for confounding by creating two strata based on the values of X.

The result of this stratification can be shown by creating a table that displays the possible response patterns based on the values of both F and X (Greenland and Poole, 1988). We can think of each *combination* of the value for F and the value for X as the exposure state. In particular, we imagine that the values of F and X can *both* be manipulated, while fixing the values of all the other factors. The response pattern for the "intervention" (F,X) now consists of a set of 4 potential outcomes instead of just 2. Table 7.15 shows the 16 possible response patterns. For example, it is possible that some people might be causal if $X = 0$ but doomed if $X = 1$ So, for these individuals, X is a causal modifier of F.

In this particular example, we denote by P_{DC} the proportion of individuals in the target population who would change from *doomed* under $X = 1$ to *causal* under $X = 0$. Proportions for other response patterns for (F,X) are similarly defined. We can now calculate the stratum-specific response-pattern distributions. The proportion of *causal* individuals when $X = 1$ is $P_{CD} + P_{CC} + P_{CP} + P_{CI}$. When $X = 0$, this proportion is $P_{DC} + P_{CC} + P_{PC} + P_{IC}$. A similar situation pertains to each of the other three response patterns. The complete distributions are shown in Table 7.16. These response-pattern distributions clearly differ

Table 7.15   Response Patterns for (F, X) when X is a Causal Modifier of F

Response Pattern	$F = E,$ $X = 1$	$F = U,$ $X = 1$	$F = E,$ $X = 0$	$F = U,$ $X = 0$	Proportion
Doomed/Doomed	D	D	D	D	P_{DD}
Doomed/Causal	D	D	D	H	P_{DC}
Doomed/Preventive	D	D	H	D	P_{DP}
Doomed/Immune	D	D	H	H	P_{DI}
Causal/Doomed	D	H	D	D	P_{CD}
Causal/Causal	D	H	D	H	P_{CC}
Causal/Preventive	D	H	H	D	P_{CP}
Causal/Immune	D	H	H	H	P_{CI}
Preventive/Doomed	H	D	D	D	P_{PD}
Preventive/Causal	H	D	D	H	P_{PC}
Preventive/Preventive	H	D	H	D	P_{PP}
Preventive/Immune	H	D	H	H	P_{PI}
Immune/Doomed	H	H	D	D	P_{ID}
Immune/Causal	H	H	D	H	P_{IC}
Immune/Preventive	H	H	H	D	P_{IP}
Immune/Immune	H	H	H	H	P_{II}

Table 7.16   Distribution of Response Patterns for F when X is a Causal Modifier of F

	$X = 1$	$X = 0$
Doomed	$P_{DD} + P_{DC} + P_{DP} + P_{DI}$	$P_{DD} + P_{CD} + P_{PD} + P_{ID}$
Causal	$P_{CD} + P_{CC} + P_{CP} + P_{CI}$	$P_{DC} + P_{CC} + P_{PC} + P_{IC}$
Preventive	$P_{PD} + P_{PC} + P_{PP} + P_{PI}$	$P_{DP} + P_{CP} + P_{PP} + P_{IP}$
Immune	$P_{ID} + P_{IC} + P_{IP} + P_{II}$	$P_{DI} + P_{CI} + P_{PI} + P_{II}$

Table 7.17   Distribution of Response Patterns for F when X is a Not a Causal Modifier of F

	$X = 1$	$X = 0$
Doomed	P_{DD}	P_{DD}
Causal	P_{CC}	P_{CC}
Preventive	P_{PP}	P_{PP}
Immune	P_{II}	P_{II}

between the two strata. Thus, conditioning on a causal modifier (or set of factors comprising a causal modifier) results in causal heterogeneity.

From this analysis, we see clearly how conditioning on a causal modifier to correct for confounding can give rise to effect modification. However, effect modification can also occur in another manner. Suppose that the study groups are exchangeable, as for example in a randomized controlled trial (RCT), and we stratify the study population based on a factor X that is *not* a causal modifier of the exposure. Then the response pattern will not depend on the value of X *for any individual*. In that case, Table 7.16 reduces to Table 7.17 and there is no effect modification attributable to causal synergy. However, it is still possible that X is statistically associated with causal factors in such a way that the resulting strata are not causally homogeneous.

In Section 3.3, we gave the example of stratification on the basis of school type (public vs. private). In that situation, the sampling mechanisms for the two school populations differed. Thus, the causal effect in the schools differed, even though the causal effect for any individual student was unrelated to which school she attended. So, effect modification can arise in a manner that is akin to sampling bias. When that happens, effect modification does not necessarily imply confounding.

Because the different populations in the strata are the result of selection forces that may be unstable, the nature of the effect modification might change over time. For example, economic conditions might worsen, making it more difficult for middle-class children to attend the private school. In that case, the causal effect in the private school might be altered, affecting the nature of the effect modification. In this sense, such "circumstantial" effect modification may be of less scientific or policy interest. On the other hand, if the selection mechanisms are stable, the effect modification could have important practical implications. For example, a public school principal would want to know whether the program being evaluated would work for her students, who might be similar to those in the public school where the study was conducted.

Confounding when the Outcome is Numerical

The conceptual issues involved with a numerical outcome are identical to those discussed in the preceding subsection. However, the mathematical representation of these issues is quite different and provides additional insight. In Chapter 5, we developed certain mathematical results to describe the nature of selection bias when the outcome is numerical and the average causal effect (ACE) is being estimated. In particular, we showed that bias associated with a nonignorable assignment mechanism can be expressed as

$$\text{Bias}_A = \frac{1}{\mu_A (1 - \mu_A)} \left[\text{Cov}_s (A, M) + \frac{1}{2}(1 - 2\mu_A) \text{Cov}_s (A, \Delta) \right] \quad (7.1)$$

For a review of the derivation of Equation 7.1 and the definition of terms, the reader is referred to Section 5.7. Intuitively, Equation 7.1 indicates that assignment bias has two components. One component is related to the statistical association (covariance) between the assignment probability A and the mean individual outcome M (the average of the individual's two potential outcomes). The second component depends on the covariance between A and the individual causal effect Δ (the difference between the individual's two potential outcomes). Confounding is absent if and only if $\text{Bias}_A = 0$.

Now suppose that confounding exists, so that $\text{Bias}_A \neq 0$. Let X be a covariate that is used to stratify the study population. If this adjustment is successful, then within each stratum there will be no bias. Let us define $\text{Bias}_{A|x}$ to be the within-stratum bias conditional on the value of $X = x$. Furthermore we will suppress the dependence on the study population, which is assumed throughout the following discussion. From Equation 7.1, we can write the conditional bias as

$$\text{Bias}_{A|x} = \frac{1}{\mu_{A|x} (1 - \mu_{A|x})} \left[\text{Cov}(A, M|x) + \frac{1}{2}(1 - 2\mu_{A|x}) \text{Cov}(A, \Delta|x) \right] \quad (7.2)$$

The covariances in Equation 7.2 are based on the joint distribution of (A,M) and (A,Δ) conditional on each x value. From Equation 7.2 we see that adjustment will be successful if $\text{Cov}(A,M|x) = 0$ and $\text{Cov}(A,\Delta|x) = 0$ for all values of X. Conversely, confounding bias will usually exist if either of these covariances is nonzero. The exception occurs if the two covariances fortuitously have nonzero values that happen to cancel out.

Now suppose that we consider the special case when the average effect does not depend on the value of X. That is, $E(\Delta|x) = E(\Delta)$ for all x. In the Appendix to this chapter, we demonstrate a necessary condition for confounding to exist ($\text{Bias}_A \neq 0$) when such causal homogeneity exists. We prove that confounding can occur only if $E(M|x) \neq E(M)$ for some value x of X. That is, confounding can occur whenever there is variability among the conditional average values of M.

In the previous section we established the general rule that if we successfully condition on a covariate, causal heterogeneity must exist. Therefore, causal *effect* heterogeneity (effect modification) would ordinarily be anticipated. However, we explained that there are special cases in which underlying causal heterogeneity does not translate into causal effect modification. In such cases, it is possible to obtain the classic pattern: equal stratum-specific effects coupled with a different unadjusted effect. The result mentioned above implies that if $E(\Delta|x) = E(\Delta)$ for all values of X, then it is possible for confounding to occur ($\text{Bias}_A \neq 0$) *without causal effect modification*. In effect, the usual presumption about confounding is seen to hold in the special case of estimating the ACE for an effect that is constant, or is at least unrelated to the covariates. Because this special case has become so predominant in statistical analyses, causal effect homogeneity is commonly regarded as an essential aspect of confounding.

In the Appendix, we show specifically how the usual applications of regression analysis and related techniques for statistical adjustment rest on the assumption of this special case. Suppose that $E(\Delta|x) = E(\Delta) = \alpha$. Then if X is sufficient for adjustment, we demonstrate that there exists a function $f(x)$ for which

$$E(W_E|x) = \alpha + f(x) \tag{7.3}$$

$$E(Z_U|x) = f(x) \tag{7.4}$$

Let Y denote the observed outcome value for any individual in the study, regardless of exposure status. Then we can express Equations 7.3 and 7.4 as

$$E(Y|x) = \alpha F + f(x) \tag{7.5}$$

For example, if we assume that $f(x)$ is linear and there is a single covariate, then

$$E(Y|x) = \alpha F + \mu + \beta x \tag{7.6}$$

Equation 7.6 is the standard form of the regression model.

In summary, standard regression modeling will generate an unbiased estimate of the ACE if the causal effect is unrelated to the covariates. On the other hand, if the causal effect is related to X, the properties of these routinely employed estimates are generally unclear. Furthermore, this entire discussion is predicated on the assumption that X is sufficient for successful control. If this is not a valid assumption, then residual bias will be present, even if the individual causal effects are not associated with X. A comprehensive discussion of the counterfactual interpretation of regression-based adjustment from a somewhat different perspective can be found in Chapter 5 of Morgan and Winship (2007).

Finally, we note that $E(\Delta|X) = E(\Delta)$ implies that we have not only causal effect homogeneity, but also effect collapsibility. We showed previously that, when adjustment is successful, confounding implies lack of *causal* collapsibility. However, *effect* collapsibility and confounding can coexist in the special case of estimating an ACE when $E(\Delta|X) = E(\Delta)$. The reason for this anomaly is that the causal heterogeneity that exists between strata pertains only to the "level" of the outcome, as embodied in M, but not to the causal effect Δ. It remains true that causal heterogeneity exists, because the distributions of response patterns actually do vary across strata. However, these differences in response-pattern distributions do not affect the average causal effects, which remain constant across the different strata.

7.2 CORRECTION: STATISTICAL CONTROL FOR CONFOUNDING

In theory, confounding bias can be eliminated by conditioning on certain covariates. However, in practice this correction of confounding is very difficult to achieve. In this section, we discuss the mechanical aspects of the problem.

Suppose that we have data on an observed set of covariates X and that conditioning on these covariates would successfully control for confounding. How can we obtain an unbiased empirical effect? Recall that successful control means that if we condition on the value of X, we can eliminate within-stratum confounding bias. In causal terms, the study groups are exchangeable and the assignment mechanism is strongly ignorable conditional on X. The critical question of how to identify a sufficient set of covariates for successful statistical adjustment is deferred to the following section.

Traditional Approach to Statistical Adjustment

A statistical adjustment technique is a method that uses data on covariates to reduce or eliminate confounding bias. The usual aim of statistical adjustment is to obtain an unbiased estimate of the empirical effect for the entire study population. Conditional effects for subpopulations based on the value of one or more covariates are also sometimes of interest. However, these interactions are typically accorded much less attention than the main effects. As discussed in the previous section, statistical adjustment can be achieved most straightforwardly by forming strata based on each possible value of X. Then we can obtain an unbiased estimate within each stratum. These conditional (stratum-specific) effect estimates can be combined by calculating an appropriate weighted average.

When there are multiple variables comprising X, the number of such strata becomes very large, and the number of individuals in each stratum very small. To address this dilemma, the analyst usually resorts to the use of a statistical modeling approach. There are many possible methods for performing statistical adjustment. Most of these techniques fall under the general rubric of *regression analysis*, which includes least-squares regression, Cox proportional hazards regression, Poisson regression, and logistic regression. For example, the logistic regression model to estimate the effect of risk factor F on the probability P of the outcome event can be expressed as

$$\text{Log}\left(\frac{P}{1-P}\right) = \mu + \alpha F + \sum_j \beta_i X_i \qquad (7.7)$$

Here the left-hand expression, called the *logit*, is the natural logarithm of the odds. The logit is expressed as the sum of a constant term μ, the causal effect (α if $F = 1$ and 0 if $= 0$), and a linear combination of the covariates. If we condition on any fixed values of the covariates, then it is well known that α is equal to the logarithm of the odds ratio (e.g., Hosmer and Lemeshow, 2000, 63). Because of this convenient mathematical property, logistic regression is commonly used to estimate the odds ratio, while "controlling" for multiple covariates.

Regression models as applied in this context embody two critical and rather restrictive assumptions. First, these models effectively assume that the causal effect is essentially uniform, or at least unrelated to any of the covariates. As usually applied, regression modeling makes no allowance for effect modification by any of the covariates. The causal effect in Equation 7.7 is always α regardless of the values of the covariates. Lifting this restriction is possible, but at the expense of greater complexity and loss of statistical power. Second, regression models assume a very specific mathematical form for the joint causal effect of multiple covariates. The total effect of the covariates is modeled as the sum of their individual effects. This modeling assumption is called *linearity* because the dependent variable or some appropriate mathematical transformation (e.g., logit) is a linear combination of the independent variables (covariates).

Although most methodologists do not believe that a regression model represents reality exactly, they accept it as a working approximation. Twenty years ago, Sander Greenland articulated clearly the limitations of the standard modeling methodology as applied in epidemiology, focusing especially on the two main underlying assumptions just mentioned. He observed that our justification for relying on these methods is rather tenuous:

> Although much research has been done examining alternative or more general forms, none of this research has directly addressed the issue of how much inferences about exposure effects will be distorted if the effects are estimated from a model whose assumptions are violated (i.e., a mis-specified model). (Greenland, 1989, 341)

This troubling state of affairs has changed little and may continue indefinitely, because there is no way to corroborate the model's working assumptions empirically. For the most part, data analysts accept the standard mathematical models uncritically, because no practical alternatives are available.

Greenland observes that techniques such as logistic regression have been enormously useful as tools for obtaining a kind of "rough approximation" to reality. But the limitations of these methods are not well understood even by methodological experts, let alone practitioners who must rely on the results generated by these methods. As he suggests, it could be useful to perform more research, both theoretical and empirical, on the extent to which standard techniques for causal inference are trustworthy. However, the limited investigation that has so far been conducted is not very encouraging. In their extensive and penetrating discussion of linear regression methods from a counterfactual perspective, Morgan and Winship observe that:

> ... much of the received wisdom on regression modeling breaks down in the presence of individual-level heterogeneity of a causal effect, as would be present in general when causal effects are defined with reference to underlying potential outcomes tied to well-defined causal states. (Morgan and Winship, 2007, 142)

Causal Perspective on Statistical Adjustment

From a counterfactual perspective, we have seen that confounding is equivalent to lack of exchangeability, or equivalently to lack of strong ignorability. In Section 7.1, we described what happens when we condition on a particular covariate. We termed the statistical adjustment for covariate(s) X successful if the study groups are exchangeable within each stratum. Such conditional exchangeability is equivalent to a constant value of the assignment variable A within each stratum. For a dichotomous outcome, this situation was illustrated in Tables 7.4 and 7.5. For any value x of X, let us now define A_x as this constant assignment probability conditional on x. For the example shown in Tables 7.4 and 7.5, A_x has two possible values: A_1 and A_0.

In Tables 7.8 and 7.9, we portrayed the special case when the assignment probability does not vary across strata, that is, $A_x = A$ for all x. In that particular situation, there is no confounding because the assignment probability *for each response pattern* is A in every stratum. Consequently, the assignment probability for each response pattern in the pooled population is also A. The study groups are exchangeable for the total study population and conditional on each stratum. Suppose that we have a large number of strata, based on different values of X and that the value of A_x can vary across the strata. If we merge some of the strata that share a common value of A_x to form a "coarser" set of strata, the study groups will be exchangeable within each of these coarser strata. Finally, suppose we aggregate *all* strata that share a particular value of A_x to construct the coarsest stratification such that the groups are exchangeable within strata. In effect, we would then be conditioning on the value of A_x itself. Consequently, A_x as a potential covariate is sufficient to permit successful adjustment for confounding. Conditioning on A_x results in the coarsest stratification that produces successful adjustment.

The practical problem, of course, is that we cannot directly observe the value of A_x. However, by virtue of conditional exchangeability, A_x has a unique value for each value x of the covariate X. Therefore, at least in principle, A_x can be expressed as some mathematical function of x. This function is called the *propensity score* (Rosenbaum and Rubin, 1983a). The propensity score is often defined simply as the conditional probability of exposure given the covariate X. This statement is technically correct. However, the propensity score's practical importance lies in its properties when conditional exchangeability occurs. Without this critical assumption, A_x would not be uniquely defined, because the assignment probability conditional on x could vary according to the response pattern. Then, conditioning on the propensity score would not produce conditional exchangeability.

Following Rosenbaum (2002), we denote the propensity score, expressed explicitly as a mathematical function of x, by $\lambda(x)$. Suppose we could somehow determine the particular mathematical form of $\lambda(x)$. Then we could calculate the propensity score for any individual based on her observed covariate value x. We could also condition on the propensity score by creating strata that are homogeneous with respect to the value of $\lambda(x)$. Within each stratum, exchangeability would prevail. In effect, the propensity score function (if known) renders the conditional assignment variable A_x *visible* and thus available for use in statistical adjustment procedures.

This approach allows the data analyst to reduce a highly multivariate statistical problem to a much simpler form. Rather than trying to adjust for many variables simultaneously, the analyst can focus on controlling for only one, the propensity score. Straightforward methods such as matching and stratification then become feasible. These methods can be used to estimate causal effects within strata that are relatively homogeneous in terms of the propensity score. These stratum-specific estimates can then be combined by averaging them to obtain an overall estimated effect. If the causal effect being estimated is the ACE, this overall effect is truly an average effect across the entire study population.

In general, the averaging must weight the strata according to their proportions in the study population, or perhaps some other target population of interest. In practice, the strata are often engineered to be of equal size, and the ACE in the study population can thus be estimated by a simple average of the stratum-specific average effects. If some other causal effect, such as the risk ratio or odds ratio, is being estimated, an average across strata can still be calculated. However, the result cannot be interpreted directly as an average individual effect.

In theory, propensity-score analysis offers the major advantage of not requiring reliance on unverifiable regression model assumptions. However, the propensity scores are not known. So, the practical value of this approach depends on our ability to estimate accurately the propensity scores. How can this estimation be accomplished? Rosenbaum and Rubin proved that, under fairly unrestrictive assumptions, logistic regression can be used to estimate $\lambda(x)$. That is, there exists a logistic model with certain parameters that can represent $\lambda(x)$. Furthermore, they showed that if this model is properly specified, its parameters can be estimated from observable data and will approach the true parameters as the sample size increases.

Unfortunately, the specific form of this logistic model is unknown. For example, a correct model may involve nonlinear functions of measured variables (e.g., powers or logarithms) or interaction terms (products of two or more covariates). So, it might seem that we face the same problem of unverifiable assumptions that bedevils traditional modeling approaches. However, Rosenbaum and Rubin proved another result that establishes a useful criterion for determining whether the estimated $\lambda(x)$ is correct. We can explain their essential logic in our notation as follows. Let $f(x)$ represent the probability distribution or density function for the covariate X. Suppose that $\lambda(x)$ is any "balancing" function such that:

$$f(x|E, \lambda) = f(x|U, \lambda) \tag{7.8}$$

Equivalently, we can express Equation 7.8 as:

$$f(x|E, \lambda) = f(x|\lambda) \tag{7.9}$$

Now in general, we have from Bayes' theorem that:

$$f(x|E, \lambda) = \frac{P(E|x, \lambda) f(x|\lambda)}{P(E|\lambda)} \tag{7.10}$$

Therefore, Equation 7.8 implies that:

$$P(E|x, \lambda) = P(E|\lambda) \tag{7.11}$$

Thus, λ must be such that A_x equals $\lambda(x)$ or is a function of $\lambda(x)$.

Thus, if the distributions of X in both study groups are identical, conditional on $\lambda(x)$, then $\lambda(x)$ must be essentially the propensity score. This "balancing" criterion can be tested by comparing the (multivariate) distributions of X in the different treatment groups after conditioning on the estimated $\lambda(x)$. In practice, it is typical to start with a relatively simple model and to test its adequacy empirically before introducing any additional complications. Based on the provisional model, the propensity score for each individual in the study population is calculated. The propensity scores are then arrayed from highest to lowest and are aggregated into a small number (usually between 5 and 10) of relatively homogeneous strata. Within each of the strata, the distributional balance of the covariates is tested. If balance appears to be achieved, the estimated $\lambda(x)$ is accepted as an adequate approximation. If these distributions are not balanced, the provisional model must be rejected. The analyst may then try introducing transformations of covariates and interaction terms to find a model able to achieve balance.

Advantages and Limitations of Propensity Scores

The analytic strategy of estimating $\lambda(x)$ using logistic regression and testing for balance can in theory solve the adjustment problem. The precise mathematical form of the logistic regression model used to estimate $\lambda(x)$ is not critically important, and it may not even be unique. What matters is whether conditioning on the estimated value of $\lambda(x)$ achieves distributional balance. If such balance is attained, then our estimated $\lambda(x)$ is a good approximation to A_x for all values of x. Conditioning on this estimated $\lambda(x)$ thus generates unbiased (conditional) causal effects. Of course, this property obtains only if the covariates are sufficient to adjust successfully for confounding. If exchangeability does not hold after conditioning on X, it can be shown that covariate balance still exists (e.g., Rosenbaum, 2002, Chapter 10). However, within-stratum exchangeability is no longer assured.

Unlike the key assumptions underlying traditional regression-based techniques, the criterion of covariate balance can, at least in principle, be validated empirically. On the other hand, there is no consensus regarding just how the test for balance should be conducted. A variety of alternative methodologies to assess balance across the groups have been proposed and debated (for a summary, see Morgan and Winship, 2007, Chapter 4). Furthermore, several different approaches to the implementation of stratification and matching techniques based on estimated propensity scores have been developed. Grappling with these practical difficulties has led to controversy among methodologists regarding whether (or at least when) propensity-score analysis really is preferable to other approaches. For example, the reevaluations of the National Supported Work Demonstration (NSW) manpower training programs discussed in Chapter 1 highlighted many of the trade-offs between traditional econometric methods and techniques based on propensity scores.

One advantage of propensity-score analysis is the ability to identify situations in which the study groups differ to an extent that any attempt at statistical control will be infeasible. From a traditional perspective, this situation might be signaled by a substantial *separation* between the covariate distributions (e.g., Anderson et al., 1980, 63–66); there may be some values of the covariate vector that appear exclusively (or predominantly) in one study group or the other. Attempting to compare groups in the case of extreme separation is referred to as *extrapolation* across groups.

Suppose that there are several covariates included in the statistical adjustment model. In the traditional analytic framework, there is no easy way to detect this problem. The mathematical algorithms applied to generate effect estimates using conventional methods do not break down because of separation between the groups. This is a strength of these methods, but it can also be a liability. These powerful statistical tools allow apparently meaningful results to be obtained with a minimum of subjective judgment. However, when there is substantial separation between groups, the results might be seriously misleading.

From the counterfactual perspective, the problem is that there are some values of x for which A_x may be very close to 0 or 1. Individuals with such x values are nearly always exposed or nearly always unexposed. Indeed, it is possible that certain individuals are *always* assigned to only one of the possible exposure states. To avoid such pathological cases, the inventors of propensity-score methodology specified the ancillary condition that $\lambda(x)$ must be strictly between 0 and 1 (Rosenbaum and Rubin, 1983a). Someone with an extremely high (or low) value of A_x has characteristics that strongly predispose her to one exposure state or the other. For example, if some women would rarely choose hormone replacement therapy (HRT) under any circumstances, then estimating the potential effect of HRT on them would be of little value, unless the factors influencing treatment decisions are likely to change in the future. Moreover, as a practical matter, it would be difficult to obtain an accurate estimate, because there would be so few exposed subjects.

Propensity-score methods provide a direct assessment of the degree of separation between the groups. After estimating the propensity score for each individual in the study population, we can easily observe how many have extreme values. If the proportion of such individuals is substantial, we may choose either to forego analysis or to eliminate those subjects who have extreme scores. One criterion that has been suggested is to restrict the analysis to individuals with propensity scores in the *region of common support* (e.g., Heckman, Ichimura, and Todd, 1997). This region is the range of propensity-score values that are represented in both study groups. Often, the lower limit of this region of support is simply the higher of the two minimal values for the study groups, and the upper limit is the lower of the two maximum values for the two groups. Any individual who has a value outside this range might be excluded from the analysis.

Focusing on the degree of separation between study groups also draws attention to the issue of external validity. Restriction of the analysis to a subset of "comparable" individuals allows estimation of a causal effect for a subpopulation that may not be representative of the target population. On the one hand, this population may be more relevant, because it excludes those for whom a change in exposure state is highly unlikely to occur. On the other hand, as we demonstrated in Chapter 5, the causal effect depends on the characteristics (and therefore response patterns) of those included in the analysis. So, the population to which the results can be generalized must be carefully considered. In effect, propensity scores can help to reveal trade-offs between internal and external validity that may otherwise go unnoticed.

Perhaps most important, but rarely recognized, stratification on the basis of propensity scores can shed light on the relationship between the effect size and the value of $\lambda(x)$. As explained in the previous section, conditioning successfully on a covariate should normally be expected to result in causal heterogeneity. Therefore, a careful interpretation of the pattern of conditional effects may be more informative than a population-wide effect. For example, suppose there is a positive correlation between the estimated effects (e.g., risk ratios) and the propensity scores. Assume also that the study population is representative of the target population observed under *natural conditions*. This finding would provide evidence of self-selection occurring in the target population, with those population members most likely to benefit also being those who are most likely to become exposed. In this situation, a weighted average effect over the entire target population, or even over the exposed population, may be of little relevance.

For example, suppose hypothetically that HRT proved highly beneficial for women who had a propensity score exceeding some cutoff, but had a slightly harmful effect for those with lower scores. An overall risk ratio for the entire target population might be close to 1.0, implying absence of any effect. However, HRT's protective effect for those currently receiving it would be much more interesting. Our hypothetical study would suggest that postmenopausal women, in consultation with their physicians, were generally capable of deciding intelligently when HRT is likely to be beneficial. Note that randomized experiments will typically not be able to identify such a phenomenon, because the natural assignment mechanism is altered by randomization. In theory, we could restrict a randomized study to individuals who were strongly predisposed to use HRT. However, such women might be very reluctant to enter a trial in which there was a possibility of being assigned to a placebo.

The relatively recent advent of methods based on propensity scores challenges the supremacy of the reigning statistical methodology. The ramifications of this alternative approach will continue to become clearer over time. At present, there is much ongoing research and debate over the proper role that traditional modeling and propensity-score methods should play. The major trade-offs between these two general approaches to statistical adjustment for confounding are fairly clear:

> The main strengths of regression analysis are its computational simplicity, its myriad forms, its familiarity, ... and the ease with which one can induce computer software to generate standard errors. ... But, ... regression models have some serious weaknesses. Their ease of estimation tends to suppress attention to features of the data that matching techniques force researchers to consider, such as the potential heterogeneity of the causal effect and the alternative distributions of covariates across those exposed to different levels of the cause. Moreover, the traditional exogeneity assumption of regression (e.g., in the case of least-squares regression that the independent variables must be uncorrelated with the regression error term) often befuddles applied researchers who can otherwise easily grasp the stratification and conditioning

perspective that undergirds matching. As a result, regression practitioners can too easily accept their hope that the specification of plausible control variables generates an as-if randomized experiment. (Morgan and Winship, 2007, 165–166)

At present, standard variants of linear regression analysis have become so entrenched that withdrawal from them has become difficult to contemplate. Like the proverbial drunk looking for his lost keys under the lamppost (because that is where the light is), researchers and methodologists feel tethered by necessity to these traditional methods. But data analysis at its best demands both deep subject-matter insight and methodological creativity. Propensity-scores analysis may foster these qualities by providing a more intuitive "feel" for the structure of the data.

7.3   DETECTION: ADEQUACY OF STATISTICAL ADJUSTMENT

How can we know whether statistical adjustment has been successful? In theory, the study groups can be considered exchangeable if we have conditioned on a set of covariates that includes all relevant causal factors, or proxies for these factors. Suppose that Factors 1 and 3 are the critical factors. Then if we were to form strata based on the possible values of these two factors, the study groups would be exchangeable within these strata. However, it might be possible to obtain such conditional exchangeability more efficiently by using a statistical modeling approach. In particular, we could condition on the propensity score.

The success of such an analysis would hinge on whether all of the causal factors that play a role in determining the different distributions of response patterns between groups have been included in the statistical model. For example, suppose we attempt to build the propensity scores based only on Factor 1. Then there exists a possibility of bias, because the distribution of Factor 3 might differ between the study groups and might result in different distributions of response patterns. In general, we do not know whether this difference actually does induce confounding, but the possibility exists. One way to be more certain would be to attempt a detailed elaboration of the causal mechanism, as in the graphical–SEM framework. However, as noted previously, this approach requires some rather stringent assumptions.

In most situations, the best we can do is to demonstrate the strong *plausibility* that all important factors have been taken into account. The remainder of this section will discuss the principal means that have been employed to establish causation from observational data. At present, there is much skepticism among methodologists about the viability of this enterprise, and it is certainly true that proof of causation is impossible. However, in scientific research on human subjects, absolute proof is an unrealistic standard. Rather, our objective must be to develop the strongest possible evidence to help reach a convincing conclusion.

Causal Criteria

Have all the relevant causal factors been taken into account? Can we interpret an empirical effect obtained from an observational study as a causal effect? The answers to these questions can almost never be established with certainty. Judgment is required, based on all the available evidence. Well-reasoned judgment applies sound logic and scientific acumen. In general, it is difficult to formalize the methods of reasoning by which any scientist reaches a conclusion about presence or absence of confounding. Ultimately, the weighing of evidence at any point in time entails a judgment:

> Evidence, even extensive evidence, does not compel belief. Rather than being forced to a conclusion by evidence, a scientist is responsible and answerable for conclusions reached in light of evidence, responsible to his conscience and answerable to the community of scientists. (Rosenbaum, 1999, 260)

Although the assessment of residual confounding is not fully amenable to systematic analysis, certain principles, or guidelines, may be helpful. In epidemiology, especially, various "criteria" for evaluation of causality have been proposed. The most famous and influential of these proposals was made by Sir Austin Bradford Hill (Hill, 1965). Hill suggested nine considerations that could be helpful in assessing the likelihood that an observed association can be attributed to causation:

1. Strength
2. Consistency
3. Specificity
4. Temporality
5. Biological gradient
6. Plausibility
7. Coherence
8. Experiment
9. Analogy

By the *strength* of a causal effect, Hill was referring to its magnitude. By *consistency*, he meant that the effect was replicated (approximately) in a variety of different settings. *Specificity* occurs if the effect (or perhaps its magnitude) appears only for the specific outcome, but not for other outcomes. *Temporality* means simply that the exposure is known to have preceded the disease. A *biological gradient* is present if increasing levels of exposure are associated with stronger effects. A putative effect has *plausibility* to the extent that it conforms to expectations based on (current) scientific theory. An effect has *coherence* if it is consistent with the available empirical evidence. By *experiment*, Hill was referring to any deliberate manipulation of the exposure states, followed by observation of any change in the outcome. By *analogy* was meant that effects for other similar exposures or situations might sometimes be supportive.

It is easy (and currently somewhat fashionable) to find fault with the idea of criteria in general, and those proposed by Hill in particular. However, most of these criticisms are directed at the use of criteria to *systematize* causal evaluation in ways that Hill did not intend. As he clearly understood, justifying a causal interpretation cannot be reduced to a mechanical exercise:

> Here then are nine different viewpoints from all of which we should study association before we cry causation. What I do not believe—and this has been suggested—is that we can usefully lay down some hard-and-fast rules of evidence that *must* be observed before we accept cause and effect. None of my nine viewpoints can bring indisputable evidence for or against the cause-and-effect hypothesis and none can be required as a *sine qua non*. What they can do, with greater or less strength, is to help us make up our minds on the fundamental question—is there any other way of explaining the set of facts before us, is there any other answer equally, or more likely than cause and effect? (Hill, 1965, 299)

Hill's "viewpoints" can be seen as *descriptions* of the perspectives often adopted by scientists (especially epidemiologists) to assess causation, not *prescriptions* that will assure a correct answer. They are simply tools and can be wielded with varying amounts of skill and appropriateness. Their utility should be appraised not in the abstract, but in terms of how expert researchers make use of them to fashion compelling arguments about the validity of causal effects.

For any comparative study, including an RCT, a scientist's willingness to infer causation depends on his being *persuaded* that no other plausible explanation exists. Causal inference by an individual scientist, or by the scientific community, is typically forged through critical dialogue and debate. A persuasive argument requires logical and rhetorical skills applied to the analysis of all available evidence and data. Various "causal criteria" can be regarded as attempts to understand better the possible elements of a strong argument that could support an inference of causation.

Perhaps the most effective way to derive and evaluate such criteria would be in the careful analysis of exactly how we have reached valid causal conclusions from observational data in the past. Almost no thinking person today doubts the harmful effects of tobacco smoking, high blood pressure, or lack of educational opportunity. Why? Methodologists might do well to examine the empirical record of what has worked and what has not. If inferences about causation must rest ultimately on informed judgment, then surely we have much to learn by studying the most successful exemplars of such judgment in action. Of course, there may be disagreement about which studies should be emulated, and what lessons learned, but even such controversy is likely to be fruitful in the long run.

No Study Is an Island

Ideally, a study would be designed to deliver a definitive conclusion regarding the validity of a specific hypothesis. However, even with well-designed randomized controlled trials, disagreement about the interpretation of findings may ensue. With observational studies, the potential for bias is typically much greater. The results of any observational study will therefore be scrutinized for methodological shortcomings, especially with respect to potential confounding. Furthermore, the study may be compared and contrasted with other research, and possibly included in a larger research synthesis to assess the overall weight of evidence.

This reality may place the study investigator in a somewhat awkward position. On the one hand, she may attempt to reach a stand-alone conclusion based on the data collected and analyzed within the study itself. On the other hand, the findings will inevitably be woven into the larger tapestry of relevant theory, practice, and opinion. The study's methodology will be evaluated by other researchers and by those professionals (e.g., physicians, lawyers, policymakers) concerned with its practical implications. In an effort to prove a hypothesis (and be publishable in a prominent journal), a researcher may be tempted to overstate her case, discounting alternative explanations without thoroughly addressing their merits.

Paradoxically, this pressure to establish causation unequivocally can weaken the evidentiary strength that might otherwise be achieved. The persuasive power of a comparative study is enhanced to the extent that *competing theories* have been explicitly considered (Rosenbaum, 1999). By a competing theory is meant an alternative explanation for an empirical effect. Such a theory may simply suggest the importance of a particular causal factor not taken into account, or it may posit an entirely new causal mechanism. A critic reading a report that describes the study might conjure up such possibilities, based on her own knowledge, experience, and personal biases. Of course, it is not possible to predict all of the objections that may be raised to a study's methodology or conclusions. However, a serious attempt to anticipate and address the most obvious and potentially damaging competing theories is always worthwhile, for at least three reasons.

First, there may be evidence based on previous research and theory that can debunk a proposed competing theory. Second, it is sometimes feasible to obtain ancillary data to test a competing theory. Third, and often overlooked, candid disclosure of potential qualifications or weaknesses would place an independent reviewer in a better position to form her own educated opinion. For example, such information might assist a practicing physician who must bring to bear her own knowledge and experience in deciding how the study results can be applied in treating a particular patient. For this purpose, a more nuanced interpretation of the findings may help this professional to assess the study's relevance.

In general, the evidentiary utility of a specific competing theory is proportional to its plausibility in light of current scientific theory and evidence. Identifying a conceivable but far-fetched confounder does little to shake our confidence in a study's conclusions. The more credible the specific competing theory in which such a factor figures prominently, the more we tend to believe that bias actually exists. Conversely, the stronger will be the corroboration of the study's causal inference if evidence *against* such a plausible confounder can be produced.

As an example, consider the observational studies of HRT. A primary hypothesis of these studies was that HRT reduced the risk of coronary heart disease (CHD). The competing theory that women who used HRT were more "health-conscious" is certainly plausible. Of course, the studies attempted to measure and control for many factors believed to be associated with health-consciousness. However, later randomized studies overturned the finding of a beneficial effect on CHD risk, suggesting that some residual confounding remained. Perhaps some critical factor related to health-consciousness was not fully reflected in the covariates analyzed.

Speculating about an unrecognized factor, or alternative causal mechanism, is two steps removed from testing its explanatory power. The first step is to operationalize what may at first be only a vague idea of how the study groups might be nonexchangeable. Transforming this abstraction into more concrete terms may suggest ways to test its plausibility. Instead of health-consciousness in general, we may be led to consider various health-promoting practices related to diet, exercise, lifestyle, etc. Is there some previously overlooked factor that might exert an impact on CHD risk? Isolating and specifying such a construct might sharpen our competing theory and shed light on its plausibility.

The second step is to obtain data that can yield evidence about the existence of residual confounding related to additional factors. In most comparative studies, there is very limited data to address this problem within the study itself. Usually, the data about possible confounders has already been exhausted in an attempt to control for confounding. Therefore, the issue of potential residual confounding is typically dealt with hypothetically. Consideration of specific competing theories is limited to logical arguments regarding their plausibility. However, some of the most persuasive observational studies are designed to collect data that help to rule out specific competing theories.

The Residue of Design

Legendary baseball executive Branch Rickey famously asserted that "luck is the residue of design." Being well positioned to deal with problems that arise unexpectedly depends on careful prior planning and preparation. When analyzing the results of a typical observational study, the statistician may often feel "out of luck" when attempting to counter a proposed competing theory. There is usually no relevant data available that captures the particular confounding factor being alleged. As in baseball, fortune smiles on those who anticipate potential threats to validity and design their research accordingly.

In Chapter 4 we discussed the basic types of research designs. In essence, a research design is a blueprint for collecting data. It defines the nature of the intervention or exposure, the outcome of interest, and various covariates. In addition, the design specifies how individuals will be selected for the study and allocated to study groups, and how the exposure, outcome, and covariate values will be ascertained. Most comparative studies include only a single unexposed group in addition to an exposed group, and a single outcome variable. In an observational study, a great deal of attention may also be paid to collecting a wide range of covariates to control for confounding.

Judicious choice of covariates can supply much protection against residual confounding. Every effort should be made to collect data on all covariates suggested by possible causal mechanisms. But since the study designer's imagination and resources are inevitably limited, the possibility of residual confounding remains. How then can a study be designed to forestall future criticisms, or provide data to address them?

Unfortunately, there is no simple formula for bulletproofing any observational study. There are, however, certain design principles that can be applied to bolster the chances of successfully defending a causal interpretation. These "design elements" have been elaborated most extensively and systematically in the applied social sciences (Campbell, 1957; Campbell and Stanley, 1963; Cook and Campbell, 1979; Shadish and Cook, 1999; Shadish et al., 2002).

Social scientists in the quasi-experimental tradition accept the reality that randomized experiments are rarely feasible for evaluating social interventions. In addition, they have categorized certain plausible threats to validity encountered frequently in program evaluation research. A high-level typology of these threats was outlined in Chapter 4. Although not being able to rule out all such threats absolutely, a study designer may be able to include design elements that can test competing theories, often referred to as *plausible rival hypotheses* by social researchers (Campbell and Stanley, 1963, 36). Designing a study strategically with an eye toward specific threats to validity is a central theme in applied social science research.

In quasi-experiments, the object of a treatment often is to modify in a desired manner the value of some individual characteristic (educational level, psychological adjustment, income, etc.). This characteristic can be measured on some meaningful scale, usually numerical rather than dichotomous. Measurements can be made at multiple times, and prior to the intervention as well as afterward. A variety of quasi-experimental design strategies involve the collection of one or more "pretest" data points.

Designs in which the outcome is a posttest (or series of posttests) are relevant for many problems outside of the social sciences as well. However, most of the questions studied by epidemiologists, and indeed some of those studied by social scientists, involve outcomes that are not "tests" and require different approaches. The task of extending the arsenal of design elements to cope with residual confounding in observational studies has not received the attention it deserves (e.g., Rosenbaum, 1999, with discussion). Much work remains to further exploit and to extend the insights and principles of the quasi-experimental design tradition (Shadish and Cook, 1999).

Informative Complexity

To detect residual confounding, research designs must strive to incorporate *informative complexity*. The potential consequences of a causal effect, or of some particular form of bias, may manifest themselves in a variety of different ways. The patterns of evidence found in the data may be more consistent with some theories than with others. Sometimes these patterns can be explored more efficiently by introducing design elements that permit multiple complementary comparisons. Each comparison can test a different competing theory. The set of comparisons may generate data that will mutually reinforce (or cast doubt on) the causal hypothesis or that may appear to be inconsistent.

> A series of designs that have different strengths and that share no directional bias allow the researcher to eliminate several competing explanations for the results of the program under study. If the direction of the potential bias is different across designs, yet the results converge on the same conclusion, one can have confidence about the program's effectiveness. If the designs do not agree in their conclusion, the researcher is faced with an empirical puzzle and must attempt to explain why this is the case. This form of multiple investigation can then lead to the elucidation of variables that moderate the effectiveness of the program. (Reynolds and West, 1987, 693)

Of course, each of the design components may themselves be so flawed that their results are hopelessly ambiguous. But the chances of reaching a persuasive conclusion (or at least inspiring a fruitful debate) will often be improved by having a complex pattern of data available. It is perhaps in this manner that the kinds of "viewpoints" proposed by Austin Bradford Hill should be utilized. For example, data pertaining to an apparent biological gradient or coherence with external data can be informative as part of a broader assessment in which the whole is more than the sum of its parts.

As suggested in the preceding quotation, the various design components of a study might not converge. In that case, the pattern of results considered in light of possible competing theories might lead to a refinement of the primary study hypothesis, rather than to its outright rejection. Thus, "failure" to confirm is transmuted into a potentially important step forward. In a real sense, all studies should be viewed as both confirmatory and exploratory. Any one study is nearly always only a link in some chain of knowledge evolution extending into the future.

> More specifically, we must increase our time perspective, and recognize that continuous, multiple experimentation is more typical of science than once-and-for-all

definitive experiments. The experiments we do today, if successful, will need replication and cross-validation at other times under other conditions before they can
become an established part of science, before they can be theoretically interpreted
with confidence. Further, even though we recognize experimentation as the basic
language of proof, as the only decision court for disagreement between rival theories, we should not expect that "crucial experiments" which pit opposing theories
will likely have clear-cut outcomes. (Campbell and Stanley, 1963, 3)

A comprehensive discussion of design methodology is well beyond the scope
of this book. In the context of residual confounding, we will highlight three
different kinds of ancillary data that are especially useful to generate informative complexity:

- Data on multiple outcome variables
- Data on multiple control groups
- Data on causal modifiers

Approaches based on such data have been developed most extensively by
social science methodologists.

In the quasi-experimental tradition, multiple outcomes are called *nonequivalent dependent variables*. Ideally, these various outcomes will be differentially
sensitive to alternative theories proposed. For example, suppose that the exposure of interest would be expected to exert a powerful effect on one particular
outcome, but no effect on two others. Alternatively, the main competing theory
posits that some other factor is the true cause and that this alternative factor
would exert a strong effect on all three outcomes. Then the pattern of effects
actually observed may help to determine which theory is more likely.

Choosing "nonequivalent" outcomes requires creativity and subject-matter
knowledge. Outcomes should be selected that are believed to respond differentially to the exposure and the suggested alternatives. An observed pattern
of results will be most informative when the expectations about these responses
are grounded in sound logic and current scientific theory. As discussed previously, the apparent reduction in cardiovascular risk found in many observational studies of HRT may have resulted from residual confounding. Those
women who used HRT may have been healthier and/or more health-conscious
in ways that were not fully measured. One way to probe such competing theories would be to collect data on health-related outcomes expected to be influenced by health-consciousness, but not by HRT therapy.

For example, being diagnosed with certain benign breast lumps might be
entirely unrelated to HRT. However, among more health-conscious individuals, greater attention to any suspicious growths might be anticipated. If we
observed a much higher rate of benign-lump diagnoses in the women exposed
to HRT than in unexposed women, the health-consciousness theory would
gain support. If the rates of such diagnoses were similar, this competing theory
would lose plausibility. Conversely, if HRT truly decreased CHD risks, then

it might also be expected to reduce the risk of ischemic stroke, which is also related to atherosclerotic and thrombotic processes. However, HRT might be expected to have little or no impact on hemorrhagic stroke, which has a different etiology. Health-consciousness, on the other hand, might help reduce the risk of hemorrhagic stroke as well as ischemic stroke, because more assiduous attention to blood-pressure control among the most health-conscious women would lower both risks. Thus, a decrease in ischemic strokes, but not in hemorrhagic strokes, might help to substantiate the benefits of HRT. However, a decrease in both types of stroke would lend support to the health-consciousness alternative.

As discussed above, nonequivalent outcomes are informative because the expected causal effects can vary across the different outcomes. *Nonequivalent control groups* can be informative for another reason: the nature or degree of certain biases might be likely to vary across the different control groups in a predictable manner. Just as with nonequivalent outcomes, the objective is to produce a set of independent comparisons that have predictable but different results under alternative causal theories. For example, an investigator might be able to locate groups of women who do not use HRT for particular reasons that correlate with health-consciousness. Some women might reject all forms of "artificial" bodily enhancement on religious or philosophical grounds, but be very educated about health issues. Arguably, this subgroup would be at least as health-conscious as the general population of women taking HRT. Suppose we compared the HRT users with this subgroup, after statistical adjustment for other suspected confounders. Then we would expect the apparent beneficial effect of HRT vs. this special comparison group to be reduced, or even eliminated. Thus, if a beneficial effect of HRT were observed, health-consciousness as an omitted covariate would not seem a plausible explanation. Now, suppose we could identify a group of women who did not use HRT and were *not* health-conscious. Then we might expect the confounding bias attributable to a "healthy-user" effect to be more pronounced, resulting in an inflated empirical effect of HRT. So, failing to observe evidence of such inflation would weigh against the health-consciousness argument.

These examples illustrate the basic idea of utilizing multiple control groups that are believed to vary predictably in their "distance" from the exposed group. By distance we mean the difference (direction and magnitude) between the unexposed group and the exposed group in terms of the potential confounding factor. Under the competing theory, larger degrees of this difference would imply larger effects of the confounding factor. This increased confounding would then result in a predictable change in the empirical effect. In the extreme, the range of control groups would "bracket" the exposed group. That is, at least one of the control groups would be expected to have a higher level of the confounder, and at least one would have a lower level. In that case, we might observe a reversal of the direction of the observed effect between the two comparisons.

Finally, the phenomenon of causal heterogeneity offers another important source of informative complexity. Suppose that if a causal effect exists, it is plausible that this effect would be more pronounced for a certain subgroup of individuals. For example, the beneficial effects of HRT might be greater for women who suffer from severe vasomotor symptoms. Perhaps these women tend to experience the most extreme hormone depletion and therefore have the most to gain from hormone supplementation. Arguably, if HRT can reduce risk of CHD, it should have an especially large effect on this subgroup. On the other hand, if the apparent benefit of HRT is actually an artifact explained by health-consciousness, analysis of this high-symptom subgroup should not magnify the observed effect.

Unfortunately, a strong focus on statistical interactions that might reveal effect modification is widely viewed as statistical heresy. Searching for meaningful patterns of effect variation is often disparaged by terms like "fishing expedition" and "capitalizing on chance." Traditional statistical wisdom tends to be extremely conservative in protecting against possible overinterpretation of apparent effects based on relatively small samples in particular subgroups. But the risk of overlooking potentially important effects is also real and should not be ignored. Ideally, an optimal balance can be struck between the equally legitimate imperatives of data exploration and hypothesis testing.

The Future of Data Analysis

The three methods of generating informative complexity by obtaining ancillary data should be regarded as building blocks. As with all design elements, their feasibility will be limited by circumstances and resources. However, there are probably many more opportunities to strengthen studies using such approaches than are currently being exploited. Many researchers, especially those outside of the quasi-experimental tradition, are simply unaware of these strategic options. However, a deeper reason for their lack of popularity may lie in the prevailing research paradigm.

Students are taught that the primary aim of a research study is to reach a "decision" whether the null hypothesis (no effect) can be rejected. Consequently, the significance level (p-value) attained is the most important aspect of the study. An extreme (i.e., very low) p-value can effectively rule out *chance* as an explanation, but it does not exclude bias as an alternative explanation. Conversely, a nonsignificant p-value (i.e., above 0.05) demonstrates that random variability cannot be excluded, but it does not prove that chance is the *best* explanation. Even in a study with apparently high statistical power to detect an effect, biases can reduce the observed effect so that it fails to reach significance. Unfortunately, the strong pressure on investigators to demonstrate statistical significance tends to preempt all other considerations. Statistical significance at the conventional 0.05 level has become a virtual threshold condition for publication in any respectable academic journal.

Ruling out random variability whenever possible is extremely important, but it is not the sine qua non of scientific research. As Campbell and Stanley remind us, "continuous, multiple experimentation" incorporating "replication and cross-validation" is the hallmark of scientific methodology. In this broader context, a study that falls short of nominal statistical significance but suggests a logical next step in the ongoing research enterprise can be very valuable. Weak or partial confirmation of a causal hypothesis is not the same as *no* confirmation. With data that generate informative complexity, the investigator may be able to illuminate why the causal hypothesis needs to be refined or why the study "failed" in a particular setting. These insights may then inform further, more definitive research.

There is a need for new methodologies that address the realities of observational data. We live now at the dawn of an information age, and opportunities to create databases of inestimable value abound. In particular, the ongoing explosion of knowledge about the human genome has stimulated quantitative research to support the rapid evolution of personalized medicine. In this dynamic context, statistical thinking must adapt to a world in which the primary challenges may entail the need to detect causal needles in the mountainous haystacks of data. In this brave new world, mining these rich data resources may require approaches that are more eclectic and realistic in coping with sources of bias.

Many statisticians and other research methodologists remain focused almost exclusively on mathematical modeling as the key to dealing with bias in general, and confounding in particular. While acknowledging the need for judgment and subject-matter knowledge in conjunction with statistical expertise, they harbor the ambition of a fundamentally quantitative solution:

> There are many individual sources of possible error. It is not sufficient to characterize a study as having or not having any of these sources of error, since nearly every study will have nearly every type of error. The real issue is to quantify the errors. As there is no precise cutoff with respect to how much error can be tolerated before a study must be considered invalid, there is no alternative to the quantification of study errors to the extent possible. Although there are no absolute criteria for assessing the validity of scientific evidence, it is still possible to assess the validity of a study. What is required is much more than the application of a list of criteria. Instead, one must apply thorough criticism, with the goal of obtaining a quantified evaluation of the total error that afflicts the study. (Rothman and Greenland, 2005, S150)

Aspiring to obtain a formal quantification of total bias is laudable, but this ideal may be a chimera. It seems doubtful that the bases for a scientist's judgment can be made fully explicit, let alone quantifiable. The weighing of evidence entails more considerations than can be captured in any purely mathematical probabilistic statement.

Statistical modeling, especially causal modeling, can potentially contribute in many ways to judgments about causal effects. However, mathematical logic can perhaps serve most usefully as a means to facilitate insight, rather than as a structure for formalizing the process of deliberation. Ideally, data-analytic tools can discover, refine, and validate meaningful patterns in the data through an iterative and dynamic process. To perform these tasks, statistical methods may need to incorporate techniques and principles from other disciplines, such as data mining, computer science, and quasi-experimental design. Statistics as a field can continue to flourish by forging data-analytic approaches that will complement and facilitate critical judgment, not supplant it.

Statistician John Tukey in his famous essay on "The Future of Data Analysis" warned against an overreliance on mathematical modeling, lest our data analyses become elegant but sterile exercises:

> Data analysis, and the parts of statistics which adhere to it, must then take on the characteristics of a science rather than those of mathematics, specifically:
>
> **(b1)** Data analysis must seek for scope and usefulness rather than security.
>
> **(b2)** Data analysis must be willing to err moderately often in order that inadequate evidence shall more often *suggest* the right answer.
>
> **(b3)** Data analysis must use mathematical argument and mathematical results as bases for judgment rather than as bases for proof or stamps of validity. (Tukey, 1962, 6)

Tukey's point (b2) is especially pertinent to the discovery and testing of causal effects. Causal inferences from observational studies are subject to both random and systematic sources of error. Teasing out valid causal effects may depend on multiple looks at the data and combining evidence from multiple studies. The *high art* of conducting exploratory analyses that "err moderately often" in order to make important discoveries and come to conclusions based on "inadequate evidence" have an essential role to play, along with the confirmatory *technology* that currently dominates statistical theory.

GUIDEPOST 7

In this chapter we have viewed the idea of confounding through the lens of counterfactual analysis. The essence of confounding can be defined at one level as nonexchangeability between the study groups. Residual confounding after conditioning occurs when there is nonexchangeability within the strata formed by conditioning on the covariate (also known as strong ignorability of the assignment mechanism). Thus, confounding and assignment bias are formally equivalent concepts. At a deeper explanatory level, the nonexchangeability is seen to result from differences between study groups with respect to the distributions of causal factors.

We distinguished between the essential aspects of confounding, which pertain to response-pattern distributions, and incidental aspects that may appear only for particular statistical models and effect measures. Certain confusing or "paradoxical" phenomena (Simpson's paradox, Berkson's bias) were seen as related to the mathematical properties of these particular models and effect measures, and not as reflections of confounding in general. Most important, the relationship between conditional (i.e., adjusted for covariates) and unconditional response-pattern distributions was explicated.

Our analysis revealed that successful control for confounding ordinarily results in different response-pattern distributions within strata (causal heterogeneity). Thus, effect modification after control for confounding is normally to be expected, except in certain special circumstances. In this sense, confounding and effect modification are not completely distinct phenomena, as is usually supposed. Effect modification can occur without confounding, but (successful) control for confounding ordinarily does imply the existence of effect modification. A notable exception occurs when the outcome is numerical and the causal effect is uniform across individuals. In that case, confounding could occur without effect modification.

The conception of confounding described in this chapter leads naturally to the centrality of the propensity score, as suggested by Rosenbaum and Rubin. The propensity score is the assignment probability as a mathematical function of the covariates. Under the assumption of conditional exchangeability (strong ignorability), the propensity score is a single covariate that can be used to (successfully) adjust for confounding. The propensity score can be used not only to obtain an overall estimated effect, but also to provide insight about possible effect modification. Of course, conditional exchangeability is a difficult assumption to substantiate in practice. The adequacy of statistical adjustment depends on being able to rule out other plausible competing theories. Evidence relevant to such alternative explanations might come from several sources and may depend critically on design elements that have been incorporated in the study.

So far, we have assumed a static situation in which the causal context remains fixed and only the exposure state can change. In the next chapter, we consider situations in which the risk factor can cause one or more of the causal factors to change. In that situation, the total effect of the risk factor might include an "indirect effect" resulting from the modified value of this intermediate variable. Counterfactual thinking sheds light on conceptual and practical issues related to estimating a causal effect in the presence of such causal mediation.

APPENDIX

In this Appendix we derive an important result referenced in Section 7.1 regarding the estimation of an average causal effect (ACE).

Theorem. Suppose that there exists assignment bias (confounding), but conditioning on a covariate X is sufficient to adjust for the bias. Then the two following conditions cannot *both* be true:

$$E(\Delta|x) = E(\Delta) \text{ for all } x$$

$$E(M|x) = E(M) \text{ for all } x$$

Proof of Theorem. From Equation 7.1 we know that because bias is assumed to occur in the absence of statistical adjustment, $Cov(A, \Delta) = 0$ and $Cov(M, \Delta) = 0$ cannot both occur. By definition, we have

$$\text{Cov}(A, \Delta) = E(A\Delta) - E(A)E(\Delta) \tag{A.7.1}$$

This can be expressed alternatively as:

$$\text{Cov}(A, \Delta) = E_x[E(A\Delta|x)] - E(A)E(\Delta) \tag{A.7.2}$$

However, because X is sufficient for adjustment, the value of A conditional on any value of X must be constant. Therefore, we obtain:

$$E_x[E(A\Delta|x)] = E_x[E(A|x)E(\Delta|x)] \tag{A.7.3}$$

Therefore, we have from Equations A.7.2 and A.7.3 that:

$$\text{Cov}(A, \Delta) = E_x[E(A|x)E(\Delta|x)] - E(A)E(\Delta) \tag{A.7.4}$$

If $E(\Delta|x) = E(\Delta)$ for all values of x, Equation A.7.4 would imply that:

$$\text{Cov}(A, \Delta) = 0 \tag{A.7.5}$$

Now we can apply the same logic, substituting M for Δ in (A.7.1)–(A.7.4) to obtain:

$$\text{Cov}(A, M) = E_x[E(A|x)E(M|x)] - E(A)E(M) \tag{A.7.6}$$

Therefore, if $E(M|x) = E(M)$ for all values of x, then we would have:

$$\text{Cov}(A, M) = 0 \tag{A.7.7}$$

Therefore, if both $E(\Delta|x) = E(\Delta)$ and $E(M|x) = E(M)$ for all values of x, then Equations A.7.5 and A.7.7 must hold, which contradicts the assumption that bias existed prior to adjustment and proves the Theorem.

Corollary. If the conditions of the Theorem hold, then $E(\Delta|x) = E(\Delta)$ for all values of x if and only if there exists a function $f(x)$ such that:

$$E(W_E|x) = \alpha + f(x)$$

$$E(Z_U|x) = f(x)$$

Proof of Corollary.   Since X is sufficient to adjust for confounding, we
have

$$E(W_E|x) = E(V_E|x) \qquad\qquad (A.7.8)$$

$$E(Z_U|x) = E(V_U|x) \qquad\qquad (A.7.9)$$

From Equations A.7.8 and A.7.9 and the definitions of M and Δ we obtain:

$$E(W_E|x) = E\left(M + \frac{\Delta}{2}\bigg| x\right) = E(M|x) + \frac{E(\Delta|x)}{2} \qquad (A.7.10)$$

$$E(Z_U|x) = E\left(M - \frac{\Delta}{2}\bigg| x\right) = E(M|x) - \frac{E(\Delta|x)}{2} \qquad (A.7.11)$$

Now, the Theorem implies that $E(\Delta|x) = E(\Delta)$ for all values of x if and only if
it is not true that $E(M|x) = E(M)$ for all values of x. Let $f(x) = E(Z_U|x)$ and
$\alpha = E(\Delta)$. Then from Equations A.7.10 and A.7.11, the Corollary is true if
$E(\Delta|x) = E(\Delta)$ and is not true if $E(\Delta|x) \neq E(\Delta)$ for some value of x.

CHAPTER 8

Intermediate Causal Factors

Heuristically, an intermediate causal factor, or *causal mediator*, is a variable that is causally influenced by the exposure and in turn causally affects the outcome of interest. Thus, the mediator is in some sense partially responsible for the total causal effect of exposure. The problem is to articulate exactly what this idea of partial causation really means. As discussed previously, classical statistical theory has no clear definition of causation in general, let alone in this more complex situation. Consequently, there has been much confusion regarding whether and how a total causal effect can be parsed into components that reflect the direct and indirect effects of exposure.

8.1 DIRECT AND INDIRECT EFFECTS

Many researchers believe that it is possible to resolve the total causal effect into direct and indirect components, using variations of ordinary statistical models (Baron and Kenny, 1986; MacKinnon et al., 2002). In essence, this is attempted by adjusting for the intermediate variable to estimate a direct (or net) effect, and then subtracting this direct effect from the total effect, to derive the indirect effect. For example, it is known that some nonsteroidal anti-inflammatory drugs (NSAIDs) can increase blood pressure. Suppose we are studying the effect of an NSAID on myocardial infarction (MI). An increase in blood pressure raises the risk of an MI. Ideally, we would like to know the direct effect of the NSAID, *as if it had no effect on blood pressure*, as well as the indirect effect operating through blood pressure as a causal mediator. But in what sense can we separate these direct and indirect effects?

The idea that an exposure can exert both a direct effect on the outcome and an indirect effect seems intuitively plausible. However, when we examine

Bias and Causation: Models and Judgment for Valid Comparisons, By Herbert I. Weisberg
Copyright © 2010 John Wiley & Sons, Inc.

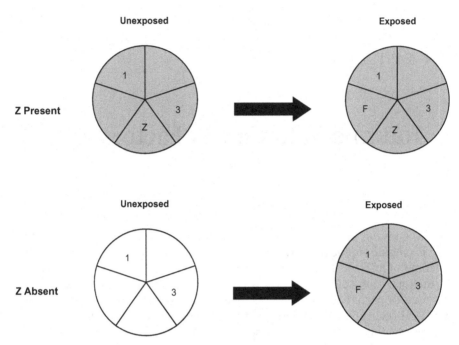

Figure 8.1. A situation in which Factor Z is a *causal modifier* of the effect of F. (The underlying causal mechanism implies that the outcome event occurs if and only if at least one of F and Z is present.) If Z is present, then the individual is *doomed,* but if Z is not present, he is *causal.*

this concept closely, we find it to be quite ambiguous. Classical statistical theory cannot adequately explain the conceptual issues involved; an explicit causal model is necessary. As we will see, the counterfactual perspective reveals that the idea of unique and well-defined components that "add up" to the total causal effect is in general not meaningful (Frangakis and Rubin, 2002; Kaufman et al., 2004; Rubin, 2004, 2006; Hafeman, 2008).

Suppose we are interested in the causal effect of a risk factor F whose causal context includes four other factors: Factors 1, 2, 3, and Z. Here Z represents a dichotomous intermediate variable that can be causally influenced by F. Let us ignore for the moment that Z is an intermediate variable. If Z is a causal modifier of the risk factor, then the response pattern can depend on the value of Z. Figure 8.1 illustrates this situation. Here a particular hypothetical individual happens to have Factors 1 and 3 present, while Factor 2 is absent. Furthermore, we will assume a simple causal model: the outcome event for this individual will occur if at least one of F and Z is present. Suppose first that Z is present. Then the outcome event occurs (indicated by shading) whether or not F is also present, so this individual is *doomed*. Now suppose that Z is absent. In that case, the event occurs if and only if the individual is exposed, so she is *causal*. The value of Z has changed the causal effect (response pattern) for this individual from *doomed* to *causal*.

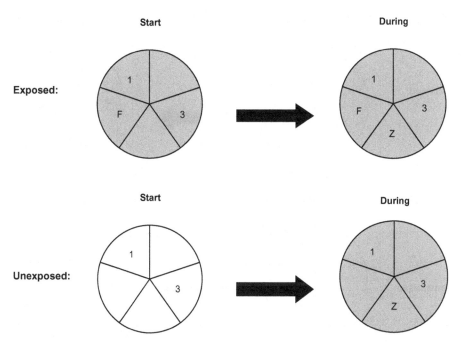

Figure 8.2. A situation in which Factor Z is a *causal mediator* of the effect of *F*, and the individual is *I-doomed*. (The underlying causal mechanism implies that the outcome event occurs if and only if at least one of *F* and *Z* is present.) For this individual, *Z* always occurs, whether or not *F* has occurred. The effect of *Z* is to change the individual's response pattern from *causal* to *doomed*.

Now suppose that *Z* is not only a causal modifier of *F*, but is also causally influenced by *F*. In other words, *F* is a cause of *Z*, given the assumed (fixed) values of the other three causal factors. We assume that the value of *F* exerts a causal effect on the value of *Z* that manifests at some point during follow-up. This modified value of *Z* then affects the study outcome.

This more complex situation is illustrated in Figs. 8.2–8.5. In this example, we assume that *Z* = 0 at the start of the study. Let us assume first that this individual is "doomed" with respect to the outcome *Z*. That is, *Z* = 1, whether or not exposure to the risk factor *F* occurs. On the left-hand side of Fig. 8.2, we portray the same individual at the outset of the study period, both with and without exposure to *F*. On the right-hand side, we show the result after *Z* has occurred. Observe that the individual is *doomed* because the outcome event always occurs, whether or not there is exposure. This response pattern describes the "total" effect of *F* on the outcome, including the "indirect effect" that is mediated by the intermediate variable *Z*.

Figure 8.2 represents only one of the four possible intermediate response patterns for the effect of *F* on the mediating factor *Z*. In the situation shown in Fig. 8.2, we will say that this individual is *intermediate-doomed*, or *I-doomed*. This means the intermediate event always occurs, regardless of the exposure

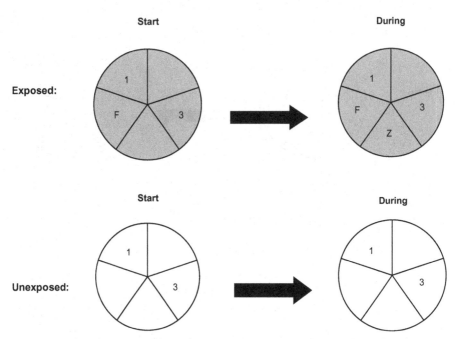

Figure 8.3. A situation in which Factor Z is a *causal mediator* of the effect of F, and the individual is *I-causal*. (The underlying causal mechanism implies that the outcome event occurs if and only if at least one of F and Z is present.) For this individual, Z occurs if and only if F has occurred. In this case, Z does not change the individual's response pattern, which remains *causal*.

state. Figures 8.3–8.5 show the other three possible scenarios for this individual. She could be *I-causal*, (Fig. 8.3), *I-preventive* (Fig. 8.4), or *I-immune* (Fig. 8.5). In each case, the total causal effect is determined by the intertwined causal influences of F and Z. But how might we define the direct effect of F on the outcome?

We have defined a causal effect for each individual as the response pattern when "everything else" is held constant. We represented "everything else" formally as the causal context. But if F changes the value of the causal modifier, then it changes the causal context. Consequently, we observe a causal effect of F that is predicated on a *modified* causal context; after Z occurs, an *I-doomed* individual would have a context in which Factors 1, 3, and Z are all present. Therefore, the response pattern for F would become *doomed*, rather than *causal*.

One way to define the direct effect would be as the causal effect of F at the outset, before its impact on Z has materialized. This is the effect that would have occurred if F did not affect Z. To obtain the indirect effect, we could then compare this direct effect with the total effect. In our example, the "difference" between the two effects would be a change from a response pattern that is *doomed* to one that is *causal*. This change represents the "influence" that

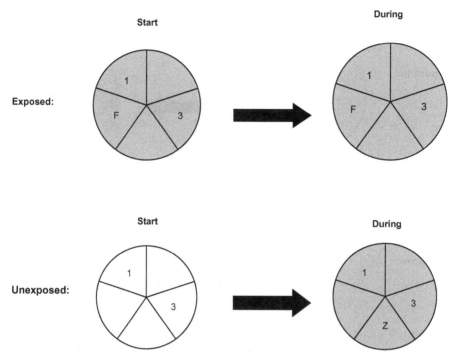

Figure 8.4. A situation in which Z is a *causal mediator* of the effect of F, and the individual is *I-preventive*. (The underlying causal mechanism implies that the outcome event occurs if and only if at least one of F and Z is present.) For this individual, Z occurs if and only if F has not occurred. The effect of Z is to change the individual's response pattern from *causal* to *doomed*.

is mediated by Z. But this influence is not a simple causal effect, that is, a difference between the potential outcomes under alternative exposure states. Rather, it is a change in the response pattern, a kind of "change in a change." Thus, we cannot simply define two component effects that sum to the total effect.

Note that if a certain individual is *I-doomed* (Fig. 8.2) or *I-immune* (Fig. 8.5), then the causal effect of F is well-defined, albeit dependent on the fixed value of Z. This effect is the total effect of F for the causal context that includes this fixed value of Z. It is a direct effect in the sense that, because the intermediate event either always occurs or never occurs, it can be considered a fixed part of the causal context. Suppose that certain individuals will experience an increase in blood pressure during the study period, whether or not they are exposed to an NSAID being investigated. For these individuals, the direct causal effect of the NSAID on coronary heart disease (CHD) will be determined in part by the causal factor of hypertension. But there is no indirect effect of the NSAID that is *specifically* attributable to increased blood pressure *caused* by this drug.

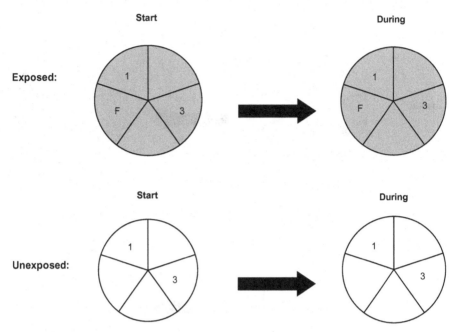

Figure 8.5. A situation in which Z is a *causal mediator* of the effect of F, and the individual is *I-immune*. (The underlying causal mechanism implies that the outcome event occurs if and only if at least one of F and Z is present.) For this individual, Z never occurs, whether or not F has occurred. In this case, Z does not change the individual's response pattern, which remains *causal*.

Now consider what happens when a person is either *I-causal* (Fig. 8.3) or *I-preventive* (Fig. 8.4). In these scenarios, changing the value of F alters the causal context. As a result, the total effect of exposure is not a simple cause, because "all else" is not being held fixed. We can conceptualize an indirect "effect" in the sense explained previously, but there is usually no way to measure this effect. For example, some individuals may experience an increase in blood pressure *caused* by the NSAID. The total effect of the drug for them includes a component that is in some sense attributable to this increase. However, to articulate, and possibly quantify, the nature of this component would require a model of the causal mechanism. For example, if we understood the science well enough to derive a structural equation model that could accurately represent the causal dynamics, then it might be possible to separate the direct and indirect effects.

8.2 PRINCIPAL STRATIFICATION

Suppose that we were somehow able to stratify the study population based on the individual's "response pattern" for the intermediate "outcome" Z as follows:

- Stratum a: *I-doomed* ($Z = 1$ if $F = 1$; $Z = 1$ if $F = 0$)
- Stratum b: *I-causal* ($Z = 1$ if $F = 1$; $Z = 0$ if $F = 0$)
- Stratum c: *I-preventive* ($Z = 0$ if $F = 1$; $Z = 1$ if $F = 0$)
- Stratum d: *I-immune* ($Z = 0$ if $F = 1$; $Z = 0$ if $F = 0$)

Let N_k represent the number of subjects in the study population who are in Stratum k. For example, N_d is the number of *I-immune* subjects. Stratification based on the intermediate response pattern has been called *principal stratification* (Frangakis and Rubin, 2002). For the case of a dichotomous mediator, there are four *principal strata*. However, the idea of principal stratification can be extended to a numerical causal mediator as well.

Interpretation of Principal Strata

A general representation of the data from a comparative study is exhibited in Table 8.1. We define P_{j1k} to be the proportion of subjects in Stratum k

Table 8.1 Principal Strata Based on Potential Occurrence of Intermediate Event: Distributions of Response Patterns within Strata

	Exposed	Unexposed
Principal Stratum a: *I-Doomed*		
Doomed	$A_{1a}N_aP_{11a}$	$(1 - A_{1a})N_aP_{11a}$
Causal	$A_{2a}N_aP_{21a}$	$(1 - A_{2a})N_aP_{21a}$
Preventive	$A_{3a}N_aP_{31a}$	$(1 - A_{3a})N_aP_{31a}$
Immune	$A_{4a}N_aP_{41a}$	$(1 - A_{4a})N_aP_{41a}$
Principal Stratum b: *I-Causal*		
Doomed	$A_{1b}N_bP_{11b}$	$(1 - A_{1b})N_bP_{10b}$
Causal	$A_{2b}N_bP_{21b}$	$(1 - A_{2b})N_bP_{20b}$
Preventive	$A_{3b}N_bP_{31b}$	$(1 - A_{3b})N_bP_{30b}$
Immune	$A_{4b}N_bP_{41b}$	$(1 - A_{4b})N_bP_{40b}$
Principal Stratum c: *I-Preventive*		
Doomed	$A_{1c}N_cP_{10c}$	$(1 - A_{1c})N_cP_{11c}$
Causal	$A_{2c}N_cP_{20c}$	$(1 - A_{2c})N_cP_{21c}$
Preventive	$A_{3c}N_cP_{30c}$	$(1 - A_{3c})N_cP_{31c}$
Immune	$A_{4c}N_cP_{40c}$	$(1 - A_{4c})N_cP_{41c}$
Principal Stratum d: *I-Immune*		
Doomed	$A_{1d}N_dP_{10d}$	$(1 - A_{1d})N_dP_{10d}$
Causal	$A_{2d}N_dP_{20d}$	$(1 - A_{2d})N_dP_{20d}$
Preventive	$A_{3d}N_dP_{30d}$	$(1 - A_{3d})N_dP_{30d}$
Immune	$A_{4d}N_dP_{40d}$	$(1 - A_{4d})N_dP_{40d}$

who would have (outcome) response-pattern j if $Z = 1$. Likewise, P_{j0k} is the proportion of subjects who would have (outcome) response-pattern j among the subjects in Stratum k if $Z = 0$ For example, P_{11a} is the proportion of *doomed* individuals among those in Stratum a (*I-doomed*), whether or not they are exposed. Note that these proportions for certain combinations of j and k are vacuous, because of how the principal strata are defined. For example, $P_{10a} = 0$, because $Z = 0$ cannot actually occur in this study for individuals who are in Stratum a (*I-doomed*).

For an observational study, the study groups might not be exchangeable within strata. Therefore, conditioning on the principal strata cannot be assumed to produce unbiased stratum-specific causal effects. Moreover, even if we could control for covariates thought to be responsible for bias in the study, these covariates would not necessarily adjust successfully for confounding within the principal strata. As a result, attempting to estimate a meaningful effect from an observational study when causal mediation occurs can be very complex.

On the other hand, when we have causal mediation in a randomized experiment, the situation is simpler. In a randomized experiment there is a constant assignment probability A that applies to all subjects. Consequently, the (expected) proportion of subjects who are exposed within each principal stratum will be A for every response pattern, as well as for the stratum as a whole. For example, there will be AN_a exposed individuals and $(1 - A)N_a$ unexposed individuals in Stratum a. However, the distributions of response patterns within the different principal strata can vary across strata, that is, the principal strata will not necessarily be causally homogeneous.

In Table 8.2, we represent the response-pattern distributions for the four principal strata. Because of randomization, the distributions of response patterns at the outset of the study would be identical for the two study groups within each of the principal strata. However, these distributions could be modified by the values of Z that arise during the study. Within Strata a and d, the study groups remain exchangeable, because the value of Z is *fixed* independently of exposure status. Therefore, an unbiased estimate of the stratum-specific causal effect can be obtained. As explained previously, this effect is conditional on $Z = 1$ (Stratum a) or $Z = 0$ (Stratum d). These can be interpreted as the direct effects of exposure when Z is fixed at these values, respectively. For Strata b and c, the situation is less easily interpretable.

From Table 8.2, we can see that the study groups within these principal strata will be exchangeable if and only if $P_{j1b} = P_{j0b}$ and $P_{j1c} = P_{j0c}$ for all values of j. However, if Z is a causal modifier of F, at least some of the response patterns would depend on whether $Z = 1$ or $Z = 0$. Therefore, exchangeability ordinarily would not hold. The empirical effect in Stratum b cannot be interpreted simply as a causal effect of exposure to F. Rather, it can be construed as a causal effect of the *combined* exposure to the presence of both F and Z. Because of randomization, the study groups within Stratum b are exchangeable with respect to the actual exposure in this stratum ($F = 1$ and $Z = 1$). Likewise, the empirical effect in Stratum c estimates the combined effect of

Table 8.2 Principal Strata Based on Potential Occurrence of Intermediate Event: Distributions of Response Patterns within Strata for a Randomized Experiment

	Exposed	Unexposed
Principal Stratum a: *I-Doomed*		
Doomed	AN_aP_{11a}	$(1-A)N_aP_{11a}$
Causal	AN_aP_{21a}	$(1-A)N_aP_{21a}$
Preventive	AN_aP_{31a}	$(1-A)N_aP_{31a}$
Immune	AN_aP_{41a}	$(1-A)N_aP_{41a}$
Principal Stratum b: *I-Causal*		
Doomed	AN_bP_{11b}	$(1-A)N_bP_{10b}$
Causal	AN_bP_{21b}	$(1-A)N_bP_{20b}$
Preventive	AN_bP_{31b}	$(1-A)N_bP_{30b}$
Immune	AN_bP_{41b}	$(1-A)N_bP_{40b}$
Principal Stratum c: *I-Preventive*		
Doomed	AN_cP_{10c}	$(1-A)N_cP_{11c}$
Causal	AN_cP_{20c}	$(1-A)N_cP_{21c}$
Preventive	AN_cP_{30c}	$(1-A)N_cP_{31c}$
Immune	AN_cP_{40c}	$(1-A)N_cP_{41c}$
Principal Stratum d: *I-Immune*		
Doomed	AN_dP_{10d}	$(1-A)N_dP_{10d}$
Causal	AN_dP_{20d}	$(1-A)N_dP_{20d}$
Preventive	AN_dP_{30d}	$(1-A)N_dP_{30d}$
Immune	AN_dP_{40d}	$(1-A)N_dP_{40d}$

the exposure in this stratum ($F = 1$ and $Z = 0$). However, we cannot separate the causal effects of F and Z. Indeed, each of these factors can be regarded as a confounder of the other's effect. From this point of view, principal stratification is a method for resolving the study population into strata in which no "confounding" by the causal mediator occurs and strata in which there is perfect confounding.

Traditional Approaches to Analysis

The theory of principal stratification can help illuminate the implications of applying straightforward methods for dealing with causal mediation. One common approach is to concentrate on the *total* effect of F on the study outcome. We ignore the particular "causal pathways" at work. This approach assumes, in effect, that the causal pathways are fairly stable, so that this total causal effect will carry over to situations in which the study's results will be

Table 8.3   Data for Full Study Population with Causal Mediation: Distributions of Pooled Response Patterns for a Randomized Experiment

	Exposed	Unexposed
Doomed	$A(N_a P_{11a} + N_b P_{11b}$ $+ N_c P_{10c} + N_d P_{10d})$	$(1 - A)(N_a P_{11a} + N_b P_{10b}$ $+ N_c P_{11c} + N_d P_{10d})$
Causal	$A(N_a P_{21a} + N_b P_{21b}$ $+ N_c P_{20c} + N_d P_{20d})$	$(1 - A)(N_a P_{21a} + N_b P_{20b}$ $+ N_c P_{21c} + N_d P_{20d})$
Preventive	$A(N_a P_{31a} + N_b P_{31b}$ $+ N_c P_{30c} + N_d P_{30d})$	$(1 - A)(N_a P_{31a} + N_b P_{30b}$ $+ N_c P_{31c} + N_d P_{30d})$
Immune	$A(N_a P_{41a} + N_b P_{41b}$ $+ N_c P_{40c} + N_d P_{40d})$	$(1 - A)(N_a P_{41a} + N_b P_{40b}$ $+ N_c P_{41c} + N_d P_{40d})$

applied. For example, we assume that any indirect effects of increased blood pressure attributable to an NSAID in a randomized controlled trial (RCT) will occur in approximately the same manner more generally.

Let N represent the total number of subjects in the study ($N = N_a + N_b + N_c + N_d$). The distribution of response patterns in each study group is shown in Table 8.3. For example, the proportion of *doomed* individuals in the exposed group is given by $(N_a P_{11a} + N_b P_{11b} + N_c P_{10c} + N_d P_{10d})/N$. However, the corresponding proportion among the unexposed is $(N_a P_{11a} + N_b P_{10b} + N_c P_{11c} + N_d P_{10d})/N$. In general, the study groups are no longer exchangeable. Hence, the empirical effect of F does not equal the causal effect. In particular, the empirical effect does not equal the causal effect that would have occurred if Z had been held fixed for all individuals. This issue can be viewed as a violation of the stable unit–treatment value assumption (SUTVA) discussed in Section 3.8. In Rubin's terminology, the variability of the "version" of the treatment received violates the SUTVA principle. From this perspective, attempting to estimate the total effect might be considered meaningless (Rubin, 1986).

On the other hand, we could define the "actual exposure" to be the nominal exposure ($F = 1$) combined with whichever value of Z happens to occur in the normal course of events. This "exposure" would then depend on the individual's principal stratum as well as his nominal exposure state. The "response patterns" defined in terms of this compound "exposure" would still be exchangeable, but the relevance of the causal effect being estimated might be in doubt. We pointed out in Section 3.7 that the total effect *may* be an accurate reflection of the total effect that would occur in practice. If it is, then from a pragmatic or public health standpoint this effect might be useful. However, it would be important to recognize that the effect could differ for a population or situation in which the "normal course of events" was not the same.

If the total effect is not considered relevant, we might be tempted to adjust for the causal mediator, treating it like an ordinary confounder. We would then (presumably) be focused only on the direct effect of exposure rather than the total effect. The direct effect is conceived as the net effect of exposure over

Table 8.4 Data for Strata Conditional on Observed Outcomes of Causal Mediator; Distributions of Response Patterns within Strata for a Randomized Experiment

	Exposed	Unexposed
	Stratum: $Z = 1$	
Doomed	$A(N_a P_{11a} + N_b P_{11b})$	$(1 - A)(N_a P_{11a} + N_c P_{11c})$
Causal	$A(N_a P_{21a} + N_b P_{21b})$	$(1 - A)(N_a P_{21a} + N_c P_{21c})$
Preventive	$A(N_a P_{31a} + N_b P_{31b})$	$(1 - A)(N_a P_{31a} + N_c P_{31c})$
Immune	$A(N_a P_{41a} + N_b P_{41b})$	$(1 - A)(N_a P_{41a} + N_c P_{41c})$
	Stratum: $Z = 0$	
Doomed	$A(N_c P_{10c} + N_d P_{10d})$	$(1 - A)(N_b P_{10b} + N_d P_{10d})$
Causal	$A(N_c P_{20c} + N_d P_{20d})$	$(1 - A)(N_b P_{20b} + N_d P_{20d})$
Preventive	$A(N_c P_{30c} + N_d P_{30d})$	$(1 - A)(N_b P_{30b} + N_d P_{30d})$
Immune	$A(N_c P_{40c} + N_d P_{40d})$	$(1 - A)(N_b P_{40b} + N_d P_{40d})$

and above the effect that flows through the "causal pathway" related to the mediating factor. The problem with this logic, however, is that a causal mediator is not just an ordinary confounding factor.

Let us consider what actually would happen if we attempted to estimate a direct effect of F by adjusting for an intermediate factor Z. Statistical adjustment would usually involve a technique that effectively *conditions* on Z. By statistical conditioning we mean stratifying on the basis of the observed value of Z, as discussed in Chapter 7. For example, suppose that we condition on $Z = 1$. Referring again to Table 8.2, we see that this stratum includes all subjects in Stratum a, the exposed subjects in Stratum b, and the unexposed subjects in Stratum c. In general, the exposed and unexposed groups within the stratum defined by $Z = 1$ will not be exchangeable, as can be observed in Table 8.4.

In theory, estimation of a direct effect is only possible within a principal stratum in which the exposure has no effect on the mediator (Frangakis and Rubin, 2002). For example, suppose we could restrict attention to those who are *I-doomed*. Then the study groups would be exchangeable. In effect, principal stratification resolves the study population into subsets (principal strata) for whom a causal effect of F is well defined (*I-doomed* and *I-immune*) and those for whom it is not (*I-causal* and *I-preventive*). For the former, the effects are well defined because the intermediate variable is *fixed* at a given value.

This fixing of the causal mediator occurs at the individual level. Being fixed in this sense is a counterfactual property of an individual; it pertains to his (intermediate) potential outcomes under different exposure states. For example, an *I-doomed* individual is one for whom the intermediate event would always occur ($Z = 1$), no matter how the value of F may be manipulated (holding all other causal factors besides Z constant). The subgroup of individuals with this property is not the same as the subgroup of individuals who happen to experience the intermediate event in the study.

Table 8.5 Data from a Randomized Study with Causal Mediation: Numerical Outcome (Sums of Outcome Values by Exposure Status and Principal Stratum)

Exposed	Unexposed
Principal Stratum a: *I-Doomed*	
$AN_a\mu_{E1a}$	$(1-A)N_a\mu_{U1a}$
Principal Stratum b: *I-Causal*	
$AN_b\mu_{E1b}$	$(1-A)N_b\mu_{U0b}$
Principal Stratum c: *I-Preventive*	
$AN_c\mu_{E0c}$	$(1-A)N_c\mu_{U1c}$
Principal Stratum d: *I-Immune*	
$AN_d\mu_{E0d}$	$(1-A)N_d\mu_{U0d}$

Estimating the Average Causal Effect

So far, we have assumed that the outcome is dichotomous. With a numerical outcome, the problems can in principle become even more complex. However, if we consider the usual special case of estimating a difference of means, the mathematical development simplifies. We will consider only the case of a randomized experiment. For an observational study, the same basic principles would apply, but the presentation would be more complicated. We can imagine that in an observational study it might be possible to stratify based on a covariate, so that the results for a randomized study would apply within each stratum.

Table 8.5 displays expressions for the totals of the observed outcome values in various combinations of exposure state and principal stratum. Each cell entry is written as the product of the mean outcome value and the number of subjects in that cell. For example, the mean value for the exposed subjects in Stratum a is μ_{E1a} and the number in this cell is AN_a. Let us define $\mu_{\Delta 1k}$ to be the average causal effect (ACE) for the members of Stratum k when $Z = 1$ and let $\mu_{\Delta 0k}$ be the corresponding ACE when $Z = 0$. We can express these as follows:

$$\mu_{\Delta ik} = \mu_{Eik} - \mu_{Uik}, \qquad i = 0, 1; k = a, b, c, d \qquad (8.1)$$

Then $\mu_{\Delta 1a}$ is a direct effect of F for the *I-doomed*, because it is the ACE for these individuals when the value of Z is held fixed at $Z = 1$. Similarly, $\mu_{\Delta 0d}$ is a direct effect for the *I-immune* when the value of Z is held fixed at $Z = 0$. Note that these two direct causal effects apply to possibly different subpopulations of the study population.

In practice, the notion of a unique direct effect is tenable only if we have reason to believe the causal effect is constant. In that special case, the direct effect could, in principle, be estimated by fixing the value of Z (at either $Z = 0$ or $Z = 1$) and calculating the ACE. More generally, if Z is a numerical variable, the direct effect could be estimated for any value of Z in the same way. As observed previously, however, fixing the value of Z does not mean simply conditioning on this value (i.e., passively observing what happens for individuals who happen to have a given value). Rather, fixing means to identify those members of the study population whose values of Z remain fixed. This may be very difficult.

In theory, it may sometimes be possible to hold the value constant by direct *manipulation*, as in a controlled experiment: "the requirement of holding the mediating variables fixed must be interpreted as (hypothetically) setting these variables to constants by physical intervention, not analytical means such as selection, conditioning, or adjustment" (Pearl, 2000, 127). For example, we could imagine a situation in which blood pressure is somehow maintained at a constant level for certain individuals, while their levels of exposure to NSAIDs vary. The practical problem, of course, would be how to exert this experimental control or even to envision how it might be accomplished. Moreover, would such hypothetical control have other "feedback" effects that would modify the causal context in unpredictable ways?

Now let us consider the consequences of trying to estimate the total and direct effects when the outcome is numerical. From Table 8.5, we can derive the following formula for the total ACE:

$$ACE_{total} = \frac{N_a\mu_{\Delta 1a} + N_b(\mu_{E1b} - \mu_{U0b}) + N_c(\mu_{E0c} - \mu_{U1c}) + N_d\mu_{\Delta 0d}}{N} \qquad (8.2)$$

The total causal effect is a weighted average of four (possibly different) causal effects. Each of these four effects corresponds to the ACE for one principal stratum. The causal effect for Stratum a is the direct effect for the individuals who fall in this stratum, under the condition $Z = 1$. The causal effect for Stratum d is the direct effect for the individuals who fall in this Stratum, when $Z = 0$. The causal effects for Strata b and c cannot be interpreted as direct effects of F net of Z. These mean differences incorporate the joint effect of changing both variables. They are confounded by Z.

Now let us consider what happens if we estimate the direct effect by simply conditioning on the observed value of Z. For example, if we condition on $Z = 1$, then we are in effect creating an exposed group that consists of the exposed subjects in Strata a and b. Similarly, we are creating an unexposed group composed of the unexposed subjects in Strata a and c. In general, the subgroups of subjects in the different principal strata will not be exchangeable. Therefore, the exposed and unexposed groups actually being compared will not be exchangeable, except under special circumstances.

From Table 8.5 we can derive an expression for the "naïve" direct effect when we condition on $Z = 1$:

$$\text{Naive ACE}_{\text{dir}} = \frac{N_a \mu_{E1a} + N_b \mu_{E1b}}{N_a + N_b} - \frac{N_a \mu_{U1a} + N_c \mu_{U1c}}{N_a + N_c} \tag{8.3}$$

The conditions necessary for this naïve estimate to be an unbiased estimate of a causal effect can be clarified using Equation 8.3. Suppose first that Strata b and c are exchangeable (or at least mean-exchangeable), so that

$$\mu_{E1b} = \mu_{E1c} \quad \text{and} \quad \mu_{U1b} = \mu_{U1c} \tag{8.4}$$

Further, assume that Strata b and c are of equal size, that is, $N_b = N_c$. Then the expression in Equation 8.3 would reduce to

$$\text{Naive ACE}_{\text{dir}} = \frac{N_a \mu_{\Delta 1a} + N_b \mu_{\Delta 1b}}{N_a + N_b} = \frac{N_a \mu_{\Delta 1a} + N_c \mu_{\Delta 1c}}{N_a + N_c} \tag{8.5}$$

That is, the naïve estimate would be a weighted average of the ACE for Stratum a and the ACE for either Stratum b or c. Furthermore, if we knew that Strata a, b, and c were all exchangeable, then the expression in Equation 8.5 would reduce to $\mu_{\Delta 1a}$ and the naïve estimate would be interpretable as a "direct" causal effect for $Z = 1$.

The theory of principal stratification has helped to clarify the essential nature of causal mediation. Many statisticians have recently been developing and testing techniques for estimating causal effects that build on this approach (e.g., Barnard et al., 2003; Griffin et al., 2008; Roy et al., 2008; Zhang et al., 2009). These approaches recognize the subtleties that are not explicitly addressed within the classical statistical framework. In so doing, they attempt to avoid fuzzy thinking and unrealistic assumptions:

> … the concepts of direct and indirect causal effects are generally ill-defined and often more deceptive than helpful to clear statistical thinking in real, as opposed to artificial, problems. A clear conceptual formulation of the relevant issues for practice must precede a correct mathematical statistical attack on these issues. (Rubin, 2004, 162)

Whether the complexities revealed by causal analysis will ultimately yield to these mathematical statistical attacks remains to be seen. It might well be the case that practical applications of the theory prove elusive. Moreover, in an observational study, the complications are compounded. In an observational study, the study groups will ordinarily not be exchangeable within principal strata as we have been assuming. Therefore, the problems of causal mediation will be superimposed upon those entailed in correcting for preexisting confounding. In any case, a clear understanding of the issues is vitally important to avoid the trap of attempting to model causal mediation without an explicit causal model.

8.3 NONCOMPLIANCE

In this section we consider a phenomenon that occurs often in randomized experiments and in certain observational studies also. Over the years, this problem has caused researchers a great deal of practical difficulty. In an RCT or quasi-experiment, the aim is usually to evaluate the causal effect of an intervention that is specified quite precisely. Among other details, the study protocol spells out exactly the nature of the treatment modalities and the conditions under which they are to be administered. The experimental subjects are expected to adhere closely to the requirements of the protocol. However, in many instances subjects may vary in terms of their degree of compliance with the study protocol.

For example, in the randomized study of supported work programs mentioned in Chapter 1, some subjects may not have been willing to participate in the program to which they were assigned. In particular, those who believed the program option offered to them was inferior to the alternative might have refused to comply. A common approach to analysis in such situations is to exclude those who fail to satisfy some prespecified standard of adequate compliance. This approach generates an *as-treated* or *per-protocol* estimate. However, those who are adequately compliant may differ in some respects from the target population, leading to sampling bias. If these differences vary according to the treatment modality assigned to the subject, the situation can entail a lack of external validity as well.

To understand the issues related to *noncompliance*, we consider a simple model that has been analyzed extensively. Imagine that two well-defined treatment modalities are being compared. Suppose that each subject can choose to comply by following the protocol requirements for the assigned modality or can opt to "defy" the study designers by choosing the "opposite" modality. Then there are four possible types of individuals in the study: compliers, always-takers, never-takers, and defiers (Angrist et al., 1996; Imbens and Rubin, 1997). *Compliers* always adhere to the assigned treatment modality. *Always-takers* adopt the experimental program, regardless of treatment assignment. *Never-takers* always adopt the control program, and *defiers* always select the opposite program.

This simplistic but illustrative formulation of the noncompliance issue can be translated into the model developed in the previous section. Let us define Z to be a causal mediator that indicates whether the subject complies with the assigned treatment ($Z = 1$) or not ($Z = 0$). Then we have the following set of equivalences:

- *I-doomed* = Compliers
- *I-causal* = Always-takers
- *I-preventive* = Never-takers
- *I-immune* = Defiers

Table 8.6 Data from a Randomized Study with Noncompliance: Numerical Outcome (Sums of Outcome Values by Exposure Status and Principal Stratum)

Exposed	Unexposed
Principal Stratum a: *Compliers*	
$AN_a\mu_{E1a}$	$(1-A)N_a\mu_{U1a}$
Principal Stratum b: *Always-Takers*	
$AN_b\mu_{E1b}$	$(1-A)N_b\mu_{U0b}$
Principal Stratum c: *Never-Takers*	
$AN_c\mu_{E0c}$	$(1-A)N_c\mu_{U1c}$
Principal Stratum d: *Defiers*	
$AN_d\mu_{E0d}$	$(1-A)N_d\mu_{U0d}$

It follows from the discussion in Section 8.2 that we could only estimate a direct effect for the compliers (*I-doomed*) or the defiers (*I-immune*). Furthermore, the estimate for the defiers would need to be "reversed" to account for the fact that all of these subjects received the opposite modality.

For illustrative purposes, it will be convenient to deal with the case of a numerical outcome and estimation of a mean difference. Table 8.6 is a version of Table 8.5 that pertains to our noncompliance model. Note that, in general, the potential outcomes can depend not only on the true exposure state, but also on the value of Z. In some situations, the assignment to a certain modality per se might have a causal effect, even if the assigned modality is resisted. For example, refusal to accept an assigned treatment modality might result in adopting some other treatment that would exert a causal effect on the outcome of interest.

Suppose, however, that each individual's potential outcomes are not affected by their compliance status except through the resulting exposure condition. This assumption is sometimes called the *exclusion restriction*. When the exclusion restriction holds, then we have

$$\mu_{E1k}=\mu_{U0k}=\mu_{Ek} \quad \text{and} \quad \mu_{U1k}=\mu_{E0k}=\mu_{Uk}, \qquad k=a,b,c,d \qquad (8.6)$$

Therefore, Table 8.6 can be rewritten as Table 8.7. In that case, the effect generated by the as-treated approach becomes

$$\text{ACE}_{\text{as-treated}}=\frac{N_a\mu_{Ea}+N_b\mu_{Eb}}{N_a+N_b}-\frac{N_a\mu_{Ua}+N_c\mu_{Uc}}{N_a+N_c} \qquad (8.7)$$

This expression will not in general be interpretable as a causal effect.

Table 8.7 Data from a Randomized Study with Noncompliance: Numerical Outcome, with Exclusion Restriction (Sums of Outcome Values by Exposure Status and Principal Stratum)

Exposed	Unexposed
Principal Stratum a: *Compliers*	
$AN_a\mu_{Ea}$	$(1-A)N_a\mu_{Ua}$
Principal Stratum b: *Always-Takers*	
$AN_b\mu_{Eb}$	$(1-A)N_b\mu_{Eb}$
Principal Stratum c: *Never-Takers*	
$AN_c\mu_{Uc}$	$(1-A)N_c\mu_{Uc}$
Principal Stratum d: *Defiers*	
$AN_d\mu_{Ud}$	$(1-A)N_d\mu_{Ed}$

The most common alternative to the as-treated approach is the *intention-to-treat* (ITT) methodology. The ITT approach estimates the total effect of exposure, essentially ignoring whether or not any particular individual actually complies with the modality assigned. The ITT approach attempts to preserve the benefits of randomization. Under an ITT regime, all randomized subjects are included in the analysis, regardless of actual compliance. An ITT analysis is often believed to approximate the way in which treatment modalities would actually be implemented in real-world applications. An RCT that estimates such a causal effect is called a *pragmatic* trial. Because such an experiment "encourages" the subjects to accept the assigned treatment, the term *encouragement trial* has also been coined (Holland, 1988).

In the case of a general causal mediator, the total effect can be represented as in Equation 8.2. In the special case of noncompliance (and assuming the exclusion restriction), the ITT effect can be expressed as

$$\text{ACE}_{\text{ITT}} = \frac{N_a\mu_{\Delta a} - N_d\mu_{\Delta d}}{N} \tag{8.8}$$

This simplification occurs because, within principal Strata *b* and *c*, subjects in both study groups have the same exposure state. For the always-takers and never-takers, noncompliance effectively nullifies the intended exposure difference.

Suppose further that it is reasonable to assume that there are no defiers ($N_d = 0$). After all, such perverse behavior would be unusual, unless the subjects were intent on sabotaging the experiment. Absence of defiers is called the *monotonicity restriction*. When we can add the monotonicity restriction, it becomes theoretically feasible to estimate the causal effect for the compliers.

This quantity has been termed the *complier average causal effect*, or CACE (Angrist et al., 1996; Imbens and Rubin, 1997):

$$CACE = \mu_{Ea} - \mu_{Ua} = \mu_{\Delta a} \qquad (8.9)$$

From Equations 8.8 and 8.9 we derive

$$ACE_{ITT} = \frac{N_a \mu_{\Delta a}}{N} = \frac{N_a}{N} CACE \qquad (8.10)$$

We can interpret Equation 8.10 to mean that the ITT effect is the CACE "diluted" by the zeroed-out (unobservable) effects for the always-takers and never-takers.

Intuitively, Equation 8.10 means that the ITT approach effectively "spreads" the sum of the individual causal effects of the compliers across the entire study population. Whether this makes sense depends mainly on the answers to two questions. First, are those who would comply in the context of the experiment likely to be similar to individuals who would comply in more general applications of the experimental treatment? That is, would the same kinds of people be induced to accept the experimental modality in real-world situations? Second, is the proportion of compliers in the study population similar to the corresponding proportion in the target population? That is, are the incentives and conditions such that roughly the same proportion would be induced to adopt the experimental modality? Utilizing the ITT approach implicitly assumes that the answers to these questions are affirmative, so that the ITT effect reflects what might be expected to occur in practice.

This discussion provides additional insight regarding the essential nature of a randomized controlled trial. On the one hand, an RCT facilitates valid causal inference by virtue of exchangeability between study groups. On the other hand, an RCT implicitly assumes that the assigned modalities will actually be implemented, both during the study and in practical applications. When compliance for some individuals is in question, the RCT's generalizability may be undermined. In a sense, randomization tries to *force* a particular modality on each individual. If such a coerced exposure status "takes" (as it does with compliers), the RCT yields valid causal inferences. But these results may not carry over into the field if the factors that affect adoption in other settings are quite different. Thus, there is often an inherent tension between the desire for tight experimental control to assure internal validity, and the need to emulate conditions that are likely to pertain in practice.

Equation (8.10) suggests the possibility of a middle course between the as-treated and ITT approaches. Suppose we knew the proportion of compliers (N_a/N) in the study population. Then it would be possible to inflate the ITT effect to "recover" the average effect for the compliers:

$$CACE = \frac{(ACE_{ITT})}{N_a/N} \qquad (8.11)$$

Under the simplifying assumptions of our compliance model, estimation of the proportion of compliers is feasible. From Table 8.7, it follows that, among all those *intended* to be exposed, the proportion *actually* exposed is $(N_a + N_b)/N$. Among those intended to be unexposed, the proportion actually exposed is N_b/N. Thus, the proportion of compliers can be obtained from

$$\frac{N_a}{N} = \frac{N_a + N_b}{N} - \frac{N_b}{N} \tag{8.12}$$

For example, suppose that 75% of those who are assigned to be exposed are found to be truly exposed (and therefore compliant), and 35% of those assigned to nonexposure are truly exposed (and therefore noncompliant). Then we would know that of the 75% truly exposed in the nominally exposed group, 35% are always-takers and 40% must be compliers.

Using Equation 8.12 we can (under the relevant assumptions) interpret the results of an RCT as estimating two parameters: (a) the CACE and (b) the proportion of individuals who could be induced to adopt the experimental modality (compliers). By considering each of these separately, we might gain a clearer picture of the full causal effect. We would also focus attention directly on the plausibility of the conditions required for generalization to a particular target population. Absence of such an explicit framing of the causal issues has been the source of much confusion in arguments about the relative merits of the as-treated and ITT approaches in various practical situations. The importance of "calibrating" the results of an RCT to the target population of interest has only recently been recognized (Frangakis, 2009).

Finally, we note that the essential logic leading to Equations 8.11 and 8.12 has recently been applied to help explicate the properties of a statistical technique that has long been cultivated by econometricians, but viewed skeptically by most statisticians. The use of *instrumental variables* has been advocated by some as a way to exploit "natural randomization" to estimate causal effects from observational data. An instrumental variable (IV) has two properties. First, the values of the IV for various individuals are determined effectively at random. Second, the IV exerts a causal effect on whether or not the individual actually adopts the treatment modality of real interest.

The variable that the IV influences is an intermediate causal factor. The IV can be regarded as providing "encouragement" to "comply with" the modality of interest. For example, suppose that educational vouchers are distributed by a random lottery to a sample of eligible families. These vouchers provide funding for tuition at certain private schools. Some of these families end up using the vouchers and others do not. This lottery may serve as an instrument that can be used to help evaluate the causal effect of a private-school education.

The causal effect being estimated in an IV analysis can be seen as analogous to the CACE from a randomized trial. This quantity has been dubbed the *local average treatment effect* (LATE). The LATE is the causal effect for individuals who would be induced to adopt the treatment (e.g., private school) if offered the opportunity defined by the IV (e.g., educational voucher). There has been much recent discussion among statisticians, econometricians, and sociologists about the practical conditions under which an unbiased estimate of a causal effect can be generated using instrumental variables (for recent reviews, see Gelman and Hill, 2007, Chapter 10; Morgan and Winship, 2007, Chapter 7).

8.4 ATTRITION

Broadly defined, lack of *adherence* pertains to any deviation from the study's protocol that could potentially affect the outcome. In the previous section, we considered the situation of noncompliance. Effectively, such a deviation changes the nature of the treatment modality actually received. However, the subjects remain in the study, so that follow-up continues and the outcome is eventually observed. But suppose that the individual completely drops out and is lost to follow-up. Such *attrition* is the most common reason for missing information on the outcome value, although *nonresponse* can sometimes occur for other reasons. In any event, missing outcome information can pose a serious problem that is akin to that of noncompliance.

Note that in general the subject can go missing at any point during the course of a study. Thus, there may be partial information about the final outcome based on observations at intermediate time points. Alternatively, if the outcome is dichotomous, we know that the subject "survived" up to the time of censoring. To understand the basic issue of attrition, we will consider the simple situation in which outcome measurement occurs only at the end of a specified follow-up period. If some of the subjects drop out of the study completely, the investigators may be faced with a Hobson's choice: flawed data or no data at all. Therefore, strenuous efforts to minimize attrition are usually made.

Attrition is similar to noncompliance in the sense that it occurs after the study has begun and can be affected by the individual's exposure state. One way to think about the problem is to define an intermediate variable R that has the value 1 if the individual is retained in the study and 0 if not. The only individuals whose outcomes could actually be observed are those who have $R = 1$. This situation is shown in Table 8.8. In Chapter 5, we considered a similar problem of retention bias in the context of our discussion of selection bias. There it was assumed that the "attrition" was related to the individual's outcome value. We can regard the retention process as a particular type of attrition in which the intermediate variable depends (either directly or indirectly) on the outcome value. Here we deal with the more general attrition problem from the causal mediation perspective developed in this chapter.

Attempting to estimate the causal effect conditional on $R = 1$ yields what we termed a naïve ACE in Section 8.2. The resulting empirical effect was given

Table 8.8   Data from a Randomized Study with Attrition (Sums of Outcome Values by Exposure Status and Principal Stratum)

Exposed	Unexposed
Principal Stratum a: *Always Retained*	
$AN_a\mu_{Ea}$	$(1-A)N_a\mu_{Ua}$
Principal Stratum b: *Retained If and Only If Exposed*	
$AN_b\mu_{Eb}$	—
Principal Stratum c: *Retained If and Only If Unexposed*	
—	$(1-A)N_c\mu_{Uc}$
Principal Stratum d: *Never Retained*	
—	—

in Equation 8.3. As explained there, this empirical effect will not in general be a causal effect. Moreover, even if it is a causal effect, this effect might differ from the causal effect in the study population.

For the naïve ACE produced by a study with attrition to be interpreted as a causal effect, two conditions must be satisfied. First, Strata b and c must be exchangeable (or at least mean-exchangeable). Second, the number of subjects in Stratum b must equal the number in Stratum c, that is, $N_b = N_c$. Intuitively, the first condition means that the individuals who remain if and only if exposed (Stratum b) are similar (at least on average) to those who remain if and only if unexposed (Stratum c). The second condition says that subjects are as likely to remain because of exposure as they are to remain because of nonexposure. Under these conditions, the effect calculated based on the retained subjects is interpretable as a causal effect. However, unless Strata b and c are also exchangeable with Strata a and d, this causal effect would not equal the ACE applicable to the study population as a whole.

GUIDEPOST 8

This chapter has attempted to unravel the conceptual intricacies of intermediate causal factors. The idea of an intermediate factor that can partially explain the total effect of an exposure is intuitively appealing but difficult to specify precisely. Intuitively, it should be possible to identify the direct causal effect, by somehow controlling for the intermediate factor and then subtracting from the total effect to obtain an indirect effect. However, this logic breaks down when analyzed carefully from the counterfactual perspective. The total effect and direct effect are not causes in the same sense, because the causal context is modified by the exposure.

A useful model to express the nature of possible bias related to an intermediate variable is based on the concept of principal stratification. A principal stratum is defined as a subgroup of subjects for whom the intermediate response pattern is constant. Within the framework provided by principal stratification, it is possible, in principle, to develop methods of analysis that avoid the misconceptions that often underlie statistical analyses of causal mediation. In particular, this framework can be applied to help understand the issues of noncompliance and attrition, and to shed light on the methodology of instrumental variables.

So far, we have dealt with the conceptual underpinnings of phenomena traditionally categorized as selection bias or confounding. We have implicitly assumed that the exposure state and the outcome value have been measured accurately. Chapter 9 will be concerned with information bias. This topic will be approached from both a traditional and a counterfactual perspective. The idea of causal mediation turns out to be useful in helping to understand the essential nature of exposure misclassification and outcome measurement.

CHAPTER 9

Information Bias

Once the subjects to be included in a study have been selected, the focus of research shifts to the collection of information about the subjects. There are two critical types of data that must be captured in any comparative study. First, for each subject we need to know the exposure status. Second, we need to record the value of the outcome variable. Errors of measurements that arise for either the exposure or the outcome can result in a biased estimate of the treatment effect.

Information bias is sometimes used also to describe the implications of measurement error in covariates. We explained in Chapters 6 and 7 how statistical adjustment techniques are based on the idea of conditioning on covariates. In theory, conditioning on these covariates can reduce or eliminate confounding bias. However, sometimes the covariates are measured with error. In the final section of this chapter, we briefly discuss how using a fallible covariate can affect the adjusted empirical effect.

Like selection bias and confounding, the concept of information bias has evolved in the absence of an explicit causal model. As a result, the essential nature of the problem has not been clearly articulated. Information bias is conceived as a discrepancy between the effect based on actual (distorted) data and the "true" effect measured without any distortion. But the true effect is simply the statistical association that would have been measured without error. Classical measurement theory offers no way to describe the relationship between the observed effect and the underlying causal effect. In this chapter we summarize the traditional measurement model and its limitations. We then discuss the issues of exposure and outcome measurement from a causal perspective.

Bias and Causation: Models and Judgment for Valid Comparisons, By Herbert I. Weisberg
Copyright © 2010 John Wiley & Sons, Inc.

9.1   BASIC CONCEPTS

For a dichotomous variable, measurement error results in *misclassification*. There are two possible types of misclassification errors. A *false negative* happens when the event of interest occurs but is not detected by the measurement process. Conversely, a *false positive* occurs when the event has not occurred but is (falsely) detected. Statistical models of misclassification attempt to explain how such misclassification errors can generate bias.

The traditional model of misclassification focuses on the error rates for these two types of misclassification errors. These error rates are conceived as conditional probabilities that describe the relationships between true and observed events in a given population. Consider any event A whose occurrence is being monitored. The probability of a false negative is the conditional probability of observing not-A (denoted $\sim A$), given that the event A has truly occurred:

$$P(\text{false negative}) = P(\sim A \text{ observed} | A \text{ true}) \qquad (9.1)$$

The *sensitivity* of a measurement process is the conditional probability that the event is in fact detected, given that it has actually occurred. That is,

$$\text{Sensitivity} = 1 - P(\text{false negative}) = P(A \text{ observed} | A \text{ true}) \qquad (9.2)$$

Similarly we define

$$P(\text{false positive}) = P(A \text{ observed} | \sim A \text{ true}) \qquad (9.3)$$

The *specificity* is then defined as the conditional probability that the event is not detected, given that it has actually not occurred:

$$\text{Specificity} = 1 - P(\text{false positive}) = P(\sim A \text{ observed} | \sim A \text{ true}) \qquad (9.4)$$

We can think of sensitivity as the probability of correctly detecting an event, and specificity as the probability of correctly identifying a nonevent.

In the context of a comparative study, the events of interest are exposure (E) and the outcome event (D). Sensitivity and specificity for the two key events (exposure and the outcome event) are the main parameters in the traditional misclassification model. Sensitivity is often imagined to be a stable property of the measurement process. However, the sensitivity found in one study might not carry over to a different study. Without an explicit causal model, there is no way to express clearly why such inconsistency across different populations might arise.

Traditional Misclassification Model

The general approach for dealing with misclassification is similar to the traditional approach for dealing with selection bias, as described in Section 5.2. We

Table 9.1   Standard 2 × 2 Table: True Data

	Disease	Healthy
Exposed	N_{ED}	N_{EH}
Unexposed	N_{UD}	N_{UH}

Table 9.2   Standard 2 × 2 Table: *Observed Data*

	Disease	*Healthy*
Exposed	O_{ED}	O_{EH}
Unexposed	O_{UD}	O_{UH}

start with the usual 2 × 2 table of cell counts from a study that is assumed to be without bias. To analyze misclassification bias, it will be convenient to rewrite the standard 2 × 2 table in a different form. In Table 9.1, the count for the exposed-case (ED) cell of the "true" table is N_{ED}. The exposed-control cell count is N_{EH}. Similarly, the unexposed-case and unexposed-control cells contain N_{UD} and N_{UH} individuals. Let us define the *observed* counts using similar notation, but with O in place of N; for example, the observed ED cell count is O_{ED}. The resulting observed table is represented as Table 9.2.

The traditional model assumes that the process of misclassification effectively "shuffles" the individuals in the four cells. For example, all individuals in the ED cell are assumed to have some fixed probability of being observed in each of the four cells. Each of these four probabilities reflects the chances of being misclassified on the exposure status, the outcome value, or both. Because there are four cells in all, there are 16 classification probabilities. However, for each cell, three of these probabilities determine the fourth, because the four must add to 1.0. Therefore, 12 *misclassification probabilities* are necessary to fully specify the measurement process.

A model with 12 parameters would be quite unwieldy. However, in many situations a much simpler form with fewer probabilities can be assumed. In the remainder of this section we consider three such simplified models that represent the main situations of practical importance:

- Misclassification of exposure status only
- Misclassification of outcome value only
- Independent misclassification of both exposure and outcome

Intuitively, independent misclassification of exposure and outcome means that the misclassification processes for exposure and outcome can effectively be separated. That is, whether or not an error is made on one of these factors does not affect the probability of making an error on the other. Nonindependent

True

Exposure Misclassified

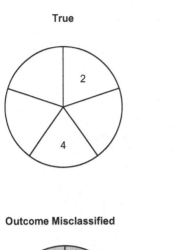

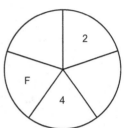

Outcome Misclassified

Both Misclassified

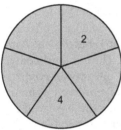

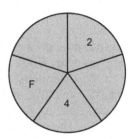

Figure 9.1. An *unexposed* individual who is subject to possible misclassification of outcome and/ or exposure. The apparent state of this individual is shown under the four possible situations that could occur.

misclassification of both factors is equivalent to the most general case with all 12 probabilities and will not be discussed further.

The remainder of our presentation follows the exposition presented in the text by Kleinbaum, Kupper, and Morgenstern (1982, Chapter 12). Besides presenting the general model, this reference reviews all of the major prior research that dealt with more specific models and their implications. More recent developments offer refinements of these results to deal with various complexities, such as misclassification when the outcome has more than two values, and bias in estimating parameters of a logistic regression model (see reviews by Höfler, 2005b; Rothman et al., 2008, Chapters 9 and 19).

Causal Misclassification Model

From a causal perspective, we focus on the individual's potential outcomes as they are observed and as they truly exist. Figure 9.1 shows a hypothetical individual with a causal context consisting of components 2 and 4 present, but components 1 and 3 absent. This person is assumed to be unexposed (*U*)

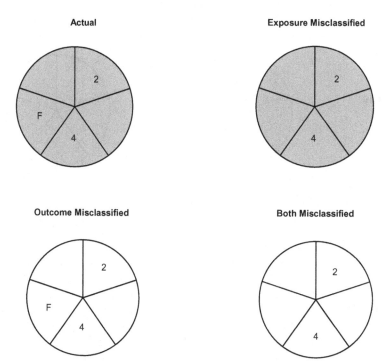

Figure 9.2. The four possible outcomes for an *exposed causal* individual who is subject to possible misclassification of outcome and/or exposure.

and healthy (*H*), but he is shown here under the three other possible scenarios. If only exposure status is misclassified, he will appear as an exposed person for whom the outcome event did not occur (*EH*). If only the outcome value is misclassified, he will seem to be an unexposed person for whom the event did occur (*UD*). Finally, if both are misclassified, he will present as an exposed individual for whom the event occurred (*ED*). Suppose now that our hypothetical individual considered previously happens to be *causal*. Then the possible observed outcomes for him, if exposed, are shown in Fig. 9.2.

Let us begin by considering what happens when there is outcome misclassification. Figure 9.3 shows the true and observed response patterns if this person is *always* misclassified; the data will represent him as *preventive* rather than *causal*. However, if he is misclassified if and only if exposed, then he will appear to be *immune,* as shown in Fig. 9.4. Conversely, if he is misclassified only when unexposed, he will appear to be *doomed*. So, the effect of outcome misclassification in this case is to transmute a *causal* response pattern into one of the other three response patterns. Similar "virtual" transformations occur if the person has one of the other true response patterns. By "changing" the response patterns in this manner, misclassification acts just like causal

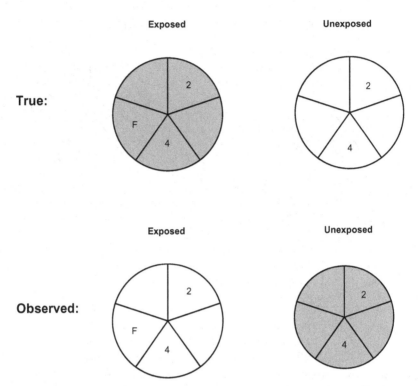

Figure 9.3. A *causal* individual whose outcome value is always misclassified, regardless of his exposure state. This individual appears as a *preventive*.

mediation. The difference is that this causal mediation is virtual, in the sense that the response pattern is only *apparently* modified.

Now consider what happens when there is misclassification of exposure. From the traditional perspective, exposure misclassification operates in the same manner as outcome misclassification. Both factors are treated completely symmetrically, ignoring the unique role each plays in a causal analysis. However, from the causal perspective the phenomenon of exposure misclassification is in reality more subtle. Suppose that the person represented by Figs. 9.1 and 9.2 is misclassified as unexposed if he is truly exposed, but correctly classified if unexposed. Then he will always appear to be unexposed, regardless of his true exposure state. However, his *outcome* will still depend on his true exposure state. Therefore, if truly exposed, he will appear to be in cell *UD* instead of *ED*, but if truly unexposed, he will appear to be in *UH* instead of *EH*.

Suppose this individual is truly exposed. Then he will not appear to be a member of the exposed study group. He will be virtually "reassigned" to the unexposed group. However, he will still be observed to experience the outcome event (because he is truly exposed and *causal*). Thus, it will be *as if* this subject

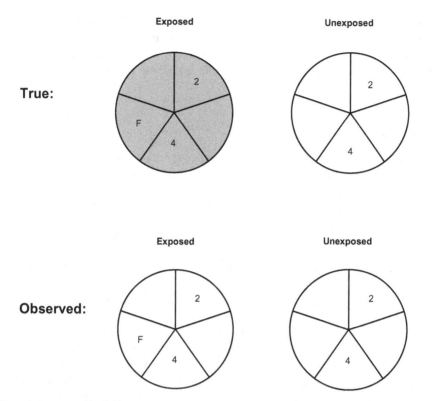

Figure 9.4. A *causal* individual whose outcome value is misclassified if and only if he is exposed. This individual appears as an *immune*.

were in fact either *doomed* or *preventive*. For practical purposes we can think of him as apparently *preventive* (Whether he is deemed apparently *preventive* or *doomed* is academic, because his observed outcome and exposure status are consistent with either). This situation can be described as involving *two* virtual transactions. First, an individual who is truly *causal* is "transferred" from the exposed to the unexposed group. Second, this individual's response pattern is modified from *causal* to *preventive*. Whenever an individual's exposure state is misclassified, there will be a virtual transfer from one study group to the other, coupled with the possibility of a virtual causal modification.

9.2 CLASSICAL MEASUREMENT MODEL: DICHOTOMOUS OUTCOME

In this section we develop the model of misclassification from a traditional perspective. Epidemiologists have been concerned about the implications of misclassification for a long time (e.g., Bross, 1954; Newell, 1962; Keys and

Kihlberg, 1963; Gullen et al., 1968; Barron, 1977; Copeland et al., 1977). Unfortunately, because no standard notation has been widely accepted, discussions in the literature can be hard to follow or compare with each other. In this section we present the main results and their practical implications in terms of the notation developed here.

Misclassification of Exposure

The concern with misclassification of exposure status derives primarily from the importance of retrospective data collection, especially in case–control studies. It is often necessary to gather exposure information long after any exposure may actually have occurred. Typically such information is obtained by accessing written records or by interviewing individuals. The potential for misclassifying exposure data can be significant. Comparative studies in the social sciences, on the other hand, are usually prospective. Thus, the identity of the treatment modality administered to each individual is typically known.

We now explain the relationship between the true data (Table 9.1) and the observed data (Table 9.2) when there is misclassification of the exposure status, but not of the outcome value. Consider an individual who is both exposed and a case (ED). As stated previously, the probability that this person will be *correctly observed* as exposed is the sensitivity, which we shall denote by C_E. Similarly, the probability that an unexposed person is correctly identified as unexposed is the specificity C_U. So, $N_{ED}C_E$ is the (expected) number of individuals in the ED cell who are correctly identified as exposed. However, there will also be some incorrectly observed (unexposed) individuals in the ED cell.

The probability that a person in the UD cell is incorrectly identified as being exposed is $(1 - C_U)$. Thus, there will be $N_{UD}(1 - C_U)$ members of the UD cell incorrectly added to the ED cell. Similar reasoning can be applied to each of the other three cells in the table. The result of these virtual "migrations" into and out of the four cells can be summarized as follows:

$$O_{ED} = N_{ED}C_E + N_{UD}(1 - C_U) \tag{9.5}$$

$$O_{EH} = N_{EH}C_E + N_{UH}(1 - C_U) \tag{9.6}$$

$$O_{UD} = N_{UD}C_U + N_{ED}(1 - C_E) \tag{9.7}$$

$$O_{UH} = N_{UH}C_U + N_{EH}(1 - C_E) \tag{9.8}$$

If we knew the values of C_E and C_U, then we could calculate the bias for any estimate of effect using Equations 9.5–9.8. Moreover, we could adjust for this bias by inverting the equations to solve for the true cell counts as functions of the observed cell counts (e.g., Kleinbaum et al., 1982). To illustrate how the relationships summarized by Equations 9.5–9.8 operate, we consider

Table 9.3 Hypothetical True Data with No Misclassification

	Disease	Healthy	Risk
Exposed	600	300	66.7%
Unexposed	400	700	36.4%
	RR = 1.83		
	OR = 3.50		

Table 9.4 Hypothetical Data: Nondifferential Exposure Misclassification
($C_E = 0.70$; $C_U = 0.90$)

	Disease	Healthy	Risk
Exposed	460	280	62.2%
Unexposed	540	720	42.9%
	RR = 1.45		
	OR = 2.19		

a numerical example. We will refer to a set of hypothetical data that has appeared in the literature (Copeland et al., 1977). The data are presented here as Table 9.3.

We can imagine that Table 9.3 represents the results of a hypothetical case–control study. We assume that all cases have been captured, but only 1% of the controls have been randomly selected. Because we have only a sample of controls, the odds ratio (OR) would become the relevant empirical effect (see Section 4.1). However, we can equally well regard Table 9.3 as the result of a cohort study and obtain qualitatively similar results in terms of the risk ratio (RR). In this example, the risk ratio is 1.83 and the odds ratio is 3.50. Recall that the odds ratio is a close approximation to the risk ratio that *would have been calculated from the entire cohort*. This can easily be verified by multiplying the values in the *EH* and *UH* cells by 100 to obtain the true RR of 3.45.

Now let us assume that the sensitivity is 0.70, meaning that there is a 30% chance of a false negative. For example, in the Yale Hemorrhagic Stroke Project (HSP) case–control study, some subjects might not have recalled their use of a product that contained phenylpropanolamine (PPA). We will assume that the specificity is 0.90, corresponding to only a 10% rate of false positives. In many studies, it may be possible to obtain ancillary information to confirm that a purported exposure actually occurred. By such means, the probability of a false positive might be minimized. In the HSP, efforts were made to verify that the subjects actually had purchased a product that contained PPA. For example, the contents of the subject's medicine cabinet were checked whenever possible.

Applying Equations 9.5–9.8 generates the results shown in Table 9.4. The odds ratio has been reduced from 3.50 to 2.19 as a result of misclassification.

Similarly, the risk ratio has been reduced from 1.83 to 1.45. The effect of misclassification in this situation is to decrease the value of a true OR or RR that is greater than 1.0. Suppose now that the data for exposed and unexposed were reversed, but values of sensitivity and specificity remained 0.70 and 0.90, respectively. Then it is easy to calculate that the perfectly classified data would yield a value of 0.29 for the odds ratio and 0.55 for the risk ratio. After misclassification, the odds ratio would be 0.46 and the risk ratio 0.69. In these examples the values of the true effect shrink closer to the null value of 1.0 by the measurement error. It turns out that under the classical measurement model summarized by Equations 9.5–9.8 this attenuation of the observed effect always occurs.

The exposure–misclassification model represented by Equations 9.5–9.8 assumes that the sensitivity and specificity have values that do not vary with the outcome value. In our example, we specified that the sensitivity was 0.70 for both the cases and controls, and the specificity was 0.90 for both groups. When the sensitivity and specificity do not vary according to outcome (disease) state, the measurement error is said to cause *nondifferential misclassification*. If these parameters did vary according to the outcome value, we would have *differential misclassification*. It has long been recognized that nondifferential misclassification always produces a bias toward the null (e.g., Bross, 1954; Newell, 1962; Keys and Kihlberg, 1963), but differential misclassification can move the empirical effect in either direction (e.g., Copeland et al., 1977). There are many possible reasons why misclassification of exposure might be differential. For example, in a case–control study a case may be more highly motivated than a control to recall risk factors possibly responsible for her illness.

To analyze differential misclassification, we must generalize Equations 9.5–9.8 by allowing the sensitivity and specificity to vary. Let C_{ED} represent the sensitivity for cases and C_{EH} the sensitivity for controls. Similarly, C_{UD} represents the specificity for cases and C_{UH} the specificity for controls. Then we can generalize Equations 9.5–9.8 to

$$O_{ED} = N_{ED}C_{ED} + N_{UD}(1 - C_{UD}) \tag{9.9}$$

$$O_{EH} = N_{EH}C_{EH} + N_{UH}(1 - C_{UH}) \tag{9.10}$$

$$O_{UD} = N_{UD}C_{UD} + N_{ED}(1 - C_{ED}) \tag{9.11}$$

$$O_{UH} = N_{UH}C_{UH} + N_{EH}(1 - C_{EH}) \tag{9.12}$$

In the article that presented the sample data in Table 9.3, the following set of misclassification parameters were used for illustration (Copeland et al., 1977):

$$C_{ED} = 0.90$$

$$C_{EH} = 0.60$$

$$C_{UD} = 0.70$$

$$C_{UH} = 0.90$$

Table 9.5   Hypothetical Data: Differential Exposure Misclassification
($C_{ED} = 0.90$; $C_{EH} = 0.60$; $C_{UD} = 0.70$; $C_{UH} = 0.90$)

	Disease	Healthy	Risk
Exposed	660	250	72.5%
Unexposed	340	750	31.2%
	RR = 2.33		
	OR = 5.82		

Inserting these values in Equations 9.9–9.12, we obtain the observed results presented in Table 9.5. The observed value of the RR is now 2.33, and that of the OR is 5.82. These are more extreme (biased away from the null value of 1.0) than the corresponding true values (1.83 and 3.50, respectively).

The specific effects of exposure misclassification are complex, depending on the correctly classified cell counts and on the misclassification parameters. To gain a more direct appreciation of the problems engendered by misclassification, it is helpful to plug various values for the cell counts and misclassification parameters into Equations 9.9–9.12 and observe the results. Such an exercise reveals that even what appear to be modest levels of misclassification can result in substantial bias. Therefore, it is essential to minimize the potential for exposure misclassification through proper design and conduct of research.

Traditional Model for Outcome Misclassification

The basic concepts developed in the previous section carry over to the situation of a misclassified outcome variable as well. A *false negative* happens when the outcome event actually occurs but is not detected by our measurement process. Conversely, a *false positive* occurs when the outcome event has not occurred but is falsely detected. An outcome measurement process has high reliability when both types of error have low probabilities. In this context the sensitivity is the conditional probability that the outcome event is detected, given that it has actually occurred. Specificity is the conditional probability that the event is not detected, given that it has not occurred. Once again, sensitivity is the probability of correctly detecting an event, and specificity is the probability of correctly identifying a nonevent.

The traditional model for outcome misclassification is virtually identical to the model for exposure misclassification. Again, Tables 9.1 and 9.2 represent the 2 × 2 tables with and without misclassification, respectively. Now, however, the exposure is assumed to be accurately observed, while the outcome is subject to misclassification. Consider the *ED* cell of Table 9.1. We denoted the sensitivity (with respect to *exposure* measurement) associated with individuals in this cell to be C_{ED}. We now define the sensitivity (with respect to *outcome* measurement) for the *ED* cell as C_{DE}. C_{DE} is the probability of correctly identifying someone in the *ED* cell as being a case rather than a control. Similarly, C_{DU} is the sensitivity for a person in the *UD* cell, and we define the specificities as C_{HE} and C_{HU}.

We can now derive the relationships between the observed and actual cell counts. For example, $C_{DE}N_{ED}$ of the individuals in cell *ED* will be correctly identified as having experienced the outcome event. In addition, $(1 - C_{HE})\, N_{EH}$ of the controls in cell *EH* will also be (incorrectly) observed in cell *ED*. Following such logic, we obtain the following relationships:

$$O_{ED} = N_{ED}C_{DE} + N_{EH}(1 - C_{HE}) \qquad (9.13)$$

$$O_{EH} = N_{EH}C_{HE} + N_{ED}(1 - C_{DE}) \qquad (9.14)$$

$$O_{UD} = N_{UD}C_{DU} + N_{UH}(1 - C_{HU}) \qquad (9.15)$$

$$O_{UH} = N_{UH}C_{HU} + N_{UD}(1 - C_{DU}) \qquad (9.16)$$

These formulas are analogous to Equations 9.9–9.12, but with the roles of the exposure and outcomes reversed. We can define nondifferential misclassification (of outcome) to mean that the sensitivity and specificity do not depend on exposure. In that special case, misclassification tends to bias the empirical effect toward the null, just as nondifferential misclassification of the exposure.

Traditional Model for Joint Misclassification

Often it is plausible that misclassification affects either the exposure or the outcome, but not both. For example, in a prospective cohort study, the exposure state is usually observed directly at the study's inception. Therefore, the possibility of substantial measurement error may be confined to observation of the outcome at a later point in time. On the other hand, in a case–control study, occurrence or nonoccurrence of an event might be evident when a potential case or control is identified, whereas the observation of the exposure state might depend on more error-prone retrospective recall.

In some situations, however, it is possible to have substantial misclassification of both the exposure state and the outcome value. For any particular individual, the observed exposure state and outcome value can depend in complex ways on the process by which information is obtained. This process therefore determines the relationship between the cell of the 2 × 2 table into which an individual *appears* to fall, and the actual cell into which she should be placed. As mentioned previously, an individual who is truly in a particular cell has some (usually unknown) probability of being observed to fall in each of the four cells. For example, an exposed case might appear to be an exposed case, an exposed control, an unexposed case, or an unexposed control. We explained how 12 misclassification probabilities are necessary to describe the full misclassification process.

Because there are so many parameters to consider, assessing or correcting for bias becomes quite complex. In statistical terms, the joint probability of being correctly (or incorrectly) classified on exposure status and of being correctly (or incorrectly) classified on outcome value can depend in unknown ways on the nature of the measurement process. The implications of joint misclassification in certain situations have been analyzed, but it is very difficult to reach any general conclusions (e.g., Copeland et al., 1977; Kristensen, 1992).

In many practical situations we can justify a simplifying assumption. Suppose that, for each individual, being correctly classified with respect to exposure state is independent of being correctly classified on the outcome value. Such *independent misclassification* is likely when the sources of information about exposure and outcome are separate. For example, in a case–control study, detection of disease should be performed by medical staff members who are blinded to exposure status, and exposure information would ideally be obtained by interviewers who are unaware of the study subject's medical condition. Separating the measurement of exposure status and outcome value in this way helps to assure that any measurement errors will be statistically independent.

In terms of our notation, the probability of correctly observing the exposure state of a person in cell ED is C_{ED}. The probability of correctly observing the outcome value is C_{DE}. Therefore, the probability that an individual in cell ED will be correctly observed as being in ED is simply $C_{ED}C_{DE}$. Consequently, $N_{ED}C_{ED}C_{DE}$ of the members of ED will be retained in cell ED of the *observed* version of the 2×2 table. In addition, $N_{EH}C_{EH}(1 - C_{HE})$ members of cell EH will migrate from EH to ED. Similarly, $N_{UD}(1 - C_{UD})C_{DU}$ members of cell UD and $N_{UH}(1 - C_{UH})(1 - C_{HU})$ members of cell UH will migrate to ED. The resulting observed count O_{ED} is the sum of these four terms. A similar chain of reasoning applies to each of the other three cells. The resulting set of relationships can be expressed as follows:

$$O_{ED} = N_{ED}C_{ED}C_{DE} + N_{EH}C_{EH}(1-C_{HE}) + N_{UD}(1-C_{UD})C_{DU} \\ + N_{UH}(1-C_{UH})(1-C_{HU}) \tag{9.17}$$

$$O_{EH} = N_{EH}C_{EH}C_{HE} + N_{ED}C_{ED}(1-C_{DE}) + N_{UH}(1-C_{UH})C_{HU} \\ + N_{UD}(1-C_{UD})(1-C_{DU}) \tag{9.18}$$

$$O_{UD} = N_{UD}C_{UD}C_{DU} + N_{UH}C_{UH}(1-C_{HU}) + N_{ED}(1-C_{ED})C_{DE} \\ + N_{EH}(1-C_{EH})(1-C_{HE}) \tag{9.19}$$

$$O_{UH} = N_{UH}C_{UH}C_{HU} + N_{UD}C_{UD}(1-C_{DU}) + N_{EH}(1-C_{EH})C_{HE} \\ + N_{ED}(1-C_{ED})(1-C_{DE}) \tag{9.20}$$

The relationships between the observed and actual cell values now involve 8 parameters rather than the full 12 necessary without the assumption of independent misclassification. In general, the bias can be large or small, and the observed empirical effect can be either more extreme or less extreme than the true causal effect. However, if an additional assumption can be justified, the situation simplifies further. Suppose that we have *nondifferential* misclassification for each factor individually. Then instead of eight misclassification parameters, there are only four: (C_E, C_U, C_D, C_H). We noted in previous sections that, for each factor individually, nondifferential misclassification always produces a bias toward the null value of the empirical effect. It has been demonstrated that if joint misclassification is both independent and nondifferential, the empirical effect will be less extreme than the corresponding causal effect (Gullen et al., 1968; Flegal et al., 1986). However, if misclassification is nondifferential and dependent, this result will not necessarily hold (Kristensen, 1992). So, when there exists joint misclassification, the only case in which the effect attenuation necessarily occurs is when both nondifferential and independent misclassification occur.

9.3 CAUSAL MEASUREMENT MODEL: DICHOTOMOUS OUTCOME

In Section 9.1 we introduced a causal perspective on misclassification. This approach focuses on the difference between the individual's characteristics as perceived and her actual characteristics. We explained how outcome misclassification is actually quite different conceptually from exposure misclassification. Outcome misclassification effectively operates like a particular type of virtual causal modification. Exposure misclassification, on the other hand, effectively entails a combination of two processes: a virtual change in the assignment mechanism coupled with a particular form of virtual causal modification. Because outcome misclassification is more straightforward, it will be considered first.

Causal Model for Outcome Misclassification

In Section 9.1 we presented a hypothetical example illustrated using Figs. 9.1 and 9.2. We explained how a *causal* individual whose outcome was misclassified could in different circumstances become perceived as if he were really *doomed*, *preventive*, or *immune*. Let us consider how misclassification can potentially generate bias. As discussed earlier, misclassification functions as a virtual causal mediator. This mediator indicates whether or not the individual's observed outcome is correct.

In Chapter 8, we introduced the concept of principal stratification. As applied to outcome misclassification, the four principal strata can be defined as follows:

- Stratum *a*: *I-doomed* = Outcome is always correct.
- Stratum *b*: *I-causal* = Outcome is correct if and only if subject exposed.

Table 9.6 Principal Strata Based on Outcome Misclassification; Distributions of True Response Patterns within Strata

True	Exposed	Unexposed
Principal Stratum a: *Outcome Always Correct*		
Doomed	$A_{1a}N_aP_{1a}$	$(1 - A_{1a})N_aP_{1a}$
Causal	$A_{2a}N_aP_{2a}$	$(1 - A_{2a})N_aP_{2a}$
Preventive	$A_{3a}N_aP_{3a}$	$(1 - A_{3a})N_aP_{3a}$
Immune	$A_{4a}N_aP_{4a}$	$(1 - A_{4a})N_aP_{4a}$
Principal Stratum b: *Outcome Correct If and Only If Exposed*		
Doomed	$A_{1b}N_bP_{1b}$	$(1 - A_{1b})N_bP_{1b}$
Causal	$A_{2b}N_bP_{2b}$	$(1 - A_{2b})N_bP_{2b}$
Preventive	$A_{3b}N_bP_{3b}$	$(1 - A_{3b})N_bP_{3b}$
Immune	$A_{4b}N_bP_{4b}$	$(1 - A_{4b})N_bP_{4b}$
Principal Stratum c: *Outcome Correct If and Only If Unexposed*		
Doomed	$A_{1c}N_cP_{1c}$	$(1 - A_{1c})N_cP_{1c}$
Causal	$A_{2c}N_cP_{2c}$	$(1 - A_{2c})N_cP_{2c}$
Preventive	$A_{3c}N_cP_{3c}$	$(1 - A_{3c})N_cP_{3c}$
Immune	$A_{4c}N_cP_{4c}$	$(1 - A_{4c})N_cP_{4c}$
Principal Stratum d: *Outcome Always Misclassified*		
Doomed	$A_{1d}N_dP_{1d}$	$(1 - A_{1d})N_dP_{1d}$
Causal	$A_{2d}N_dP_{2d}$	$(1 - A_{2d})N_dP_{2d}$
Preventive	$A_{3d}N_dP_{3d}$	$(1 - A_{3d})N_dP_{3d}$
Immune	$A_{4d}N_dP_{4d}$	$(1 - A_{4d})N_dP_{4d}$

- Stratum c: *I-preventive* = Outcome is correct if and only if subject unexposed.
- Stratum d: *I-immune* = Outcome is never correct.

As before, let N_k represent the number of subjects in the study population who are in Stratum k ($k = a, b, c, d$). Then Table 9.6 displays the number of individuals with each combination of principal stratum and response pattern. Here P_{jk} represents the proportion of subjects who have response pattern j in Stratum k. Table 9.6 represents the situation in which the assignment probabilities can potentially vary across response patterns and across the principal strata.

In the special case of a randomized control trial (RCT), there would be a constant assignment probability both within and across strata. The data would then simplify to the form shown in Table 9.7. Tables 9.6 and 9.7 represent the study data as it truly is, but not as it is actually observed. For example, consider the subjects who are in Stratum c. For those in the unexposed study

Table 9.7   Principal Strata Based on Outcome Misclassification; Distributions of True Response Patterns within Strata for a Randomized Experiment

True	Exposed	Unexposed
Principal Stratum a: *Outcome Always Correct*		
Doomed	AN_aP_{1a}	$(1-A)N_aP_{1a}$
Causal	AN_aP_{2a}	$(1-A)N_aP_{2a}$
Preventive	AN_aP_{3a}	$(1-A)N_aP_{3a}$
Immune	AN_aP_{4a}	$(1-A)N_aP_{4a}$
Principal Stratum b: *Outcome Correct If and Only If Exposed*		
Doomed	AN_bP_{1b}	$(1-A)N_bP_{1b}$
Causal	AN_bP_{2b}	$(1-A)N_bP_{2b}$
Preventive	AN_bP_{3b}	$(1-A)N_bP_{3b}$
Immune	AN_bP_{4b}	$(1-A)N_bP_{4b}$
Principal Stratum c: *Outcome Correct If and Only If Unexposed*		
Doomed	AN_cP_{1c}	$(1-A)N_cP_{1c}$
Causal	AN_cP_{2c}	$(1-A)N_cP_{2c}$
Preventive	AN_cP_{3c}	$(1-A)N_cP_{3c}$
Immune	AN_cP_{4c}	$(1-A)N_cP_{4c}$
Principal Stratum d: *Outcome Always Misclassified*		
Doomed	AN_dP_{1d}	$(1-A)N_dP_{1d}$
Causal	AN_dP_{2d}	$(1-A)N_dP_{2d}$
Preventive	AN_dP_{3d}	$(1-A)N_dP_{3d}$
Immune	AN_dP_{4d}	$(1-A)N_dP_{4d}$

group, the distribution of response patterns is not distorted in any way. On the other hand, for those who are exposed, the response-pattern distribution is (virtually) altered. The nature of this causal modification is displayed in Table 9.8.

Based on the information in Table 9.8 we can represent the virtual response-pattern distributions for various combinations of principal stratum and exposure state. The results are displayed in Table 9.9. For example, a *preventive* individual in Stratum c who is truly exposed would be observed as if *doomed*. From Table 9.8 we see that the only virtual *doomed* individuals will in fact be the true *preventives*. Therefore, the number of virtual *doomed* among those who are exposed (and therefore misclassified) in Stratum c is $A_{3c}N_cP_{3c}$. Similar relationships hold for each of the other response patterns, and each of the principal strata.

From Table 9.9, it is apparent that the total (virtual) response-pattern distribution in each study group has been changed from the corresponding true distribution. Furthermore, the distributions within all of the principal strata

Table 9.8   Virtual Causal Modification Induced by Outcome Misclassification

True Response Pattern	Exposed Value	Unexposed Value	Observed Response Pattern	Exposed Value	Unexposed Value
Stratum a: *Outcome Always Correct*					
Doomed	D	D	Doomed	D	D
Causal	D	H	Causal	D	H
Preventive	H	D	Preventive	H	D
Immune	H	H	Immune	H	H
Stratum b: *Outcome Correct If and Only If Exposed*					
Doomed	D	D	Causal	D	H
Causal	D	H	Doomed	D	D
Preventive	H	D	Immune	H	H
Immune	H	H	Preventive	H	D
Stratum c: *Outcome Correct If and Only If Unexposed*					
Doomed	D	D	Preventive	H	D
Causal	D	H	Immune	H	H
Preventive	H	D	Doomed	D	D
Immune	H	H	Causal	D	H
Stratum d: *Outcome Always Misclassified*					
Doomed	D	D	Immune	H	H
Causal	D	H	Preventive	H	D
Preventive	H	D	Causal	D	H
Immune	H	H	Doomed	D	D

except Stratum *a* will be modified. The resulting information bias depends on three sets of parameters:

- Response-pattern distributions in each principal stratum
- Assignment probabilities in each principal stratum
- Proportions of study population in each principal stratum

Furthermore, it is impossible to disentangle the effect of each factor from those of the others. The best way to gain insight into the inner workings of exposure misclassification is to create a spreadsheet and experiment with different values for the parameters. Such an exercise engenders a healthy appreciation of the potential complexity of outcome misclassification.

If we are dealing with misclassification in a randomized experiment, the situation is somewhat simpler, as represented in Table 9.10. Note that for Stratum *a* (individuals who are always observed correctly), the empirical effect

Table 9.9   Principal Strata Based on Outcome Misclassification; Distributions of Virtual *Response Patterns within Strata*

Virtual	Exposed	Unexposed
	Principal Stratum a: *Outcome Always Correct*	
Doomed	$A_{1a}N_aP_{1a}$	$(1 - A_{1a})N_aP_{1a}$
Causal	$A_{2a}N_aP_{2a}$	$(1 - A_{2a})N_aP_{2a}$
Preventive	$A_{3a}N_aP_{3a}$	$(1 - A_{3a})N_aP_{3a}$
Immune	$A_{4a}N_aP_{4a}$	$(1 - A_{4a})N_aP_{4a}$
	Principal Stratum b: *Outcome Correct If and Only If Exposed*	
Doomed	$A_{1b}N_bP_{1b}$	$(1 - A_{2b})N_bP_{2b}$
Causal	$A_{2b}N_bP_{2b}$	$(1 - A_{1b})N_bP_{1b}$
Preventive	$A_{3b}N_bP_{3b}$	$(1 - A_{4b})N_bP_{4b}$
Immune	$A_{4b}N_bP_{4b}$	$(1 - A_{3b})N_bP_{3b}$
	Principal Stratum c: *Outcome Correct If and Only If Unexposed*	
Doomed	$A_{3c}N_cP_{3c}$	$(1 - A_{1c}) N_cP_{1c}$
Causal	$A_{4c}N_cP_{4c}$	$(1 - A_{2c}) N_cP_{2c}$
Preventive	$A_{1c}N_cP_{1c}$	$(1 - A_{3c}) N_cP_{3c}$
Immune	$A_{2c}N_cP_{2c}$	$(1 - A_{4c}) N_cP_{4c}$
	Principal Stratum d: *Outcome Always Misclassified*	
Doomed	$A_{4d}N_dP_{4d}$	$(1 - A_{4d})N_dP_{4d}$
Causal	$A_{3d}N_dP_{3d}$	$(1 - A_{3d})N_dP_{3d}$
Preventive	$A_{2d}N_dP_{2d}$	$(1 - A_{2d})N_dP_{2d}$
Immune	$A_{1d}N_dP_{1d}$	$(1 - A_{1d})N_dP_{1d}$

would be correct. In Stratum *b*, on the other hand, the virtual response-pattern distribution for the exposed subjects differs from the distribution for the unexposed. So, the empirical effect would be biased. The situation in Stratum *c* is similar. In Stratum *d*, the response-pattern distributions for the study groups are the same, but they have been distorted. The resulting empirical effect does represent a causal effect but measures the "reverse effect" of exposure on nonoccurrence of the outcome event.

Table 9.10 reveals that the only causal effect of *F* that could in principle be estimated correctly would be that which occurs in Stratum *a*. If we could somehow identify those individuals who are never misclassified, we could then restrict the analysis to this subgroup. For example, it might in some situations be feasible to follow certain individuals much more closely, to be certain their outcomes are recorded accurately. Aside from the smaller sample size that would result, the main potential issue would then be whether this subgroup is in fact exchangeable with the remainder of the study population.

Table 9.10 Principal Strata Based on Outcome Misclassification; Distributions of *Virtual* Response Patterns within Strata for a Randomized Experiment

Virtual	Exposed	Unexposed
Principal Stratum a: *Outcome Always Correct*		
Doomed	AN_aP_{1a}	$(1-A)N_aP_{1a}$
Causal	AN_aP_{2a}	$(1-A)N_aP_{2a}$
Preventive	AN_aP_{3a}	$(1-A)N_aP_{3a}$
Immune	AN_aP_{4a}	$(1-A)N_aP_{4a}$
Principal Stratum b: *Outcome Correct If and Only If Exposed*		
Doomed	AN_bP_{1b}	$(1-A)N_bP_{2b}$
Causal	AN_bP_{2b}	$(1-A)N_bP_{1b}$
Preventive	AN_bP_{3b}	$(1-A)N_bP_{4b}$
Immune	AN_bP_{4b}	$(1-A)N_bP_{3b}$
Principal Stratum c: *Outcome Correct If and Only If Unexposed*		
Doomed	AN_cP_{3c}	$(1-A)N_cP_{1c}$
Causal	AN_cP_{4c}	$(1-A)N_cP_{2c}$
Preventive	AN_cP_{1c}	$(1-A)N_cP_{3c}$
Immune	AN_cP_{2c}	$(1-A)N_cP_{4c}$
Principal Stratum d: *Outcome Always Misclassified*		
Doomed	AN_dP_{4d}	$(1-A)N_dP_{4d}$
Causal	AN_dP_{3d}	$(1-A)N_dP_{3d}$
Preventive	AN_dP_{2d}	$(1-A)N_dP_{2d}$
Immune	AN_dP_{1d}	$(1-A)N_dP_{1d}$

Our analysis suggests that the phenomenon of misclassification bias may be extremely complex. This is true even for a randomized experiment, and the complexity increases dramatically for an observational study. A causal model, such as that presented here, *might* become the basis for a realistic solution under certain conditions. However, experience has shown that even the relatively simple traditional model of misclassification has been of very limited practical utility. Consequently, the main value of the causal model may be in underlining the importance of implementing data collection procedures that assure very high levels of measurement accuracy (e.g., Michels, 2001).

Causal Model for Exposure Misclassification

In Section 9.1, we introduced the idea that exposure misclassification is conceptually distinct from outcome misclassification. Outcome misclassification entails a virtual causal modification for some subjects, but the study groups remain intact. Exposure misclassification involves not only a form of virtual causal mediation, but also changes the composition of the study groups

Table 9.11 Principal Strata Based on Exposure Misclassification; Distributions of *Virtual* **Response Patterns within Strata Arranged by** True **Exposure State**

True	Exposed	Unexposed
Principal Stratum a: *Exposure Always Correct*		
Doomed	$A_{1a}N_aP_{1a}$	$(1 - A_{1a})N_aP_{1a}$
Causal	$A_{2a}N_aP_{2a}$	$(1 - A_{2a})N_aP_{2a}$
Preventive	$A_{3a}N_aP_{3a}$	$(1 - A_{3a})N_aP_{3a}$
Immune	$A_{4a}N_aP_{4a}$	$(1 - A_{4a})N_aP_{4a}$
Principal Stratum b: *Exposure Correct If and Only If Exposed*		
Doomed	$A_{1b}N_bP_{1b}$	$(1 - A_{1b})N_bP_{1b}$
Causal	$A_{2b}N_bP_{2b}$	$(1 - A_{3b})N_bP_{3b}$
Preventive	$A_{3b}N_bP_{3b}$	$(1 - A_{2b})N_bP_{2b}$
Immune	$A_{4b}N_bP_{4b}$	$(1 - A_{4b})N_bP_{4b}$
Principal Stratum c: *Exposure Correct If and Only If Unexposed*		
Doomed	$A_{1c}N_cP_{1c}$	$(1 - A_{1c})N_cP_{1c}$
Causal	$A_{3c}N_cP_{3c}$	$(1 - A_{2c})N_cP_{2c}$
Preventive	$A_{2c}N_cP_{2c}$	$(1 - A_{3c})N_cP_{3c}$
Immune	$A_{4c}N_cP_{4c}$	$(1 - A_{4c})N_cP_{4c}$
Principal Stratum d: *Exposure Always Misclassified*		
Doomed	$A_{1d}N_dP_{1d}$	$(1 - A_{1d})N_dP_{1d}$
Causal	$A_{3d}N_dP_{3d}$	$(1 - A_{2d})N_dP_{2d}$
Preventive	$A_{2d}N_dP_{2d}$	$(1 - A_{3d})N_dP_{3d}$
Immune	$A_{4d}N_dP_{4d}$	$(1 - A_{4d})N_dP_{4d}$

actually being compared. For example, the *apparently* exposed group consists of those individuals who are counted as exposed after any measurement errors have occurred.

As we did for outcome misclassification, we can define an intermediate variable that indicates whether or not a correct exposure state is observed. The principal strata for this exposure mediator are as follows:

- Stratum *a*: *I-doomed* = Exposure state is always correct.
- Stratum *b*: *I-causal* = Exposure state is correct if and only if subject exposed.
- Stratum *c*: *I-preventive* = Exposure state is correct if and only if subject unexposed.
- Stratum *d*: *I-immune* = Exposure state is never correct.

Then Table 9.11 shows the data broken down by *true* exposure state and principal stratum. The observed data, though, will reflect the effect of the

Table 9.12 Principal Strata Based on Exposure Misclassification; Distributions of *Virtual* Response Patterns within Strata Arranged by *Virtual* Exposure State

Virtual	*Exposed*	*Unexposed*
	Principal Stratum a: *Exposure Always Correct*	
Doomed	$A_{1a}N_aP_{1a}$	$(1 - A_{1a})N_aP_{1a}$
Causal	$A_{2a}N_aP_{2a}$	$(1 - A_{2a})N_aP_{2a}$
Preventive	$A_{3a}N_aP_{3a}$	$(1 - A_{3a})N_aP_{3a}$
Immune	$A_{4a}N_aP_{4a}$	$(1 - A_{4a})N_aP_{4a}$
	Principal Stratum b: *Exposure Correct If and Only If Exposed*	
Doomed	N_bP_{1b}	
Causal	$N_b [A_{2b}P_{2b} + (1 - A_{3b})P_{3b}]$	
Preventive	$N_b [A_{3b}P_{3b} + (1 - A_{2b})P_{2b}]$	
Immune	N_bP_{4b}	
	Principal Stratum c: *Exposure Correct If and Only If Unexposed*	
Doomed		N_cP_{1c}
Causal		$N_c [(1 - A_{2c})P_{2c} + A_{3c}P_{3c}]$
Preventive		$N_c [(1 - A_{3c})P_{3c} + A_{2c}P_{2c}]$
Immune		N_cP_{4c}
	Principal Stratum d: *Exposure Always Misclassified*	
Doomed	$(1 - A_{1d})N_dP_{1d}$	$A_{1d}N_dP_{1d}$
Causal	$(1 - A_{3d})N_dP_{3d}$	$A_{3d}N_dP_{3d}$
Preventive	$(1 - A_{2d})N_dP_{2d}$	$A_{2d}N_dP_{2d}$
Immune	$(1 - A_{4d})N_dP_{4d}$	$A_{4d}N_dP_{4d}$

misclassification. For example, a *causal* person in Stratum *b* who is truly unexposed would be misclassified as exposed. Such an individual's outcome would be *H*, because he would be truly unexposed. But because he is observed to be exposed, he will effectively register as if *immune* or *preventive*. So, we can think of this subject as being virtually transmuted from *causal* to *preventive*. Note that we could equally well have considered him to be *immune*, because all calculations would be the same. Likewise, a *preventive* person can be considered to be virtually changed into a *causal*. On the other hand, *doomed* and *immune* individuals would not be modified.

Table 9.12 displays the data as actually observed. Subjects have been reorganized and reclassified to indicate the number in each combination of principal stratum, observed exposure state, and virtual response pattern. For example, Stratum *b* contains those subjects who are correctly classified when they are truly exposed. The "exposed" group within Stratum *b* includes all of the subjects who are truly exposed, but also those who are truly

Table 9.13 Principal Strata Based on Exposure Misclassification; Distributions of *Virtual* Response Patterns within Strata: Randomized Experiment Arranged by *Virtual* Exposure State

Virtual	Exposed	Unexposed
	Principal Stratum a: *Exposure Always Correct*	
Doomed	AN_aP_{1a}	$(1-A)N_aP_{1a}$
Causal	AN_aP_{2a}	$(1-A)N_aP_{2a}$
Preventive	AN_aP_{3a}	$(1-A)N_aP_{3a}$
Immune	AN_aP_{4a}	$(1-A)N_aP_{4a}$
	Principal Stratum b: *Exposure Correct If and Only If Exposed*	
Doomed	N_aP_{1a}	
Causal	$N_b\left[AP_{2b} + (1-A)P_{3b}\right]$	
Preventive	$N_b\left[AP_{3b} + (1-A)P_{2b}\right]$	
Immune	N_bP_{4b}	
	Principal Stratum c: *Exposure Correct If and Only If Unexposed*	
Doomed		N_cP_{1c}
Causal		$N_c\left[(1-A)P_{2c} + AP_{3c}\right]$
Preventive		$N_c\left[(1-A)P_{3c} + AP_{2c}\right]$
Immune		N_cP_{4c}
	Principal Stratum d: *Exposure Always Misclassified*	
Doomed	$(1-A)N_dP_{1d}$	AN_dP_{1d}
Causal	$(1-A)N_dP_{3d}$	AN_dP_{3d}
Preventive	$(1-A)N_dP_{2d}$	AN_dP_{2d}
Immune	$(1-A)N_dP_{4d}$	AN_dP_{4d}

unexposed. However, for the misclassified individuals, the *causal* and *preventive* effectively switch roles, resulting in the numbers of virtually *causal* and *doomed* that are displayed. Similar logic leads to the other entries in Table 9.12.

Now suppose that we are dealing with a randomized experiment instead of an observational study. Table 9.13 shows the data that would be observed in this special case. Randomization implies that the assignment probability A is constant, both within and across the strata. Even in this relatively simple situation, an empirical effect (e.g., risk ratio) does not have a simple interpretation. Its value depends on the response-pattern distributions in various principal strata, the relative sizes N_k of the strata, and the assignment probability A. Once again, our model highlights the paramount importance of accuracy in measurement.

The main lesson from the counterfactual causal formulation of exposure misclassification is that the effects of the assignment mechanism and exposure misclassification are inextricably intertwined. In Chapter 5, we derived mathematical models for the bias resulting from the assignment mechanism. In principle, these models could be applied within each stratum and used to derive a complex algebraic expression for the overall bias. This expression would not, except perhaps in special cases, resolve neatly into a component of bias attributable to the assignment mechanism and a component attributable to exposure misclassification. Therefore, the effect of misclassification on the overall bias can depend not only on properties of the misclassification mechanism itself, like sensitivity and specificity, but also on how misclassification interacts with the assignment mechanism.

9.4 CLASSICAL MEASUREMENT MODEL: NUMERICAL OUTCOME

When the outcome is numerical, the general linear model is typically used for estimation of causal effects. As discussed in the context of selection bias (see Section 5.7), this approach brings to bear powerful mathematical machinery, but at the cost of some strong assumptions. The issue of measurement error in a general linear model has been considered extensively in the methodological literature, especially by econometricians. The problem of *errors-in-variables* has been attacked in many different ways. This work is aimed at finding practical solutions to estimating a true set of relationships among variables based on observed variables that are subject to measurement error. These approaches might depend on theoretical assumptions alone, or make use of supplementary data, such as multiple measurements, validation by a gold standard, known reliability in other populations, etc. Our primary purpose here is to describe the conceptual issues involved when there are errors in measuring the treatment and/or outcome.

Traditional Model for Misclassification of Exposure Status

The traditional linear model was introduced in Section 5.7:

$$Y_i = \mu + \alpha F + e_i \tag{9.21}$$

Here Y_i represents the outcome value of subject i, and e_i is the "error" term. The error term is assumed to be a random variable with means μ_{eE} and μ_{eU} in the two study groups. The errors are assumed to be statistically independent across individuals. That is, the value of e_i for any individual is unrelated to the value for any other. Furthermore, we assume that the means of the error terms *within each group* are zero. Under these assumptions, the constant treatment effect α can be estimated by the difference between the means for the two groups.

Now suppose that the exposure is measured with error. Then each individual has some probability of being correctly classified. In general, this probability of correct classification could depend on both the exposure state and the subject's value of the error term (or equivalently the outcome value). Misclassification of exposure that depends on the outcome is termed *differential misclassification*. The implications of differential misclassification can be quite complex.

If misclassification of the exposure state is nondifferential, the situation becomes simpler. Under nondifferential misclassification, we can define C_E as the probability of correctly classifying an exposed individual and C_U as the probability of correctly classifying an individual who is not exposed. Let N_E be the total number of subjects in the exposed group, and let N_U be the number in the unexposed group. Misclassification virtually "transfers" a proportion $1 - C_E$ of the exposed subjects into the unexposed group and a proportion $1 - C_U$ of the unexposed subjects into the exposed group.

Assume that we are estimating the constant causal effect α. After a bit of algebra, we obtain the following expression for the observed effect:

$$\text{Observed ACE} = \alpha(\text{PPV} + \text{NPV} - 1) \tag{9.22}$$

In this expression

$$\text{PPV} = \frac{N_E C_E}{N_E C_E + N_U(1 - C_U)} \tag{9.23}$$

$$\text{NPV} = \frac{N_U C_U}{N_U C_U + N_E(1 - C_E)} \tag{9.24}$$

where PPV stands for *positive predictive value* and NPV for *negative predictive value*. The PPV and NPV are "inverse probabilities" of the sensitivity and specificity, respectively. For example, rather than the conditional probability of being observed as exposed given that the individual is truly exposed (sensitivity), the PPV is the conditional probability of being truly exposed given that she is observed to be exposed.

Intuitively, Equation 9.22 implies that, under nondifferential exposure misclassification, the observed effect will be reduced, or *attenuated*. The amount of this attenuation depends on the proportions of misclassified subjects who are misclassified. Note that if PPV = NPV = 1 (i.e., no misclassification), there is no bias, and if PPV = NPV = 0.5, the estimated effect is reduced to zero. In this case, a person observed to be in a particular group (exposed or unexposed) has a 50% chance of actually being in the other group. Thus, the data provide no information about the true assignment. For values between these extremes, the effect of nondifferential misclassification is to shrink the empirical effect toward the null value of zero.

Classical Model for Outcome Measurement Error

The classical model for measurement error, developed primarily by psycho-metricians, assumes that an observed "score" can be expressed as the sum of two components: the "true score" T and the error ε. Thus, for any individual i we have

$$Y_i = T_i + \varepsilon_i \tag{9.25}$$

Here the error variable ε is assumed to have a mean value of zero, *conditional on the value of the individual's true score*. Each individual is envisioned as having a distribution of errors that would occur in repeated sampling from a hypothetical population of different "occasions" of testing. The score Y_i that is actually observed is the fixed value of T_i plus a randomly selected observation from the error distribution. Furthermore, it is usually assumed (often implicitly) that these error distributions are identical in form for all subjects. Indeed, the normal distribution is often assumed, at least as an approximation.

In the context of a comparative study, the observed outcome value in the absence of measurement error is the individual's true score. Thus, we can rewrite Equation 9.21 inserting T_i for Y_i:

$$T_i = \mu + \alpha F + e_i \tag{9.26}$$

Therefore, the observed score Y_i becomes

$$Y_i = \mu + \alpha F + e_i + \varepsilon_i \tag{9.27}$$

Here ε_i is the error that results from the measurement process. Because the classical measurement model assumes these (individual) error distributions are identical and have (individual) means of zero, the mean values $\mu_{\varepsilon E}$ and $\mu_{\varepsilon U}$ of the measurement errors in the exposed group and unexposed group must both be zero. However, in general, it is possible that these means could differ. This could occur, for instance, if the observations of a program being evaluated are made by a researcher who is not impartial. If this researcher believes that the intervention is effective, she might either deliberately or unconsciously tend to rate outcomes in the exposed group more favorably.

If such systematic distortions can be ruled out, the classical model holds, then $\mu_{\varepsilon E}$ and $\mu_{\varepsilon U}$ are both zero, because the errors would represent a random sample from the same distribution (i.e., the hypothetical distribution presumed to determine each individual's errors in repeated sampling). This "purely random" error thus constitutes a kind of "noise" that averages out to zero in each study group. In this case, the two error terms can be subsumed into a single error term, say μ_i, that includes both:

$$Y_i = \mu + \alpha F + u_i \tag{9.28}$$

So, the measurement error vanishes, as it becomes folded into the overall "error" term.

However, in some situations, the measurement-error distributions might not be identical for all individuals. For example, the magnitude of errors might be related to the absolute level of the true outcome value. Alternatively, the errors could depend on the values of certain covariates, either measured or unmeasured. Such nonclassical error structures would require appropriate models that are more complex than the classical measurement model.

For the most part, more complex error models have not been tackled by research methodologists. As a result, it has become an article of faith that random outcome measurement error does not create bias in a comparative study. This conclusion may seem paradoxical when we consider that nondifferential misclassification for a dichotomous outcome always attenuates an effect. If such "random" measurement error in the context of a dichotomous outcome invariably produces bias, why is bias absent when the outcome is numerical? The answer lies in the strong, but usually unacknowledged, assumptions underlying the classical measurement model. To the extent that these assumptions are dubious, we cannot rest assured that bias is absent.

9.5   CAUSAL MEASUREMENT MODEL: NUMERICAL OUTCOME

In principle, the problem of measurement error could become much more complex when the outcome is numerical instead of dichotomous. The resulting information bias could depend on the process that generates the measurement errors and on the particular causal effect being estimated (e.g., difference of means, medians, percentiles, variances). For most such measures, the mathematical modeling to represent the consequences of measurement error would be intractable.

In practice, most studies focus on the difference of means, which we have denoted by $\mu_{WE} - \mu_{ZU}$. This empirical effect is meant to estimate the corresponding causal effect $\mu_{VE} - \mu_{VU}$. We have also described this difference as the average causal effect (ACE), or μ_Δ, because it is mathematically equal to the average of the individual causal effects (ICEs), $\Delta_i = V_{iE} - V_{iU}$. When we restrict attention to estimating the ACE, the situation is somewhat easier to handle, at least conceptually.

Causal Model for Exposure Misclassification: Numerical Outcome

The problem of exposure misclassification arises primarily in the context of retrospective studies, especially case–control studies. A causal model of exposure misclassification, even when the ACE is the preferred measure of effect, would therefore be extremely complicated. Because retrospective studies are by nature observational, a general model would need to account for possible confounding (i.e., nonexchangeability of the study groups). Moreover, the form of the nonexchangeability could vary across the principal strata (i.e., values of the assignment probabilities corresponding to each response pattern might differ across strata).

Of course, it is possible that simplifying assumptions can be justified in some practical situations. Then it might become feasible to develop methods to adjust for exposure misclassification. In particular, suppose that the assignment mechanism is known (or believed) to be effectively random. In a randomized study, the assignment probability is constant both across and within principal strata. Furthermore, the study groups are exchangeable within strata. As in Chapter 8, let μ_{Ek} and μ_{Uk} represent the mean outcome values for Stratum k.

This special situation is portrayed in Table 9.14. Even with randomization, the empirical effect is a complex expression that depends on the following:

- The assignment probability
- The relative sizes of the different principal strata
- The mean outcome values in each stratum for exposed subjects
- The mean outcome value in each stratum for unexposed subjects

Table 9.14 Observed Data from a Randomized Study with Exposure Misclassification: Numerical Outcome (Sums of Outcome Values by Principal Stratum and *Virtual* Exposure State)

Exposed	*Unexposed*
Principal Stratum a: *Exposure Always Correct*	
$AN_a\mu_{Ea}$	$(1 - A)N_a\mu_{Ua}$
Principal Stratum b: *Exposure Correct If and Only If Exposed*	
$N_b\left[A\mu_{Eb} + (1 - A)\mu_{Ub}\right]$	
Principal Stratum c: *Exposure Correct If and Only If Unexposed*	
	$N_c\left[(1 - A)\mu_{Uc} + A\mu_{Ec}\right]$
Principal Stratum d: *Exposure Always Misclassified*	
$(1 - A)N_d\mu_{Ud}$	$AN_d\mu_{Ed}$

Recall that, from the traditional perspective, the observed effect resulting from exposure misclassification when the causal effect is assumed to be a constant α is given by Equation 9.22. Suppose furthermore that the strata are all mean-exchangeable. Then it follows from Table 9.14 that Equation 9.22 holds, because the predictive probabilities are:

$$\text{PPV} = \frac{A(N_a + N_b)}{A(N_a + N_b) + (1 - A)(N_b + N_d)} \tag{9.29}$$

$$\text{NPV} = \frac{(1 - A)(N_a + N_c)}{(1 - A)(N_a + N_c) + A(N_c + N_d)} \tag{9.30}$$

In this sense, the traditional model can be viewed as a special case of the causal model, but one that is unlikely to be valid in general.

Causal Model for Outcome Measurement Error: Numerical Outcome

When the outcome is numerical and is measured with error, we are faced with a situation that can potentially be highly complex. The true potential outcomes for an individual are V_{iE} and V_{iU}. Corresponding observed potential outcomes can be denoted by O_{iE} and O_{iU}. For each individual, the relationship between the true response pattern (V_{iE}, V_{iU}) and the observed response pattern (O_{iE}, O_{iU}) can depend on a variety of individual characteristics. The distortion caused by measurement error can be regarded as an intermediate variable that (virtually) changes the outcome values. However, this causal mediator is a numerical variable, rather than the dichotomous mediator discussed in Section 9.3.

In most practical situations, measurement error is defined as a simple arithmetic difference between the observed and true outcome values. Furthermore, the ACE is usually the relevant measure of causal effect. Under these circumstances, the situation greatly simplifies. For any individual, we can define the *potential errors* ε_{iE} and ε_{iU} as

$$\varepsilon_{iE} = O_{iE} - V_{iE} \tag{9.31}$$

$$\varepsilon_{iU} = O_{iU} - V_{iU} \tag{9.32}$$

Here ε_{iE} represents the measurement error for individual i if she is exposed, and ε_{iU} represents her error if not exposed. We can think of measurement error as a second outcome variable. The difference $\varepsilon_{iE} - \varepsilon_{iU}$ represents the virtual causal effect of F on this secondary outcome. The observed ICE is the sum of the true ICE and this measurement-error effect:

$$O_{iE} - O_{iU} = V_{iE} - V_{iU} + \varepsilon_{iE} - \varepsilon_{iU} \tag{9.33}$$

Furthermore, the observed empirical effect is the sum of the true ACE and the measurement-error ACE:

$$\mu_{OE} - \mu_{OU} = \mu_{VE} - \mu_{VU} + \mu_{\varepsilon E} - \mu_{\varepsilon U} \tag{9.34}$$

Let us first consider what happens in a randomized experiment. Because of randomization, the exposed and unexposed groups are exchangeable with respect to the outcome variable, so the empirical effect would equal the causal effect under perfect outcome measurement. But randomization insures that the groups are exchangeable also with respect to the errors. Therefore, the mean error difference must be a *causal effect* of the exposure. This causal effect is the measurement bias.

There are two possible conditions that can result in absence of bias. First, there may be no causal effect at all on the errors. That is, the sharp null hypothesis ($\varepsilon_{iE} = \varepsilon_{iU}$) holds for all subjects. For example, in a well-conducted RCT with blinding of the subjects and investigators, each individual's treatment modality is completely hidden. Therefore, any error caused by the measurement process should not depend on which modality is being received. Alternatively, some of the individual error effects may be nonzero, but there may be no average causal effect of F on the errors. That is, the ACE for the error outcome would be zero ($\mu_{\varepsilon E} = \mu_{\varepsilon U}$). Note that it would not be necessary for the average bias in each group to be zero; it is possible for the mean outcome in each study group to be equally distorted. For example, if the measuring device registered five points too high for every individual, the mean in each group would also be five points too high, and there would be no bias.

In a randomized study, measurement error can lead to bias if and only if F causes a nonzero average error difference between study groups. This would typically occur when the exposure state is known to the subjects or observers, and influences their behavior. For example, consider a randomized experiment of supported work programs like the National Supported Work Demonstration (NSW). It would be difficult to mask the nature of the treatment modality received by the participant. If the researchers recorded a subjective assessment of work satisfaction by interviewing subjects, the responses given might be affected by which treatment modality was received.

Now let us consider the case of an observational study. It is possible that measurement errors are causally related to exposure state, just as in a randomized experiment. However, in addition, bias could result from lack of exchangeability with respect to the error "outcome." Therefore, the error effect might also depend on the assignment mechanism. Consider a study of supported work performed without random assignment. Once again, we assume that a subjective measure of job satisfaction is obtained. It is possible that certain individuals will tend to overstate (or understate) their true levels of satisfaction. Because the study is observational, these errors might not balance out across the study groups. For example, the workers assigned to the control group might tend to have more negative perceptions than those receiving the intervention.

The relationship between bias and the assignment mechanism for a numerical outcome is detailed in Section 5.7. The theory developed there can be applied to the errors as well as to the true outcomes. We can define response patterns and corresponding assignment probabilities for the error outcome. The results in Section 5.7 can then be applied. For example, suppose that for measurement error, the sharp null hypothesis held true, that is, $\varepsilon_{iE} = \varepsilon_{iU}$ for all subjects. In a randomized study, there would be no bias. However, in an observational study, there may be a correlation between the error and the assignment probability. Suppose that subjects with positive errors tend to fall into the exposed group, whereas those with negative errors tend to be unexposed. Then measurement error will clearly tend to inflate the observed effect. Equation 5.50 provides a precise expression for the magnitude of the bias in this situation. Conversely, the bias will be zero if there is no correlation between the size of the error and the probability of being exposed.

We mentioned earlier the common belief that measurement errors do not induce bias when the outcome is numerical. This notion is predicated on the restrictive assumptions underlying the classical measurement model. In this section, we have seen that measurement bias is possible, even in a randomized experiment. The explicit causal model presented here clarifies the conditions under which bias will occur. Once again, the most important practical implication is that we rarely have sufficient information to correct such bias, so a strong emphasis on prevention is warranted.

9.6 COVARIATES MEASURED WITH ERROR

In observational studies, it is almost always necessary to control in some manner for confounding factors. As discussed at length in Chapters 6 and 7, most statistical adjustment strategies are based on the idea of conditioning on covariates. After adjustment, it is hoped that the lack of exchangeability between study groups has been eliminated. For example, if we employ a version of stratification, the goal is to achieve exchangeability within each stratum.

So far, we have implicitly considered the covariates to be perfectly measured. Obviously, this may not always be the case. When errors of measurement are suspected, we may be concerned that bias will result. When the covariate is dichotomous, measurement error takes the form of misclassification. Traditional approaches to address this issue are in the same spirit as those described for exposure and outcome misclassification in this chapter. These analyses are rather complex and difficult to apply in practice (for a review, see Rothman et al., 2008, Chapters 9 and 19).

When the covariate is numerical, nearly all discussions have assumed the general linear model, the simplest form of which is Equation 9.28. When there is a covariate X, this model becomes

$$Y_i = \mu + \alpha F + \beta X + u_i \tag{9.35}$$

The traditional approach to this problem applies the classical measurement model to the covariate. Thus, we assume that instead of the true X, we observe a "fallible" covariate X' that contains an error term, say ε':

$$X_i' = X_i + \varepsilon_i' \tag{9.36}$$

Then it has long been a standard statistical result in the social sciences, especially econometrics, that regression coefficients become distorted (biased in the statistical sense) when independent variables are measured with error (e.g., Johnston, 1963, Chapter 6). In the context of statistical adjustment in comparative studies, this bias leads to an "underadjustment" that attenuates the effect toward zero (e.g., Cochran, 1968a; Elashoff, 1969).

This underadjustment phenomenon is a particular instance of what is broadly termed the *regression effect*, or a *regression artifact*. If we condition on a fallible covariate, then we are effectively stratifying on the basis of the random error component as well as the true X value. Suppose we match two subjects, one from each study group, on the basis of X' and observe their outcomes. Assume that the average true-score value in the exposed group is higher. Then under the classical model assumptions, the expected outcome value for the subject from the exposed group will be higher than that for the subject from the unexposed group.

This artifact happens because the exposed subjects with a given covariate score will, on average, have higher true-scores than unexposed subjects with the same observed pre-score. For concreteness, consider the common situation in which the covariate is a pretest, and the outcome a posttest. In particular, suppose the true-score mean for the exposed study group is 125, and the true-score mean for the unexposed group is 75. Consider the individuals with a pretest score of 100. The exposed subjects who have a pre-score of 100 will have an average error value that is negative, while the average for the unexposed subjects with a pre-score of 100 will be positive. However, on the posttest, the average measurement error for each of the two study groups is expected to be zero. Thus, exposed individuals with pre-score 100 will, on average, move somewhat closer on the posttest to the true-score mean of 125 for the exposed group, whereas the unexposed will tend to be closer to the true-score of 75 for the unexposed group. This "regression toward the mean" would not occur if there were no measurement errors. For a comprehensive and readable reference on the subject of regression artifacts, see Campbell and Kenny (1999). Of course, the logic expounded above is based on the classical measurement model. As noted previously, this model rests on some strong assumptions.

From the causal modeling standpoint, matters are much less clear-cut. Once again, suppose we assemble a matched pair, based on a pretest score. For each individual, we can regard a test score as an imperfect measure of some underlying trait (e.g., verbal reasoning). The individual's true score for this trait can be interpreted as a causal factor. If we could match based on this causal factor, then the response pattern for each member of the pair would be identical. The problem in practice is that we cannot condition on the true score, because we can only measure a fallible observed score.

The value of this observed covariate may be determined by various causal factors in addition to the trait measured by the true score. Some of these other factors can reasonably be viewed as *transitory*. There are various physical factors (e.g., head cold, inadequate sleep) and psychological factors (e.g., breakup with boyfriend, family argument) that might impact test performance on a given day. Most of these effects may be fairly minor and may have a "random" quality. The classical model assumptions may reflect such effects quite accurately.

On the other hand, there may also be factors that are quite stable, at least over the course of the program. For example, some subjects may suffer consistently from a disability such as test anxiety or dyslexia. Such a condition could affect the test score but would not be related directly to the underlying construct (e.g., verbal reasoning) the test is intended to assess. A stable factor would be a determinant of the observed test score, but not of the true score, unless we circularly define the true score to be the expected value of the test score in repeated testing. So, the stable factor would be a measurement error, but it will not induce regression toward the group mean and will not contribute to underadjustment bias.

In theory, it would be correct to control for these stable error factors in a comparative study. For example, suppose we were somehow able to condition on the true verbal reasoning score for an exposed subject who regularly suffered from severe test anxiety. This individual is matched with an unexposed subject who does not have test anxiety. If the exposure actually has no causal effect, the observed posttest score for the exposed subject will tend to be lower than the posttest score for the unexposed. On the other hand, conditioning on the observed pretest would have yielded the correct result.

From a causal perspective, we should not be concerned primarily with whether the covariate is measured with error relative to some hypothetical true score. Rather, the critical issue is whether the study groups are exchangeable after conditioning on the covariate. In this regard, stability of the error is an important consideration, because a stable component of measurement error effectively becomes part of the causal context for an individual, and therefore a potential confounder:

... viewed in terms of model specification, the fallibility of the covariate is not in itself the problem. Having a perfectly reliable covariate will not guarantee correct adjustment, and having a fallible covariate does not necessarily result in bias. (Weisberg, 1979, 1160)

The idea that reliability of measurement per se is essential to successful adjustment for bias has caused particular confusion in studies of growth and development. The most famous illustration of the difficulty was presented by psychometrician Frederic Lord and has come to be known as Lord's paradox (Lord, 1967). Lord describes a hypothetical study of weight gain at a university during the course of a semester. The study aims to determine whether gender is related to the increase in weight. Two statisticians analyze the resulting data. The first simply calculates the average difference between the preweight and postweight, and finds no average gain for either group. The boys are of course initially heavier on average than the girls, but the increase is identical. He concludes that there is no effect of gender. The second statistician performs an analysis of covariance, controlling for initial weight as a covariate. He finds "that the boys showed significantly more gain in weight than the girls when proper allowance is made for initial differences between the sexes" (Lord, 1967, 305).

Note that in Lord's example, there is no pure measurement error in the usual sense, because the weights are presumably ascertained with perfect accuracy. The "measurement" problem is that the preweight may not be a perfect predictor of weight gain for each individual. To unravel the "paradox" in such a situation, we would need a model of individual growth (Bryk and Weisberg, 1977; Weisberg, 1979; Rogosa et al., 1982). We can imagine the weight gain as being potentially determined by two causal factors: the preweight and the "natural" growth rate (i.e., weight gain rate under normal conditions).

The first statistician implicitly assumes an equal *average* natural growth rate for the students of each gender. Furthermore, he assumes that the growth rate for each individual is stable, but the preweight reflects transitory characteristics that are not stable with respect to the increase in weight. So, he believes that the two gender groups are exchangeable and computes a simple (unadjusted) difference in gains as his estimated effect.

The second statistician believes the preweight for each individual is an index of stable characteristics that determine his or her expected gain in weight. As a result, the average natural growth rate for the boys is expected to be be higher than the girls' rate. Because preweight is a confounder in this conception, the two gender groups are not exchangeable. However, this analyst assumes that the preweight is the only stable causal factor. Therefore, conditioning on the preweight will produce conditional exchangeability. In particular, if the preweight and growth rate are perfectly correlated, the analysis of covariance will adjust successfully for the initial difference in weight between the sexes (Bryk and Weisberg, 1977).

The two growth models implicit in Lord's scenario represent extremes. More realistically, the statistician is faced with situations in which both transitory and stable error factors may come into play. In principle, the analyst must aim to identify and adjust for the stable factors, but not to adjust for the transitory factors that vary randomly across testing occasions.

GUIDEPOST 9

This chapter has discussed information bias. There are two main problems that fall under this rubric: exposure misclassification and outcome measurement error. When the outcome is dichotomous, exposure misclassification and outcome misclassification are traditionally analyzed using a model that treats them symmetrically. When the outcome is numerical, the conventional measurement model for exposure is similar to that for a dichotomous outcome. However, the model for outcome measurement treats the errors as random additive perturbations. All of these conventional models are probabilistic, essentially regarding individuals as indistinguishable and characterized by the same values of sensitivity and specificity.

From a counterfactual perspective, exposure misclassification is quite distinct from outcome misclassification. Exposure misclassification can be conceptualized as a virtual change in the assignment mechanism coupled with a form of virtual causal mediation. Outcome misclassification does not change the assignment mechanism, but involves another type of virtual causal mediation. For a numerical outcome, when the ACE is estimated, measurement error can be seen as the result of adding a secondary (virtual) outcome value to the true outcome value. These counterfactual formulations are deterministic. Whether the exposure state or outcome value is assessed correctly is assumed to be determined by various characteristics and circumstances, usually unknown. The apparent randomness of these errors in some situations is attributable to our ignorance of the processes that give rise to these errors.

Finally, we briefly discussed measurement error in a covariate. From the traditional perspective, the covariate is seen as a fallible measure of the true covariate of interest. The relationship between the true and observed covariate values is specified by the classical measurement model. This perspective leads to the conclusion that errors in covariates tend to attenuate the observed effect by failing to adjust fully for the effects of confounding. From a causal modeling perspective, the issue is seen as more complex. The errors in the covariate are viewed as emanating from deterministic but unknown causal factors that may be either transitory or stable. Transitory factors tend to be random and should not be controlled. Stable factors can be confounders for which statistical control would be appropriate.

Chapters 5 through 9 have developed a theoretical framework for understanding the nature of bias. In the next chapter, we switch to more practical concerns. How can possible biases in a particular study be identified? Chapter 10 attempts to delineate the main *sources of bias*. A source of bias is some aspect of a study's design or implementation that could possibly lead to bias. Awareness of these various sources can help researchers in their planning and conduct of comparative studies.

CHAPTER 10

Sources of Bias

Previous chapters have alluded to various types of bias that can occur in comparative studies. These biases result from imperfections in the study design or implementation. In the current chapter, we attempt a more comprehensive summary of the various sources of bias. We have defined a source of bias to be any aspect of a comparative study that results in a discrepancy between the empirical effect generated by the study and the causal effect of interest.

We will consider the real-world conditions that can affect a study and potentially give rise to bias. These sources of bias have been described informally in many epidemiological and social science publications, often in the context of particular studies or areas of research. Our purpose here is to describe and organize these sources under the umbrella of causal theory developed throughout this book. In doing so, we are not attempting an exhaustive compendium or proposing a checklist that can be used to validate a study's methodology in a mechanical fashion. Rather, we attempt to provide a fairly high-level overview of the landscape of bias, focusing primary attention on the most important sources that arise in practice.

Our ability to interpret an observed effect in causal terms depends on being able to rule out competing theories. In practice, this entails making a convincing case that various possible sources of bias are unlikely to account for the observed effect. This idea was elaborated in Section 7.3, in the context of confounding. The general principle holds true for all types of bias. In practice, the strength of a causal argument depends on the plausibility of the competing theories that are advanced as possible alternative explanations.

As explained in Chapter 4, specific *types* of bias can be described in many different ways. Identifying and cataloging every conceivable type of bias, even if feasible, would be of limited utility. On the other hand, understanding the main *sources* of bias could help to structure a search for plausible alternative

Bias and Causation: Models and Judgment for Valid Comparisons, By Herbert I. Weisberg
Copyright © 2010 John Wiley & Sons, Inc.

Table 10.1 Tentative List of Sources of Bias

Aspect of Study	Potential Source of Bias
Sampling	Participation in the study is voluntary.
	The target population is not well defined.
	Certain potential subjects are difficult to reach.
	The study's eligibility criteria are restrictive.
Assignment	The subject can influence assignment.
	A service provider can influence assignment.
	Social stratification can influence assignment.
	Natural forces can influence assignment.
	There are concomitant causal factors.
	The "exposure" is an immutable characteristic.
Adherence	The subject finds the study requirements onerous.
	The subject experiences a change in circumstances.
	The subject dies or becomes incapacitated.
Exposure ascertainment	Exposure information reported by the subject is inaccurate.
	Exposure information recorded by a third party is inaccurate.
	Exposure actually occurred after the outcome event.
Outcome measurement	The outcome value reported by the subject is inaccurate.
	The outcome value recorded by a third party is inaccurate.
	The instrument used to measure the outcome is inaccurate.
	The measured outcome is a surrogate for the true outcome.

explanations. The limited attention usually paid to competing theories is partly attributable to the lack of a comprehensive lexicon for bias that would facilitate communication by methodologists with each other, and with scientists and practitioners.

In this chapter we identify 20 general sources of bias. A complete listing of these bias sources appears in Table 10.1. This list should be regarded as provisional, offered in the hope of stimulating more attention to the systematic exploration of bias sources. Although our list is meant to encompass the great majority of bias sources encountered in practice, there may well be others that can occur in various particular contexts. Furthermore, we have not paid special attention to sources of bias that are specific to case–control studies. As discussed in Chapter 4, the case–control design can introduce particular types of bias that have not been considered in detail in this book. Similarly, we have not attempted to deal with issues that pertain specifically to time-dependent outcomes, or to particular areas of research.

We have organized these 20 sources of bias into 5 broad categories, as first described in Chapter 4:

- Sampling of subjects
- Assignment to exposure states

- Adherence to nominal exposure states
- Exposure ascertainment
- Outcome measurement

Each category represents one of the main components of a study's design or implementation that may be susceptible to problems leading to bias.

10.1 SAMPLING

Ideally, a comparative study would be based on a random sample from the target population of interest. Techniques for obtaining a simple random sample or a more complex type of representative sample (e.g., stratified sampling, cluster sampling) are well developed within the statistical specialty area of *survey sampling* (e.g., Cochran, 1965a; Kish, 1965). Some methodologists have advocated the wider use of such approaches in epidemiological and social science research. However, there are practical limitations on the extent to which survey sampling methods can be implemented.

The sampling mechanism is of concern primarily because the study population might not be exchangeable with the target population. As a result, the study might lack external validity; the causal effect in the study population might not equal the causal effect in the target population. In Section 5.4, we introduced a model to represent the sampling mechanism when the outcome is dichotomous. This model clarified how the risk ratio and risk difference are affected by the sampling probabilities associated with various types of response patterns. For example, the risk ratio tends to be inflated (farther from 1.0) if *doomed* individuals tend to be screened out. A corresponding sampling model for a numerical outcome was presented in Section 5.7. We found that the sampling bias depended on the covariance between the sampling probability and the individual causal effect (ICE).

There are many specific reasons nonrandom sampling of subjects can occur. Most of these result from one of the following four situations:

- Participation in the study is voluntary.
- The target population is not well defined.
- Certain potential subjects are difficult to reach.
- The study's eligibility criteria are restrictive.

Voluntary Participation

In many studies, the prospective study participants can decide whether or not to be included. In particular, participation in a randomized experiment is nearly always a voluntary activity. Therefore, certain individuals will inevitably be more motivated than others to take part. Some of these predisposing characteristics are unrelated to the study outcome, and self-selection on such incidental factors generally does not lead to bias. In causal terms, the distribution of response patterns will be similar to the distribution in the target population. However, there are at least three common ways in which attitudes toward study participation can result in selection bias:

- General reluctance to participate in an experiment
- Desire to obtain potential benefits from participation
- Unwillingness to experience an undesirable treatment modality

Many studies entail considerable time, effort, and even expense, on the part of subjects. Some individuals may be unwilling or unable to make such a commitment. For example, some workers might not be able to take time off to appear at scheduled follow-up visits for medical monitoring. Moreover, there may be intangible costs, such as perceived invasion of privacy. Whether exclusions for such personal reasons pose any concern depends largely on the nature of the intervention being evaluated. Thus, employment status may be a relevant selection factor in a study of an adult education program, because workers might differ from nonworkers in terms of prior experience and motivation. On the other hand, employment status might be irrelevant to a study of a new antibiotic.

Despite the possible burdens of participation, some individuals might actively seek to become included in a research study. This situation typically occurs when the intervention being evaluated is thought to offer some valuable benefit. In a randomized control trial (RCT), individuals who believe they are good candidates for an experimental intervention may be highly motivated to volunteer for the study. For example, some *causal* responders may be capable of correctly determining that the intervention would work for them. On the other hand, some *immune* responders may opt to participate because they have failed to achieve success with previously tried solutions. Studies of smoking cessation programs offer a classic example; some smokers seeking entry to a study may have previously attempted unsuccessfully to quit using several approaches.

The flip side of positive self-selection aimed at *obtaining* a potential benefit is negative self-selection to avoid possible assignment to an undesirable treatment modality. The randomized trials of hormone replacement therapy provide an excellent example of this phenomenon. In the Women's Health Initiative (WHI) study, women who had previously been taking hormones were required to abstain from hormone replacement therapy (HRT) for 3 months prior to randomization, and then to risk being allocated to placebo. Women who felt that HRT was important to their health and well-being were unlikely to accept these conditions. Consequently, the study sample contained an unusually high percentage of women who had not previously used HRT.

Defining the Target

In principle, a comparative study is designed to estimate the causal effect in some well-defined target population. In reality, the precise nature of this target population is often rather vague. As discussed in Section 3.1, researchers tend to focus primarily on the problem of establishing internal validity and to worry much less about external validity. This attitude is sometimes justified by a belief that the effect is fairly uniform across individuals. However, when there is effect variation, the causal effect found in a study might not extend to the target population.

For external validity, the study population should ideally be representative of the future target population. This target population is often the general population of individuals who meet certain broad criteria. For instance, a new antidepressant medication might be aimed at all adults who are diagnosed with severe depression. However, often the appropriate target population is defined much more narrowly. Sometimes, identifiable subgroups of the target population might be deliberately excluded from the study for a variety of reasons.

Often, the efficacy (or lack thereof) might already be known for certain types of individuals. Thus, use of low-dose aspirin for those with established heart disease (secondary prevention) is well established, whereas extension to various lower-risk subgroups (primary prevention) currently remains controversial (Antithrombotic Trialists Collaboration, 2009). In such a situation, the relevant target population would be the more restricted class of individuals who are not *already* deemed eligible for treatment. As long as such a restriction is explicit, the results of the study can be valid—and properly interpreted. Unfortunately, a study's findings are often construed to apply more broadly to the general population. Bias can occur when there exists a mismatch between the actual population to which generalization is *truly* intended and a nominal target population to which generalization is *believed* to be intended.

Difficulty in defining the target population can be exacerbated when background conditions are changing. The actual population of potential interest may be a "moving target" because the circumstances and characteristics prevailing in the general population are not fixed. In particular, interventions studied in many experiments address problems for which one or more solutions have already been tried. As a result, the nature of potential subjects available at a given point in time may depend on the availability and success of these prior treatments. Individuals who remain at risk for the outcome of interest are a subset of those who *would* have been at risk in the absence of these previous preventive measures. Alternatively, an intervention or risk factor may be similar to ones that have previously been found to have had adverse consequences. In either case, the pool of individuals willing to try a new intervention may be reduced. The existence of such processes for winnowing the potentially eligible population has been called the *depletion of susceptibles* (Moride and Abenhaim, 1994).

For example, suppose we are testing the efficacy of a new therapy to assist smokers to quit. Because smoking rates have already declined dramatically in many developed countries, current smokers include many who have previously tried to quit but failed. Therefore, the effectiveness of a new approach should ideally be evaluated in the context of this prior experience. A program that would have been successful 20 years ago may now be less effective, because the population has been largely depleted of those who would have been susceptible to its impact.

Accessibility

Some candidates for study participation may be difficult to reach, for a variety of possible reasons. Recruitment of prospective subjects may be facilitated by various means, including promotion in the media, social networking, and channeling by professionals. In the context of clinical trials, referrals by family physicians and other medical providers usually play a major role. For any individual, the likelihood of coming in contact with the study's intake process can depend on several factors, including geographic proximity, ethnicity, occupation, and education. Selection based on such characteristics may lead to bias.

In addition to having a direct impact, the referral process can also amplify the effects of self-selection. Those who are potentially motivated to participate might well be those most likely to be aware of the study. For example, a health-conscious individual is more likely to have access to medical providers who can then channel her into a clinical trial that may be beneficial. It has been documented that referrals by physicians to clinical trials are often highly selective, representing an informal layer of restriction beyond that imposed by the stated eligibility criteria (Rothwell, 2005).

Accessibility is a particular concern in case–control studies. As explained in Section 4.1, a case series is typically obtained from a particular set of health-care facilities. Because cases are relatively rare, all qualifying patients are commonly included. The controls, on the other hand, are sampled from the population of hospital patients or from the neighboring communities. Ideally, the controls will be representative of the study base, which consists of individuals who *would* have been identified as cases had they contracted the outcome disease. Because this definition is predicated on a counterfactual condition, it is impossible to identify such individuals with perfect accuracy. Moreover, controls will often be much less motivated than cases to participate. For these reasons, access constraints for the controls might impair internal validity as well as external validity.

Eligibility Criteria

Eligibility criteria are often subclassified into *inclusion criteria* and *exclusion criteria*, although the dividing line is somewhat arbitrary. Generally speaking,

inclusion criteria specify the characteristics of individuals to whom generalization is intended (i.e., the target population). For example, patients in a clinical
trial are often meant to be representative of future candidates for the drug
being tested. Most medications are only suitable for patients with certain
demographic (e.g., gender, age) and medical (e.g., diabetes, hypertension)
characteristics. These characteristics dictate inclusion criteria intended to
produce a study population that closely resembles the target population.
Specifying exclusion criteria, on the other hand, can occur for many possible
reasons. In clinical trials, patients are excluded primarily for various "medical
reasons" tied to ethical considerations, or for "scientific reasons" related to
study methodology (Britton et al., 1999).

There is often a cogent rationale based on ethical considerations for excluding certain classes of individuals from a study. Participation in the study may
entail unacceptable cost, inconvenience, or risk. There are two main sources
of ethical concerns. First, investigators must avoid exposing any subjects to an
unacceptably *high level of risk*. Second, they should avoid exposing someone
to a treatment modality already known to be *suboptimal*.

Suppose first that the intervention is suspected of greatly increasing the risk
of suffering an adverse event. It would clearly be unethical to include such
high-risk patients in the study if there is any possibility of exposing them to
serious harm. The second common motivation for exclusion on ethical grounds
is related to the potential efficacy of the intervention. A basic principle of
experimentation on humans is the need for clinical *equipoise*. Equipoise means
that a priori we believe the treatment options to be equally effective. Suppose,
however, that a subgroup can be identified for which the intervention is
believed to be more effective than the comparison modality. It would be
unethical to expose these individuals to the risk of receiving the less effective
comparator.

In principle, exclusions for ethical reasons should mimic the decisions made
by physicians and other decision-makers (teachers, therapists, etc.) in the real
world. For example, a treating doctor would avoid prescribing a drug likely to
induce a serious side effect in her patient. Ideally, the selection criteria in the
experiment and in the field would be perfectly aligned. In reality, identifying
individuals susceptible to adverse effects and/or to lack of efficacy may be
difficult. Consequently, the study's criteria may result in a population that does
not faithfully represent the target population.

Furthermore, some criteria may be related to nonspecific risk indicators,
such as advanced age, low educational level, socioeconomic status, or preg-
nancy (Van Spall et al., 2007). These global exclusions are often undertaken
in the spirit of an abundance of caution. However, the effects of such
practices can be counterproductive; exclusion from the trial does not necessarily protect the excluded individuals, who may remain subject to the same
(or even worse) risk. Moreover, these exclusions may contribute to sampling
bias, by restricting the study sample much further than necessary (Rothwell,
2005).

Suppose that a clinical trial attempts to exclude a class of potential subjects who are believed to be at high risk for a certain adverse event. However, the criteria employed might not distinguish individuals who are *doomed* from those for whom the treatment is in fact *preventive*. After all, prior to treatment the *preventive* and *doomed* individuals might both appear to be at high risk. The counterfactual models for sampling discussed in Section 5.3 have been used to explore the implications of such situations (Weisberg et al., 2009). In particular, it is possible that overselection of low-risk patients can reduce the *absolute* event rates in the study but artificially inflate the *relative* risk observed in the study.

Exclusion criteria can also be based on methodological considerations. If the outcome is numerical, potential benefits can derive from the creation of a more *homogeneous* study population. The focus of this book has been on systematic errors, ignoring the more traditional statistical concerns about random variability. However, considerations of statistical efficiency play a prominent role in the design of actual studies with limited sample sizes. Statistical efficiency refers to the extent to which a high level of precision (e.g., narrowness of a confidence interval or of a Bayesian credible interval) can be achieved with a given number of experimental subjects. Because obtaining data has costs (both financial and personal), sample sizes are constrained. Standard statistical theory implies that analyzing data on a restricted sample that has less variability can improve statistical efficiency.

If the outcome is dichotomous, the statistical efficiency is mainly a function of the number of outcome events observed. Thus, a restricted sample will be efficient if it generates a higher frequency of events than would occur in the target population. For example, eligibility criteria might deliberately favor individuals more likely to experience the event. This is usually accomplished by implementing an *enrichment strategy* that preferentially selects individuals who possess known risk factors. (e.g., Amery and Dony, 1975; Brittain and Wittes, 1990). Enrichment strategies enhance statistical efficiency by increasing the total number of outcome events in the study. However, the enriched sample might have a distribution of response patterns that differs from that in the target population. For example, suppose the high-risk subjects recruited for enrichment purposes include a high proportion of *doomed* individuals. Then, as we explained in Section 5.4, the risk ratio could be attenuated.

Sample restriction is sometimes also employed as a means of reducing the potential for assignment bias. In a randomized study, the samples are exchangeable *in expectation*, but they can still be quite different in actual realization if the sample of subjects is not very large. When the outcome is numerical, this "random bias" depends on both the sample size and the variability of the outcome score. So, a more homogeneous sample reduces the probability that a meaningful imbalance between study groups can arise by chance.

Finally, there is an additional methodological criterion that can affect selection, whether explicitly or implicitly. During the follow-up period, a subject may fail to comply with the study's principal requirements, either by failing to adhere to the treatment modality prescribed or by dropping out of contact entirely. To minimize such eventualities, a study may try to screen out individuals who appear likely to become noncompliant. This screening process can be either informal or designed into the run-in phase prior to randomization. For example, sometimes those who fail to achieve a prespecified compliance rate (e.g., at least 80% of pills taken) during a placebo-only run-in period will not qualify for randomization.

10.2   ASSIGNMENT

In a randomized study, the assignment mechanism is controlled by the investigator and effectively guarantees exchangeability between the study groups. In an observational study, the groups have been formed by a nonrandom process. This process may reflect a more "natural" assignment mechanism, thus avoiding the "artificial" restrictions necessarily imposed on most RCTs. However, this possible advantage is purchased at a very high cost: possible absence of exchangeability between the study groups. Lack of exchangeability can be attributed to a variety of possible sources.

The lack of exchangeability that characterizes most observational studies is a critical disadvantage relative to a randomized experiment. Indeed, some methodologists are highly skeptical of almost all results obtained from nonrandomized comparative studies. Recent analyses comparing results obtained from RCTs and observational studies, most notably in the context of supported work programs and HRT, have fueled pessimism about the ability to correct for confounding successfully. On the other hand, most research studies are necessarily performed without benefit of randomization. Therefore, it is important for researchers (and study reviewers) to understand just how nonrandom assignment can precipitate bias. In particular, knowledge of the plausible sources of "assignment bias" can suggest competing hypotheses for consideration.

In Section 5.5 we introduced a model that explicitly represented the assignment mechanism when the outcome is dichotomous. This model clarified how the risk ratio and risk difference are affected by the assignment probabilities for various types of response patterns. In general, we found that, when the event is relatively rare, these measures both increase to the extent that *causal, preventive*, and *doomed* individuals are preferentially assigned to the exposed group. However, the assignment probability for *doomed* subjects is most important, because placing a *doomed* individual in the exposed group, rather than the unexposed, both reduces the event frequency in the unexposed group and increases the event frequency in the exposed group (see Section 5.5). A corresponding model for a numerical outcome was presented in Section 5.7. We found that the assignment bias depends primarily on the covariance between the potential outcome mean (M) and the assignment probability (A).

It would, of course, be impossible to identify all the specific aspects of biomedical and social reality that might give rise to nonexchangeability between groups in various circumstances. However, most of these sources of bias can be ascribed to one or more of the following factors:

- The subject can influence assignment.
- A service provider can influence assignment
- Social stratification can influence assignment.
- Natural forces can influence assignment.
- There are concomitant causal factors.
- The "exposure" is an immutable characteristic.

The Subject

In observational studies, a particular subject's exposure state can sometimes be determined, at least to some degree, by the subject herself. In the most extreme case, a group of volunteers for a new intervention can be compared against a group of nonvolunteers. This approach has the advantage of including an exposed group that may be representative of people who would actually accept (or even seek out) the intervention under normal conditions. However, the volunteers and nonvolunteers are unlikely to be exchangeable. The same logic applies when the exposure is a risk factor, rather than a potentially beneficial intervention. For example, there are many "lifestyle" exposures (smoking, high-fat diet, alcohol consumption, high-risk sports, etc.) that are at least partially voluntary. Isolating the causal effects of such factors is difficult, in large part because those who engage in a particular behavior tend to differ in many respects from those who do not. In causal terms, the participants may have a different distribution of response patterns from the abstainers.

In Chapter 1, we discussed the research undertaken a century ago to evaluate an antityphoid vaccine that had recently been developed. This "primitive" study, performed before the technique of randomization had even been formulated, recruited volunteers from the British Army. Soldiers who were inoculated could certainly not be considered equivalent to those who refused or were otherwise unavailable. Therefore, this study would not be considered probative by modern standards. On the other hand, the study produced a wealth of relevant evidence, both statistical and nonstatistical, that seemed to confirm the vaccine's efficacy.

An important consideration in the debate at the time was whether the use of volunteers invalidated the comparison between the inoculated volunteers and the controls. To answer this question, we would need to know much more about characteristics of the volunteers and the process of recruitment. This information could shed light on the distribution of response patterns among the volunteers. For example, if soldiers who would have contracted typhoid fever whether or not inoculated (*doomed*) tended to become volunteers, the result would be biased against the vaccine. Because so little was known about the disease at that time, it might be argued that the volunteers were likely to be exchangeable with the nonvolunteers. On the other hand, a more detailed understanding of the process by which volunteers were selected might suggest reasons for doubt.

Service Providers

Many interventions are evaluated under the normal conditions in which they are implemented. Consequently, individuals may be chosen to receive the intervention, rather than some other type of treatment, based on apparent need. That is, the assignment mechanism is deliberately designed to enrich the exposed group with individuals who are likely to respond positively. In the medical context, symptoms that signal heightened risk of a particular negative outcome are termed *indications* for treatment. In general, patients who display such indications have relatively high probabilities of being assigned to the intervention. Consequently, the exposed and unexposed groups cannot be considered exchangeable. In epidemiological research, this problem is termed *confounding by indication* or *channeling bias*.

In the social sciences, the same phenomenon is usually subsumed under the broad category of *selection*. This class of threats to validity encompasses all issues related to what we have termed nonrandom sampling and nonrandom assignment. Many social or educational interventions are intended to benefit relatively disadvantaged individuals. Therefore, it may be deemed unethical to assign the intervention other than according to need. In that case, the evaluation of benefits is necessarily based on a comparison between nonexchangeable groups.

What kind of bias might we anticipate in a comparison of groups formed by such purposeful intervention? Consider a hypothetical educational program aimed at reducing the number of high school dropouts. The program would target those students who would otherwise fail to graduate (*preventives*). A favorable effect would imply that the risk ratio (for the outcome of dropping out) would be substantially less than 1.0. However, the ability to distinguish the *preventives* from *doomed* individuals would undoubtedly be imperfect. So, the exposed group may be "enriched" in terms of both *doomed* and *preventive* individuals. The intentional channeling of *preventives* toward the intervention would make the program appear worse by removing them from the unexposed control group, thereby reducing the drop-out rate in that group. The unintentional channeling of *doomed* would make the program appear worse by increasing the drop-out rate in the program group and decreasing the drop-out rate in the control group.

Social Stratification

In many observational studies, it is necessary (or at least convenient) to compare study groups that are formed by complex social forces. Individuals are affected by such sorting into social strata (broadly defined) throughout their lives. People gravitate toward various schools, occupations, recreational activities, social groups, neighborhoods, etc. The determinants of this "social stratification" are many and complex. In the social sciences, many studies attempt to evaluate the impact of membership in various strata. For example, the comparison of alternative educational programs and settings is an active area of research with important public policy implications. Anthony Bryk and colleagues performed a landmark study comparing the educational and social effects of Catholic parochial schools and public schools (Bryk et al., 1993). Obviously, this research had to contend with a variety of factors related to the process by which children are sorted into different types of schools.

In epidemiology, social stratification is important primarily in relation to risk factors under study. Those exposed to a particular risk factor, such as an environmental or occupational hazard, may not be exchangeable with those who are unexposed. Becoming exposed, or perhaps the degree of exposure, often results from a set of individual characteristics and prior experiences that position a person in a particular time and place. As a result, certain causal factors may be more prevalent among the exposed than among the unexposed.

Consider a risk assessment of some suspected toxin used in a manufacturing process. Suppose we compare workers in a large factory who are exposed to this chemical with co-workers who are not exposed. The employees in contact with the chemical may be similar in many respects to their counterparts who are not, but different in other respects. Sometimes a group of exposed workers is compared with a group of unexposed individuals drawn from the general population. In this case, the workers may tend to be unusually healthy or physically fit. This phenomenon has been called the *healthy worker effect*.

Natural Forces

In many observational studies, especially in epidemiology, the study groups are formed by natural forces. These forces are effectively random in the sense of not being guided by any purpose or intelligence (religious and philosophical considerations aside). However, the subpopulation of individuals who are exposed to a risk factor may not be random in the statistical sense. For example, the exposure might be the result of a biological or environmental catastrophe that affects everyone within some large geographical area. This area may happen to be characterized by a high prevalence of other risk factors that could explain an apparent causal effect.

Suppose we are investigating the relationship between the incidence of cancer and proximity to a certain hazardous waste site. The incidence rate of various cancers in the surrounding community might be compared with an overall national or regional rate. Assume that a highly significant elevation of the rate for bladder cancer is observed, even after making proper statistical adjustment for the fact that multiple waste sites and multiple cancer types have been investigated. Then the epidemiologist must try to determine whether the community is exchangeable with the control population. For example, an unusually high level of smoking or the existence of a chemical plant nearby might be considered confounding factors.

Concomitant Causes

In many observational studies, exposure may be accompanied by other identifiable factors that can affect the outcome. If these concurrent causal factors affect all exposed individuals, they can be considered as part of the definition of the exposure. For example, suppose we are evaluating the efficacy of a new antibiotic on a certain type of infection. If it is known that this experimental treatment tends to cause gastrointestinal distress, then a concomitant antacid may be prescribed for everyone in the study. On the other hand, a concomitant medication may be given only to a subset of the study population who are identified as especially susceptible to stomach problems. In that case, the concomitant medication may be considered to be either an effect modifier or a confounder.

In the observational studies of HRT, many women with prior cardiovascular disease might have taken low-dose aspirin for secondary prevention. Suppose that, in a certain study, the rate of aspirin use was higher among those using HRT than among nonusers. Then low-dose aspirin could be considered a confounding factor for which statistical adjustment would be warranted. Alternatively, we could focus on the role of aspirin as a causal modifier. In that case, separate analyses for each stratum (aspirin vs. nonaspirin) might reveal whether the causal effect of HRT varies according to whether low-dose aspirin is also being taken.

As discussed in Section 7.1, confounding and effect modification are traditionally regarded as distinct phenomena. But we explained how these are actually two sides of the same coin: different aspects of what transpires when we condition successfully on a covariate. The critical issue is whether the concomitant factor is considered to be a causal factor within our conceptualization of the causal mechanism. Suppose that taking aspirin actually had no effect on coronary heart disease (CHD) for the women in the HRT studies. Then it would still be possible that the causal effect of HRT on CHD would differ between the aspirin and nonaspirin subgroups. But what if aspirin were a causal factor? Then aspirin use could be a confounder. For simplicity, let us assume it was the only confounder. Then, if we conditioned on aspirin use, we would obtain an unbiased estimate of the HRT effect. However, based on the discussion in Section 7.2, we would expect to find effect modification, that is, the HRT effect would differ for the aspirin and nonaspirin strata.

Immutable Characteristics

Sometimes the potential causal factor is a fixed characteristic that cannot actually be modified, such as gender or ethnicity. Some methodologists argue that it is improper to speak of causes in such situations, that there can be "no causation without manipulation" (Holland, 1986). Others believe that it is meaningful to *imagine* what would change if an individual, or one similar to her, were identical in all respects except the characteristic of interest. This hypothetical (though impossible) manipulation may correspond to what we mean by a cause in ordinary conversation: "How much more successful would Mary have been if she had been born a boy?" What exact meaning can be attached to such a statement (if any) can be debated by philosophers, but most people would have some idea of what is being asked.

Suppose we are trying to assess the causal effect of gender on wages. We obviously cannot assign people randomly to gender. Gender is an attribute that is fixed for any individual; so the counterfactual scenario of becoming male while holding everything else fixed is somewhat nonsensical. A counter to this argument is that in discussing the causal effect of gender or race, we are really thinking in the abstract of a person who has a certain constellation of characteristics. The causal effect of interest might be interpreted as the difference in wages that a woman with these characteristics would receive relative to a man who possesses the same characteristics. This causal effect could be studied by comparing groups of men and women, adjusting for covariates that capture all of the relevant factors that might affect the outcome.

The feasibility of such an exercise depends on how far our imaginations can be stretched in this way. Certain attributes might be so strongly associated with gender that comparing an "equivalent" man and woman would be virtually impossible. Any actual groups of men and women would be qualitatively different in terms of relevant causal factors precluding successful statistical adjustment. When the risk factor is "bundled together" with such factors, the implicit "assignment" probability for many configurations of causal factors may be very close (or equal) to either 1 or 0. That is, certain sets of characteristics can "predict" a person's gender with virtually 100% accuracy.

The study of death sentencing discussed in Chapter 1 illustrates this problem. Recall that a primary aim of the research was to find out if the victim's race played a causal role in determining whether a death sentence was imposed. But the complex circumstances of the population (convicted murderers) displayed very limited overlap between the race-of-victim groups. After adjustment for many legitimate explanatory factors based on statistical modeling, the death-sentence rates remained different. This could be interpreted to mean that race itself, or other illegitimate factors correlated with it, affected the sentencing decisions. Alternatively, the legitimate factors (or model specification) might have been inadequate to overcome the qualitative differences between the two groups of crimes. In the end, the Supreme Court ruling appeared to say that, even after conditioning on many relevant variables, the groups of black-victim and white-victim defendants could not be assumed exchangeable.

10.3 ADHERENCE

Ideally, the subjects in a comparative study would adhere perfectly to the requirements of the research design. In an RCT, and sometimes in a quasi-experiment, these requirements are usually laid out formally in the study's *protocol.* The protocol describes in detail the clinical and statistical methodology, including precisely how the treatment modality is to be administered and the data collected. In particular, it prescribes how the subject is to participate (e.g., take one pill twice daily, attend classes 3 days per week) and to provide data (e.g., come in weekly for blood tests, take a written examination every 2 weeks).

In reality, there are a number of reasons certain people could deviate from these requirements. Most of these reasons fall into three general categories:

- The subject finds the study's requirements onerous.
- The subject experiences a change in circumstances.
- The subject dies or becomes incapacitated.

In general, we can say that *something* happens during the follow-up period that causes the subject to deviate from the protocol's specified requirements. The exact nature of this discontinuation can range from a minor lapse in cooperation to a complete termination of the treatment, or even to withdrawal from the study entirely (loss to follow-up). The impact of nonadherence will obviously depend on exactly how the participant's behavior changes. From a methodological perspective, nonadherence in general can be framed as a problem of causal mediation, as discussed in Chapter 8.

In most studies, we are interested in the "direct effect" of an exposure under full adherence to certain specified conditions. Substantial lack of adherence presents a conundrum: the fully compliant subset of the subjects is selected nonrandomly. The empirical effect that results from analyzing this nonrandom subset may be biased. The discussion in Chapter 8 delineates the conditions that must be met for this as-treated effect to be unbiased. Alternatively, assuming that outcome data are available on all subjects, an intention-to-treat (ITT) analysis can be performed. Because the ITT approach includes all subjects who are randomized, regardless of adherence, the advantage of randomization is retained. However, the causal effect being estimated will be the total effect (including the "indirect" effects), rather than only the direct effect. An RCT using the ITT approach is often regarded as a pragmatic study, but this interpretation is valid only if factors leading to lack of adherence in the study are similar to those that exist in the real world.

Finally, it is possible that some subjects will be completely lost from the study. In that event, the ITT strategy is not an option, because the outcomes are unknown for these subjects. The analysis must be restricted to those who remain in the study. Whether or not "attrition bias" results then depends on essentially the same considerations that apply to an as-treated analysis to address noncompliance.

Voluntary Noncompliance

The requirements of the study may be unpleasant, inconvenient, or demanding in various ways. Consequently, some subjects may decide to "adapt" the treatment modality or simply to stop participating altogether. In an RCT, the person may take the prescribed medication less frequently than required or may simply refuse to take any medication at all. The former situation has been termed one of *partial compliance*; the latter has been described as *simple noncompliance* (e.g., Jin and Rubin, 2008) In Section 8.3, we discussed a special case of simple noncompliance in which the individual either fully complies with the modality assigned or elects to adopt the alternative modality.

When the subject decides to alter or terminate the assigned modality, ascertaining the reasons can be important. Consider an RCT comparing a new cough medicine with a placebo. Suppose that subjects on the new drug tend to consume lower doses than subjects on placebo, because the medication has an unpleasant taste. Because this source of noncompliance is an intrinsic property of the medication, we might expect that future consumers would respond similarly. If so, then a pragmatic study based on ITT would seem appropriate. On the other hand, suppose that noncompliance occurs because of particular experimental conditions unlikely to be replicated in routine use. Then ITT might seriously underestimate efficacy, because the future users would be expected to comply more readily.

Inability to Continue

Especially if the follow-up period is prolonged, there can be many reasons certain individuals become unable to maintain the prescribed regimen. Unexpected events in the subject's life might interfere with participation. Such idiosyncratic problems would ordinarily be distributed "at random" to the different study groups and would not necessarily lead to bias. On the other hand, the complications leading to lack of continuation might be specific to the experimental setting. Thus, it may be unclear whether the same kinds of problems would interfere with longer-term use in a natural environment. If not, then the causal effect in the study might differ from the corresponding effect that would be achieved in actual use. Of particular concern is whether the reasons for stopping treatment were caused by being exposed, or possibly by not being exposed.

One particular type of involuntary discontinuation is endemic to RCTs and is explicitly contemplated in every protocol. This is the problem of *adverse events*, or side effects. To protect the subjects, contingencies requiring termination of treatment are spelled out in advance. For example, in a double-blind, placebo-controlled trial to evaluate a new type of pain reliever, a sustained increase in blood pressure or serious bout of gastrointestinal discomfort might require withdrawal from the treatment. Such adverse events are often analyzed as secondary endpoints. The efficacy of the primary medication being studied would need to be weighed against any causally related side effects. However, the loss of subjects because of adverse events must also be considered as a source of bias potentially affecting the analysis of the primary endpoint (e.g., pain relief).

With respect to the primary endpoint, an adverse event can be regarded as an intermediate causal factor. Section 8.3 discussed the issue of attrition generally. We defined a causal mediator R that specified whether or not the individual was retained in the study or dropped out. For each type of adverse event, we could define a particular causal mediator. The overall effect of attrition because of adverse events would then be the joint effect of all these intermediate factors. In practice, this issue is extremely complex. It has been explored most extensively in the special case of death as the endpoint of interest.

Mortality

In many studies of medical interventions for potentially life-threatening conditions, the endpoint is literally death, or possibly some debilitating condition (e.g., a major stroke) that curtails further follow-up. If an individual dies (or becomes seriously incapacitated) during the study, the primary cause of the event will nearly always be known. For example, deaths in which lung cancer is the primary cause would be identified with high accuracy. So, it may be feasible to define lung-cancer mortality as the study endpoint, as in fact is commonly done. However, subjects can die for a variety of other reasons. Such other specific causes of death (or at least removal from observation) are called *competing risks*. Suffering a fatal traffic accident or experiencing a fatal myocardial infarction (MI) would be examples of competing risks.

When cause-specific mortality is the outcome, the potential for bias depends on the frequency and nature of these competing risks. Suppose we are studying the relationship between passive smoking and lung-cancer mortality. A large cohort of nonsmokers is monitored for 20 years, and their exposure to secondhand smoke is recorded. The exposed and unexposed groups are ultimately compared with respect to incidence of mortality from lung cancer. In this example, those who were killed in traffic accidents during the follow-up period might effectively comprise a random sample. Arguably, their removal from observation would not create bias.

What about those who died from an MI? Tobacco smoke (from active smoking) is well known to increase the risk of an MI. Therefore, it is plausible that some of the MIs were in fact caused by the passive smoking. An MI might thus be a marker for a higher level of exposure or susceptibility to secondhand tobacco smoke. In that case, the loss of the MI victims from the study could have produced bias, because this subgroup would probably not be random. Rather, it might have contained a relatively high proportion of individuals who were exposed and *causal*. Such individuals *would* have died of lung cancer if they had not been killed by the MI. Their removal would therefore lead to an underestimation of the causal effect.

One possible way around this problem is to define the study endpoint to be *all-cause mortality*. That way, we are concerned only with the "bottom line" of whether the exposure increases mortality. Because we ignore the particular mechanism by which the outcome occurs, every subject's outcome is observed. This approach is very much in the spirit of the ITT strategy for dealing with noncompliance. In an RCT, the study groups remain exchangeable (barring other types of attrition besides mortality), because follow-up is always complete. So, the study would yield an unbiased estimate of a causal effect. But a causal effect of what?

In a sense, the total effect of an exposure on all-cause mortality is always mediated by a large number of intermediate factors. For example, the effect of passive smoking would operate through a variety of disease processes that can be promoted by tobacco smoke. So, using all-cause mortality as the outcome yields an estimate of causal effect that may be unbiased, but perhaps of little scientific relevance. Moreover, the indirect effects that comprise the total effect might vary across different studies, because the prevalence of particular competing risks might depend on the nature of the particular study population.

Our discussion here has just scratched the surface of this complex topic. Moreover, we have ignored the time factor that is a central element of mortality studies. Rather than comparing simple frequencies, as in the risk ratio, we typically compare incidence rates or hazard rates (e.g., events per unit of time). As discussed in Section 3.6, special methods are necessary to deal with time-dependent measures of effect. At the present time, a widely accepted framework for dealing with competing risks has not been fully elaborated. Various complex statistical methods have been proposed to adjust for competing risks by estimating "the counterfactual risk of the study disease if the competing risks were removed," but this formulation is ambiguous: "How are the competing risks to be removed?" (Rothman, Greenland, and Lash, 2008, 56). For example, if all MIs were suddenly preventable, deaths from lung cancer would presumably increase substantially. But under this counterfactual, what other specific causes of mortality would also provide increased "competition" to at least partially offset the decrease in fatal MIs.

10.4 EXPOSURE ASCERTAINMENT

In a prospective study, the exposure status for each subject will nearly always be known very accurately at the study's outset. So, exposure misclassification is primarily an issue in retrospective studies. In particular, it is a major concern in case–control studies. The extent of the problem depends on how the information regarding exposure has been obtained. In general, we can identify three main sources of potential distortion in the process of data collection:

- The exposure information reported by the subject is inaccurate.
- The exposure information recorded by a third party is inaccurate.
- Exposure actually occurred after the outcome event.

Reporting by Subjects

In many retrospective studies, the information about outcomes is obtained by interviewing the subjects. In the Yale Hemorrhagic Stroke Project (HSP) study of phenylpropanolamine (PPA) and hemorrhagic stroke, specially trained interviewers asked the cases about various products used during the 2-week period prior to the onset of the stroke. Controls were asked about the same products during a presumably comparable 2-week period. Exposure to PPA was defined as any consumption of a PPA-containing product (cold-cough remedy or diet pill) during the 2 weeks.

In such a situation, there are various reasons why the recall of past exposure might be fallible. Especially if substantial time has elapsed since the event, memory of details may be unclear. Moreover, some risk factors may be emotionally loaded, so the subject may be uncomfortable being totally candid. Interviewing skill and training are often necessary to get around such challenges. Furthermore, training is necessary to make sure the *interviewers* are objective. Ideally, the interviewer would not know if the subject is a case or a control. However, with a serious disease like stroke, such blinding might be virtually impossible to maintain.

Sometimes data about exposure cannot be obtained directly from the subject, because he has died or is otherwise unable to respond. For example, stroke victims may be too cognitively impaired to provide reliable information. In these situations, a proxy respondent close to the subject (spouse, caregiver) may be available. The data may then be less reliable, because it is elicited at one remove. On the other hand, the proxy may be not only more clearheaded, but perhaps more objective as well. These factors must be weighed in evaluating the quality of the resulting data.

As shown in Chapter 9, the implications of reporting errors for bias depend on the nature of the misclassification. A general model of exposure misclassification was presented in Table 9.11. This model is quite complex. Moreover, the model does not take into account the issue of "coherence" between the case series and the controls. As explained in Section 4.1, a precise definition of coherence that has been widely adopted has not yet been promulgated. What is clear, however, is that the rates of exposure (i.e., assignment probabilities) in the populations from which the cases and controls are drawn should be equal. Since exposure misclassification can distort the proportion *apparently* exposed among the controls, it can generate bias. In the HSP study, it was possible that some truly exposed controls were embarrassed about admitting their use of weight-control products that contained PPA. The observed odds ratio may have been inflated, because these truly exposed controls (*EH*) were counted as unexposed (*UH*). In causal terms, an exposed control must be either a *preventive* or an *immune* individual. Assuming that PPA was never preventive for hemorrhagic stroke, these diet-drug consumers who did not reveal their exposure would be *immune* individuals who had been virtually "reassigned" from exposed to unexposed. Of course, a similar effect may have occurred for the cases as well. However, a case might be expected to be much more motivated to identify a possible explanation for her stroke.

Administrative Records

In many case–control studies, the exposure status is ascertained by reviewing extant files containing previously recorded information. For example, an environmental study of a toxic chemical might assess exposure by proximity to a source of contamination, such as a well or toxic waste dump. Public information from such potential sources as the Federal Census, local land court files, and school records might provide relevant data. Such administrative records are not designed specifically for research purposes and may not be complete and accurate. However, a potential advantage is that the data were recorded contemporaneously and in a presumably objective fashion. So, any exposure misclassification might well be considered random.

Of course, there is still potential for error on the part of the clerical personnel who extract the data from the historical record. Like interviewers, these file reviewers must be trained to collect data accurately and without an agenda. In particular, it may be unwise for the investigators to be personally involved in data collection, except in a supervisory role. Also, if possible, the disease status (case or control) of the person whose exposure state is being traced should be unknown to the data collector.

Post Hoc Exposure

In most case–control studies, a case is identified very soon after the time when the outcome event has occurred. Information about possible exposure is then obtained by interview or by searching through historical administrative records. Because the exposure generally has occurred well before the outcome, the temporal ordering is obvious. However, for some diseases, there may be a substantial lag between the time of onset and the time of detection. It is then possible for exposure to occur during this "dark period" before the outcome event is observed.

One type of bias that can result from such apparently post hoc exposure has been termed *reverse causation*, or *protopathic bias*. Reverse causation occurs when an early symptom of the actual outcome event is the cause of exposure, but the event is not recognized until after exposure. For example, in the HSP study, there appeared to be several instances in which a full-blown hemorrhagic stroke was preceded by a so-called "sentinel headache" that was really an early stroke symptom. But PPA, the drug under study, might sometimes be taken to relieve sinus headaches. So, it was possible that some exposure to PPA actually happened because a stroke had already taken place (Weisberg, 2004).

Reverse causation is frequently confused with confounding by indication. Confounding by indication involves exposure assignment based on some indication for treatment that is *predictive* of the outcome. Some methodologists emphasize the need to distinguish clearly between confounding by indication and reverse causation (Salas et al., 1999). However, it is not surprising that these concepts are often confused, especially in studies of long-term chronic disease processes. Part of this confusion is a consequence of ambiguity in how the endpoint is defined. For example, we often use cause-specific mortality as an outcome, because it is easy to obtain reliably, but this may really be a proxy for "contracting a fatal disease." If *having* a stroke was the real outcome in the HSP, as opposed to *being diagnosed* with a stroke, then the sentinel headache was not an indication, but actually a symptom.

Studies of the benefits of smoking cessation provide a classic example. In these studies it is common to determine retrospectively a former smoker's duration of abstinence, and then to monitor his experience prospectively for some time period. The rates of disease-specific mortality for former smokers are then compared against the rates for smokers during the follow-up period. It is now well known that, for the first few years after quitting, a smoker's apparent mortality incidence from lung cancer is actually higher than the corresponding incidence for similar continuing smokers (e.g., Knoke et al., 2008). This paradox is attributed to the "quitting-ill" effect; many smokers quit because they have been diagnosed with the early stages of lung cancer. Thus, the exposure (smoking cessation) did not precede the real outcome of interest (contracting an ultimately fatal lung cancer), although it obviously preceded the person's actual death.

This instance of reverse causation could be distinguished from a very similar situation of confounding by indication. Suppose that the former smoker actually quit because of being diagnosed with lung cancer at a very early stage of development. Because the eventual course of the disease would not necessarily be certain, this indication for cessation could be regarded as a risk factor for lung cancer mortality, rather than as an unambiguous indication that the outcome had already occurred.

From a causal perspective, the critical issue is whether the defined outcome event truly occurred prior to exposure. If so, then we can best describe the problem as a type of exposure misclassification. The individual is actually unexposed (e.g., still a smoker) at the time of the event (lung cancer). But he *appears* to have been exposed (e.g., having quit smoking) prior to the outcome, because he became exposed by the time the observed outcome (lung-cancer fatality) was eventually observed. As explained in Section 9.3, a person who is misclassified in this manner will be virtually "transferred" from the unexposed group to the exposed group. The impact will be to attenuate the apparent effect of the exposure. For example, the empirical effect of smoking cessation (e.g., relative risk less than 1.0) would become closer to 1.0, thus underestimating the benefits of quitting.

10.5 OUTCOME MEASUREMENT

Outcome measurement error has been viewed traditionally in two very different ways, depending on whether the outcome is dichotomous or numerical. If the outcome is dichotomous, the traditional model for outcome misclassification is based on the probabilities of correct classification (sensitivity and specificity). This model implies that we can say very little about bias unless we make the assumption that outcome misclassification is unrelated to the individual's exposure state (i.e., nondifferential). In that special case, we can say a lot, namely that the bias tends to attenuate the empirical effect. But there is rarely any solid evidence to justify this assumption.

If the outcome is numerical, the classical theory of measurement is similarly dependent on strong assumptions that are rarely made explicit or seriously challenged. The measurement error for each individual is posited to be an additive increment that is randomly selected from a hypothetical probability distribution. This hypothetical distribution is usually assumed to be normal and to be identical for all individuals. These conditions imply that measurement error can be seen as random noise. It increases variability, but does not generate bias.

From a causal perspective, we developed in Chapter 9 a model in which errors arose because of a form of (virtual) causal mediation. Whether the outcome is dichotomous or numerical, this same idea is applied, although the mathematical expression of the causal mediation is quite different in these cases. This causal model provides a conceptual framework within which the problem can be better understood. For example, it becomes evident that the nature of bias can depend on whether the study is an RCT or an observational study. In an observational study, the effects of a nonrandom assignment mechanism are intertwined with those attributable to the measurement process. Therefore, parameters that describe the measurement process alone, like sensitivity and specificity, would not normally be sufficient to quantify the impact of information bias.

Dealing with measurement error is very difficult, because the nature of the measurement process is generally not understood. As our causal model indicates, the implications of measurement errors are complex, unless special assumptions are plausible. So, it is important to minimize the potential for measurement error through proper study design. This requires an awareness of the main aspects of the data collection that might lead to bias:

- The subject's reported outcome value is inaccurate.
- The outcome value recorded by a third party is inaccurate.

- The instrument used to measure the outcome is inaccurate.
- The measured outcome is a surrogate for the true outcome.

Subjective Reporting

Especially in social science research, outcome values are often ascertained by self-report. The subject may be interviewed or may complete a structured questionnaire. Obviously, there are many potential reasons the subjects might be unable or reluctant to provide accurate information. Broadly, these errors might be characterized as cognitive or emotional. Cognitive issues include recall of facts, interpretation of questions asked, and linguistic and cultural misunderstandings. Emotional issues pertain to such factors as apprehension, embarrassment, and desire to please. In designing a study, careful attention to such issues is necessary to minimize potential for error.

Accuracy of reporting may or may not be related to exposure state. In an observational study, the study groups are generally not exchangeable. Therefore, a bias can result, as discussed in Chapter 9. In an RCT, the groups are exchangeable, but misreporting may be influenced by the subject's awareness of (or belief about) his exposure state. For example, a subject receiving the intervention might be more likely than one in the control group to perceive a significant improvement. To avoid such problems, the tactic of *blinding* is utilized whenever feasible. The experimental subjects are said to be blinded when the exposure state is masked. This is usually accomplished by engineering the exposed and unexposed conditions to appear virtually identical. For example, in testing a new medication, the control group may be given a placebo or a competitive drug designed to look and taste just like the drug being tested. That way, the subjects have no basis on which to discriminate between the intervention and control in a way that might influence their responses.

Blinding of the subjects is often possible, but it may not be completely successful; there may be subtle cues that can "break the blind" for some subjects. Imperfect blinding does not necessarily lead to bias, but the potential exists. The actual extent of bias depends on whether or not awareness of one's exposure state translates into a difference in outcomes. For example, in the 1970s a number of observational studies suggested that vitamin C (ascorbic acid) might prevent or ameliorate the common cold. The main endpoint in these studies was a subjective evaluation of cold symptoms. In due course, several RCTs were performed to settle the issue.

At first, the benefits of vitamin C appeared to be confirmed. However, these results were overturned when analyses were stratified on the basis of whether or not the subjects *believed* they were receiving vitamin C rather than placebo. It turned out that many of the subjects could accurately identify the vitamin C by its distinctive taste. The apparent benefits were restricted to the subgroup of individuals who believed they were getting the vitamin C (Chalmers, 1975). When analyses were performed on subjects who remained blinded to their actual treatment modalities, no meaningful difference between the vitamin C and placebo groups occurred. Although controversy persists, the absence of effect concluded by Chalmers is now generally accepted by the medical research community.

Power of suggestion resulting from awareness (or belief) that an inactive treatment modality is actually an effective agent is commonly referred to as a *placebo effect*. The placebo effect can in some situations be an objectively measurable physical effect, but more often it is an alteration of perception. A placebo effect can operate in a variety of specific ways, depending on the study design (e.g., characteristics of treatment modalities, type of outcome variable). In an RCT, the main source of bias is the *differential* placebo effect that could occur if more subjects in one study group than in the other think they are receiving the active treatment. Then the observed effect would include a component attributable to this differential placebo effect.

Observer Bias

In many comparative studies, the outcome values are ascertained by third-party observation. Consider a hypothetical RCT of antidepressant medication in which the endpoint of suicidality is assessed in part by a psychologist viewing therapy sessions from behind a one-way mirror. Ideally, the observer would be completely dispassionate, and also blinded to the type of medication (e.g., SSRI or placebo) being taken by the subject. However, the observer might have a professional "bias" in favor of the drug, and this bias could color her judgment. If she suspects that a particular patient is taking a placebo rather than active treatment, she might (consciously or unconsciously) be somewhat less vigilant in looking for evidence of suicidal tendencies.

From a statistical viewpoint, the issues for observer bias are very similar to those entailed in fallible subject reporting. The simplest solution, when it is feasible, is to have the observer(s) blinded. However, this may be impractical in many contexts, because the nature of the modalities being compared cannot be masked. Observation biases can be particularly potent, because a single observer can affect the data for many subjects. On the other hand, it may be easier to train observers to be aware of their tendencies and avoid subjectivity. In some cases, methods based on independent observation by two or more observers may enhance accuracy.

Instrument Accuracy

Outcome values are often obtained by utilizing an appropriate measuring instrument. This might be a piece of physical equipment, such as a scale to measure weight, or a mental "test" of ability or performance. To measure a physical quantity, such as distance, weight, or voltage, a reading is taken on some conventional scale using a measuring device. This measurement is subject to "random" deviations that are presumed to fluctuate around the "true" value, often with a known standard deviation. These "errors" arise because the device is affected by minor physical forces that can vary from occasion to occasion, even assuming that the underlying value remains constant.

In psychology, educational research, and much other behavioral research, the instrument is often a questionnaire or test in which a series of items captures the subject's preferences, perceptions, or knowledge. The responses are combined to generate an overall score. This test score then serves as the outcome value. The classical theory of measurement developed by psychologists seems to be motivated by the metaphor of a physical device. Psychometric theory assumes that the individual possesses a "true" underlying score that reflects some stable characteristic of interest (e.g., mathematics achievement, IQ). As discussed in Chapter 9, this metaphor of an intrinsic true-score around which errors fluctuate randomly may be overly simplistic for many psychological and educational tests.

Even within the classical test measurement framework, there are two major issues that have been identified as potential threats to validity. *Testing* effects occur when the performance on a test is modified simply because of having taken a pretest. This can result from increased familiarity with the testing demands or because of "reactivity" on the individual's part. For example, taking a pretest might help the subject to become "test-wise" when later tested again. Alternatively, a student might react by working harder in response to a poor initial performance, or perhaps lose interest because of boredom and do worse on a later test. In a comparative study, bias can result if the impact of these factors differs between study groups.

Instrumentation refers to distortions that arise from the structure of the measurement device. An observed difference between study groups may actually be an artifact of the technology. For example, some tests of intellectual ability or academic performance are suspected of being culturally biased. In this case, two individuals from different ethnic backgrounds might score differently, even if they share an identical value of the underlying construct. If the ethnic distributions differ across study groups, a bias could result.

Surrogate Outcomes

In many studies, the ultimate outcome event of interest can take a long period to manifest. Consider a large-scale RCT studying the effect of low-dose aspirin for primary prevention of myocardial infarctions. Even in a very high-risk population, the incidence of MIs is very low in absolute terms. It generally requires thousands of subjects and several years of surveillance to garner enough events for meaningful statistical analysis. An increasingly popular alternative is to define the endpoint to be a surrogate variable that reflects some aspect of the disease process leading to the ultimate outcome of interest. For example, coronary calcification by electron beam computed tomography has been employed with some success as a marker of progression toward a possible MI or other CHD event (Valabhji and Elkeles, 2000).

The use of surrogate endpoints offers a number of attractive methodological advantages, but it can entail some major drawbacks as well (Lonn, 2001). Often, the surrogate is an intermediate factor "along the causal pathway" that describes a disease process. The description of possible causal interrelationships between the exposure, intermediate factor, and endpoint in particular situations can be extremely complex, and bias can result in various ways (Fleming and DeMets, 1996). At one level, we can think of the surrogate (e.g., calcification) as a fallible measure of an underlying latent outcome variable (e.g., CHD progression). This approach is based on the idea that the entire *disease process* is the real object of scientific interest. From this perspective, an MI is regarded as a "random" occurrence that depends on many factors besides those we can hope to control.

Suppose that coronary calcification represents a perfect indication of the degree of CHD progression within a 5-year period. Then an RCT that found a substantial decrease in the level of calcification during a 1-year period would be clinically important. On the other hand, if the measure of coronary calcification is only weakly associated with actual CHD progression, then the RCT's results might have little relevance to clinical practice. The utility of surrogate endpoints in various areas of research is a topic of current interest and debate.

One of the arguments for using surrogates, apart from their potential cost-effectiveness, is the ability to help elucidate specific mechanisms of disease. This argument may continue to gain force as knowledge of biological mechanisms, especially at the molecular level, continues to evolve (Bell, 1998). Although major clinical *events* may remain the ultimate objects of interest, insight regarding complex disease *processes* may be very valuable. For example, learning how to reduce the progression of coronary calcification may turn out to be important for certain types of individuals, under certain conditions. So, the impact of such a reduction may *by itself* have a small impact on MIs but may contribute much as a piece of some larger puzzle.

GUIDEPOST 10

This chapter has attempted to organize and describe the main sources of bias. Each of these sources represents a situation that has the potential to generate bias. These situations can create flaws in a particular aspect of study design or implementation. There are five main aspects of a comparative study:

- Sampling of subjects
- Assignment to exposure states
- Adherence to nominal exposure states
- Exposure ascertainment
- Outcome measurement

Each specific source of bias affects the degree of experimental control that is possible when implementing one or another of these study components. A preliminary list of 20 sources has been offered as a tentative categorization (see Table 10.1). For each source of bias, we have sketched its nature and potential implications. Ideally, this list will be expanded and refined.

The following chapter brings this book to a close. We have attempted to place the subject of bias solidly within an explicit causal framework. We have paid little attention to methods for "solving" the problems of bias. There are numerous texts and articles that deal with the mechanics of statistical correction. In the concluding chapter, the main conventional approaches are surveyed. An alternative perspective on data analysis is then proposed as a possible expansion of statistical methodology to situations in which biases are a main consideration. This new paradigm would be aimed more at sharpening quantitative reasoning, and less at attempting to offer precise solutions to well-defined mathematical problems.

CHAPTER 11

Contending with Bias

Bias presents a conundrum. On the one hand, the presence of bias is viewed as a flaw that invalidates the results of a comparative study. On the other hand, nearly every study is affected to some extent by bias, and most scientific progress has been based on such imperfect research. This chapter discusses the ways that bias has traditionally been addressed and suggests the need for a broader perspective.

We have divided the sources of bias into five general categories:

- Sampling of the study subjects
- Assignment to exposure states
- Adherence to nominal exposure states
- Exposure ascertainment
- Outcome measurement

Assignment bias is widely considered to be most critical and has received by far the greatest attention. Internal validity is often accepted as the sine qua non of causal inference. Conversely, other forms of bias are often viewed as much less serious concerns. While recognizing the possibility of these other forms of bias, investigators tend to downplay their importance. As a result, many methodologists have come to adopt a very skeptical attitude toward conclusions reached in the absence of randomized experimentation but to be somewhat more cavalier about other threats to validity.

The causal modeling perspective suggests that traditional attitudes may be shortsighted. Bias is an almost unavoidable fact of research life that must be taken into account in evaluating a study's methodology. Rather than an attempt to engineer perfection, the research process may be better regarded as attempting to solve a complex puzzle in light of ambiguous and sometimes misleading

Bias and Causation: Models and Judgment for Valid Comparisons, By Herbert I. Weisberg
Copyright © 2010 John Wiley & Sons, Inc.

hints and clues. In this light, randomization is not a panacea, and observational studies are not always defective. The scientist's job is to see through a fog of incomplete evidence and misdirection, and attempt to discern the underlying causal pattern. Toward this end, she must employ every tool of inference at her disposal and ultimately must make a convincing argument to support a credible, albeit often tentative, conclusion.

Many of the researcher's tools are aimed at the *prevention* of bias. Sampling bias can often be prevented by carefully specifying the target population and selecting the subjects. Assignment bias can in principle be completely avoided when randomization is feasible. In observational studies, building informative complexity into the study design can be helpful. Problems of adherence can be minimized through various procedures to encourage the subjects' ability and willingness to adhere fully to the study requirements. Exposure and outcome measurement errors can be minimized through procedures to assure accuracy, including training, motivation, and skillful instrument construction.

Despite such efforts, certain sources of bias may remain operative. In this common situation, there are two main approaches. In most cases, the researcher deals with the subject of bias informally. He may discuss briefly some possible sources of bias, and assess for each one its likelihood and potential impact. This qualitative approach is not very precise and often not especially persuasive. Such subjective assessment can sometimes become little more than an exercise in rationalization. This state of affairs has motivated some methodologists to seek more rigorous methods of accounting for bias. The tools that have been developed are based on explicit mathematical–statistical models.

In the following section, we review the various solutions that have been proposed to deal with bias within this statistical modeling framework. The primary focus of these methods is on assignment bias, because that is usually considered the most critical threat. All of these techniques share a common basic premise about the general nature of the problem posed by bias. They all assume that the main objective of a comparative study is to estimate the causal effect that would have been estimated in a perfectly executed randomized experiment. This causal effect is usually conceived as a uniform effect that applies, at least approximately, to all members of the study population. In subsequent sections we suggest that this perspective is quite limited and sketch an alternative view that may be more fruitful when individual causal effects are variable.

11.1 CONVENTIONAL SOLUTIONS

From the traditional statistical point of view, the main problem lies in the different distributions (especially the mean values) of confounding factors in the study groups being compared. Statisticians tend to define confounding factors, somewhat circularly, as variables whose control would reduce bias. Epidemiologists tend to view confounding factors as variables that differ between groups and exert a "causal" effect on the outcome, but are not "on the causal pathway" between exposure and outcome. In most situations, several (often many) candidate confounders can be identified. To adjust for these confounders, an empirical effect [e.g., average causal effect (ACE), risk ratio (RR)] is calculated after conditioning on the confounders.

There are then two main analytic problems to solve. First, what statistical technique will be used to adjust for the effects of the relevant confounding variables? Second, which covariates need to be included in the calculation? Residual bias can remain after adjustment if either of these problems is inadequately addressed. For reasons of mathematical tractability, the choice of technique is usually quite limited. In most cases, a version of linear modeling is adopted, although various matching and stratification techniques are growing in popularity. Attention then focuses primarily on the issue of whether any "unobserved variables" would be necessary to specify fully a model in which the causal effect can be identified. But how can we ever be certain that there is no additional covariate necessary to avoid the possibility of "hidden bias" in our estimate of effect?

Collapsibility

In Chapters 6 and 7, we explained how conditioning on a set of covariates is the primary method applied to adjust for confounding. The covariates might be variables considered to be confounders, or perhaps surrogates for the confounders. In principle, we form strata that are homogeneous in terms of the values of all the covariates being considered. As a practical matter, a statistical model is usually employed, especially when there are several covariates. The causal effect can be represented mathematically as a parameter within the model. Setting aside the credibility of the model's mathematical form itself, the question of *which* covariates to include then becomes the central analytic concern.

The usual approach, in simplified form, can be summarized as follows. First, a provisional statistical model is fitted to the data. The model includes all of the covariates that are deemed relevant on theoretical or empirical grounds. The adequacy of this provisional model is then tested by adding other candidate variables suggested by alternative competing theories. If the parameter that represents the effect in this model changes only negligibly, then the provisional model is deemed adequate; if not, the expanded model is used. In practice, the final model chosen may be the culmination of several iterations, and other factors besides the size of the change in effect may be taken into account.

The idealized decision-rule just described has been called the *collapsibility criterion*. There are two serious problems in relying on collapsibility. Most important, the full set of potentially available covariates might not be adequate to achieve successful adjustment. That is, even if we could condition perfectly on the entire set, the study groups might not be exchangeable within the resulting strata. However, even if we can justify the assumption of exchange-ability (equivalently, strong ignorability), there is another issue to consider. As shown in Section 7.3, collapsibility in terms of the model parameter does not necessarily imply *causal* collapsibility. What this means in practical terms is that this criterion depends strongly on the assumption of a correct statistical model. In particular, most models effectively assume that the causal effect is uniform across different individuals, or at least that the within-stratum ACEs do not vary across strata. If this basic assumption is incorrect, then it is unclear exactly what is being estimated by the model, even if collapsibility holds.

Propensity Scores

Collapsibility is a criterion to help decide when inclusion of additional covari-ates is warranted. The need for such a guideline arises from the potential availability of many confounders and the resulting practical difficulties of adjusting for them all. In theory, invention of propensity scores has gone a long way toward solving this practical problem. As explained in Section 7.2, the propensity score is a mathematical function of the covariates. Under the assumption of strong ignorability (conditional exchangeability), this function represents the individual's assignment probability. In principle, propensity scores can solve the problem of *which* measured covariates to include. We simply calculate the propensity score based on all useful covariates, and condi-tion on its value. Here, a covariate is considered useful if it contributes to achieving balance of the covariates across study groups.

The derivation of propensity scores was motivated directly by an explicit causal model. Consequently, this approach disclosed the key insight that a focus on the *assignment mechanism* is central to adjustment for confounding. Since data relevant to the assignment mechanism can be *observed*, the problem of selecting from among measured covariates can, in principle, be addressed empirically. Of course, whether strong ignorability, conditional on the observed covariates, actually holds is not an empirical question. Even perfect balance across groups in terms of all measured covariates would not guarantee that hidden bias has been eliminated.

Sensitivity Analysis

Sensitivity analysis addresses the problem of bias by focusing on a simple question: if a hidden bias exists, how much of a difference could it possibly make? Answering this question would place a bound on the causal effect. In particular, can we at least be fairly certain about the direction of a causal effect, even if the precise magnitude cannot be determined? If a particular interven-tion can be found to have *some* beneficial effect, further studies might be able to establish its size. Accordingly, sensitivity analysis usually asks whether an apparent effect could be eliminated by adjustment for any plausible source of bias. This question can be addressed by determining how "large" the bias would need to be to completely account for the observed effect. If this bias is implausibly large, then at least a useful qualitative conclusion can be reached.

To carry out this program, an estimate of the causal effect is first calculated, based on the available covariates. Then, a mathematical model is developed that includes not only the measured covariates, but also a parameter to reflect the impact of possible bias. Most commonly, the bias parameter represents the effect of an unobserved covariate. The bias parameter can be varied, and the resulting value of the empirical effect is observed. In this way, the analyst determines how extreme the bias would need to be in order to account for the observed effect, in the sense that adjustment for this bias would nullify the apparent effect.

Sensitivity analysis can be undertaken using a traditional formulation of models to represent bias (see Rothman et al., 2008, Chapter 5) or an explicit causal modeling perspective (see Rosenbaum, 2002, Chapter 4). In either case, the critical issue is how to decide when a particular assumption about the unobserved covariate is too extreme to be plausible. Although the formal structure of sensitivity analysis is helpful to structure thinking about possible competing theories, it does not obviate the need for judgments about plausibility.

Partial Identification

Sensitivity analysis effectively trades precision for reduced uncertainty. For example, we might be willing to settle for a qualitative inference (i.e., direction of the effect) in exchange for greater confidence that our conclusion is correct. An alternative approach developed primarily by econometricians makes this trade-off even more explicit. Recall that a parameter of a model can be *identified* if the data are adequate to allow determination of its value *precisely* by obtaining a sufficiently large sample of observations. Standard statistical theory allows us to quantify the random error within the context of a specified model, and to reduce this error to any level by increasing the sample size. Bias can arise because of nonrandom (systematic) error that is to some degree (possibly completely) unknown.

Sensitivity analysis makes various assumptions about the possible bias and judges their plausibility. A qualitative inference [e.g., selective serotonin reuptake inhibitors (SSRIs) increase the rate of suicidality] would be sustained if it remains true in the face of the most extreme plausible hidden bias. An alternative strategy is to consider a *range* of assumptions that vary in terms of the evidentiary or theoretical "strength" that seems to be justified. These assumptions may pertain directly to a bias parameter, or to some other relevant aspect of the study. For example, the plausible values for an outcome may be restricted to some specified range. Then, the causal effect may be narrowed to an interval of possible values, even if it cannot be fully identified. This situation has been dubbed *partial identification* (Manski, 2007). The core concept underlying partial identification has been articulated by econometrician

Charles Manski as the *law of decreasing credibility*: "The credibility of inference decreases with the strength of the assumptions maintained." Weaker (less stringent) assumptions generate wider ranges of possible values, but these inferences possess greater credibility. Stronger (more stringent) assumptions generate narrower ranges, but have less credibility. Ultimately, it is up to the decision-maker to act based on her own convictions regarding the "strength" of assumptions and the trade-offs between precision and uncertainty.

Partial identification is an elegant and fruitful idea. In principle, it allows a disciplined and explicit accounting for systematic uncertainty. Especially appealing is the largely overlooked notion that different assumptions vary in terms of how strongly they warrant serious consideration. Therefore, the role of judgment is made explicit, without forcing a commitment to one or another assumption. The main practical problem concerns whether strong enough assumptions to generate useful bounds are plausible.

Probabilistic Bias Analysis

Both sensitivity analysis and partial identification focus attention on what could be inferred under various explicit assumptions regarding the extent of bias. An alternative approach is to quantify the uncertainty about the bias parameter. Rather than placing a plausible maximum on the bias parameter (sensitivity analysis) or assessing the relative strength and implications of a few alternative assumptions (partial identification), it may be possible to place a probability distribution on the bias parameter. We could formulate a probability distribution for the value of a bias parameter, based on theoretical considerations or on data from previous research.

The result of a probabilistic bias analysis is a probability distribution for the causal effect that incorporates the uncertainty about bias parameters. Obtaining such a distribution entails complex mathematical methods that are in the spirit of Bayesian "subjective probability" analysis (Greenland, 2005b). The necessary calculations often require simulation methods, because the effect distribution cannot be expressed in a closed form. This is especially true in a "multiple-bias" analysis that accounts for two or more sources of bias, such as confounding and outcome measurement error. So, although probabilistic bias analysis is a theoretically attractive idea, it has so far proven difficult for applied researchers to implement.

Instrumental Variables

The benefits of random assignment may sometimes be attainable without formal randomization. In these cases, assignment bias can be avoided by exploiting "natural experiments." As explained in Section 8.3, an instrumental variable (IV) is a variable (1) whose values have been determined effectively at random and (2) that is indirectly related to the outcome through a causal mediator. In our illustrative model, the IV was conceptualized as a form of encouragement to adopt a particular exposure of interest. The causal mediator was a dichotomous variable that indicated whether or not the exposure of interest actually occurred. Under certain assumptions, it is possible to obtain an unbiased estimate of the causal effect.

An especially fruitful application of this idea has been developed recently, based on genomic research. Suppose we are interested in the relationship between a certain risk factor and a particular disease. It may be possible to identify a genetic marker that is randomly distributed throughout the general population and causally related to the risk factor. That is, having the marker increases one's probability of having the risk factor. Then the genetic marker can serve as an instrumental variable, or *instrument*, for estimating the relative risk associated with the risk factor.

In honor of the pioneering geneticist Gregor Mendel, this approach has been termed *Mendelian randomization* (Smith and Ebrahim, 2003). For instance, Mendelian randomization has recently been used to identify a causal relationship between lipoprotein (a) and myocardial infarction (MI), using a particular polymorphism in a certain gene (the LPA gene) as the instrument (Kamstrup et al., 2009). This polymorphism is randomly distributed and is indirectly related to having an MI through its effect on the level of lipoprotein (a). In terms of the causal model, the polymorphism is the "encouragement" factor and the level of lipoprotein (a) is the causal mediator.

Theoretically, instrumental variables provide a powerful alternative to observational studies in those situations where it is applicable. However, for the vast majority of research questions, there are no instruments readily available. Furthermore, it may be difficult to determine whether the required assumptions supporting this methodology hold true.

Structural Equation Modeling

In this book, we have drawn a distinction between ordinary *variables* and causal *factors*. A variable is *any* characteristic that can be measured and can vary across individuals. A factor for a particular individual is a particular variable that can potentially affect that individual's response. The causal effect of a factor depends on the individual's causal context. Because of this dependence, the statistical association between variables, including factors and nonfactors, is theoretically quite complex.

Structural equation modeling (SEM) is predicated on the critical simplifying assumption that a factor's causal effect is independent of the causal context. Under this assumption, it becomes feasible, under certain conditions, to identify and estimate causal effects. In particular, it becomes possible to estimate causal effects from comparative studies. The recent work by Pearl and others related to Bayesian networks has clarified the precise conditions that are required for such identification (Pearl, 2000).

11.2 STANDARD STATISTICAL PARADIGM

The various methods for dealing with bias summarized in Section 11.1 are certainly valuable on a *conceptual* level and can sometimes be applied in particular situations. For example, a sensitivity perspective can help structure thinking about the extent of bias. Although these approaches are becoming more widespread, attempts to apply them rigorously in practice have generally been rather limited. One limitation stems from their mathematical complexity. Another constraint is that these techniques remain grounded in what we will refer to as the standard statistical paradigm.

The standard statistical paradigm for analyzing a comparative study can be characterized as *procedural*. The basic idea is that rigid adherence to a set of procedures guarantees validity. These procedures rule out any discretion on the part of the subjects or researchers, and any reliance on qualitative information. These strictures are put in place in order to maintain tight experimental control. As in a laboratory, the procedures assure that the estimated effect is free of all systematic influences that cannot be controlled. Random errors alone remain, and these can be managed effectively using classical statistical techniques.

This procedural approach implicitly adheres to certain principles that govern the conduct of the study:

- The purpose is to reach a decision about a causal effect.
- The study must be essentially self-contained.
- The proposed causal effect can be fully articulated in advance.
- Individual effect variability is assumed to be negligible.
- Chance must be categorically ruled out as a plausible explanation.
- Internal validity is paramount and external validity optional.
- The stable unit–treatment value assumption (SUTVA) must hold.

This paradigm and methods based upon it evolved throughout the first half of the twentieth century, in response to important research questions of that era. At the heart of this paradigm was R. A. Fisher's innovation of randomization. Equality of the study groups, and a probabilistic basis for statistical inference, could be guaranteed by *removing* all systematic influences over treatment allocation. The causal effect that would be obtained in a properly conducted randomized controlled trial (RCT) has become, in effect, the *counterfactual* that has virtually defined the "true" causal effect.

The principles just listed may seem obvious and relatively innocuous. Together, they lead inevitably to the RCT as a gold standard, along with certain other restrictions, such as blinding of subjects and observers, and the desirability of including a placebo comparison. This precise and logical system has relegated all observational studies to a much lower methodological caste. However, considering that most studies are necessarily observational rather than experimental, it is worthwhile to consider the nature and implications of these precepts.

Standard statistical methods are often viewed as providing information that will assist a hypothetical decision-maker. The results are usually presented in the form of a point estimate, along with a measure of uncertainty, such as a significance level, a confidence interval, or perhaps a Bayesian credible interval. This information can be valuable for *deciding* how to act: Should we reject or accept the null hypothesis? Should we act as if there is a large effect or a small one?

This decision-oriented perspective implicitly assumes that all of the relevant data can be generated *within the study* itself. Background scientific theory or results of previous related research may be referenced, but they are rarely incorporated directly in the analysis (except in a meta-analysis). When this external information is formally included, as in the conventional approaches described in Section 11.1, it must be reduced to a very specific mathematical form (e.g., prior probability distribution, limited parameter range). This information is ordinarily a highly "processed" summarization, possibly distilled from a complex web of prior theory and research. This summary then *conditions* the analysis of data within the study, but this analysis is essentially self-contained.

The decision is often framed as a choice between a "null hypothesis" and various possible alternative hypotheses. The null hypothesis must be clearly *specified in advance*. This is done mainly so that a well-defined probabilistic model for the data can be articulated. This model forms the basis for obtaining the estimate of effect and calculating a measure of uncertainty associated with random error. This prespecified effect is an overall summary statistic, such as the ACE or RR, often termed the "main effect" of the study.

The possibility of substantial *effect modification* is generally downplayed on both practical and theoretical grounds. Most theories are framed in terms of specific causes that are assumed to affect everyone in roughly the same manner. This assumption was often reasonable for the kinds of simple exposures or interventions of interest when the standard paradigm was being developed. As a practical matter, analyzing effect modification requires stratification of the study sample, effectively reducing the relevant sample size. For example, recent speculation has focused on a possible relationship between the effects of hormone replacement therapy (HRT) and a woman's age (Rousseouw et al., 2007). To test this theory, it is necessary to obtain estimates of effect for different age strata. Whether by stratification or by using a statistical model, the random sampling variability of an age-specific estimate will typically be substantially greater than that of the main effect.

Accounting for *random variability* is the primary concern of traditional statistical methodology. Concepts such as tests of significance, statistical "power," and confidence intervals (either conventional or Bayesian) were developed to answer an important question. Can the observed results be explained by chance alone? If random variability offers a plausible competing theory for the results, then causation is not even a consideration. In particular, the level of statistical significance, (p-value) has taken on tremendous psychological importance, and the value of 0.05 has often been interpreted as a de facto threshold for "rejecting" the null hypothesis. Literally billions of dollars can sometimes ride on whether or not a study achieves the iconic 0.05 level of significance.

Internal validity of a comparative study is usually thought to be an absolute requirement for causal inference. The enormous emphasis placed on randomization reflects the belief that internal validity is an absolute criterion for causal interpretability. We first ask whether there exists any causal effect at all. Then the generalizability of this effect becomes a secondary consideration. This prioritization is motivated by the notion that there exists a "true" causal effect that is relatively independent of individual characteristics and circumstances. Therefore, the causal effect found in the study is often *presumed* to generalize, absent some specific reason to the contrary.

In Section 3.8, we introduced Rubin's *stable unit–treatment value assumption* (SUTVA). SUTVA is usually assumed implicitly, and it effectively requires that individual response patterns do not depend on the particular way in which treatment modalities have been assigned to individuals in the study. So, the treatment modalities are well defined and do not depend on the particular conditions that exist in the study. For example, a person's causal effect should not depend on whether she is familiar with any other individuals in the study. SUTVA is considered important because it insures that the intervention's effect is related to the intervention per se and not to the particular combination of the intervention and the experimental conditions. SUTVA can thus be regarded as a threshold condition for external validity.

11.3   TOWARD A BROADER PERSPECTIVE

We noted in Section 11.2 that the standard statistical paradigm is procedural. The aim is to ensure tight experimental control of systematic influences that could engender bias. The procedures are effective to the extent that they rule out any discretion on the part of the subjects or researchers, and any reliance on qualitative information. However, in the vast majority of studies, even many RCTs, the researcher's ability to enforce experimental control is limited. Consequently, inferring causation becomes to some degree a matter of *judgment*. From the traditional perspective, subjectivity is a methodological transgression, and judgment is precisely what the standard paradigm is designed to eliminate. But judgment is an essential element in nearly all comparative studies, and it has played a critical role in some of the finest examples of applied statistical research. Thus, Susser urges us to:

... recognize that the difference between types of evidence is relative, and not absolute, and apply all available criteria of judgment to any particular instance. Perhaps we may advance beyond our present limitations by systematizing our criteria of judgment, and by expanding the number and type of available criteria. (Susser, 1977, 14)

In other words, judgment is inevitable, so we must learn how to sharpen our capacity to exercise it wisely.

It is common to regard observational studies as inferior to randomized experiments, primarily *because* of their dependence on judgment. Indeed, under conditions in which the standard paradigm is optimal, judgment can and should be minimized. But in many other situations, it may not be realistic, or perhaps even desirable, to eliminate judgment entirely. In those circumstances, our methodological arsenal may need expansion to exploit fully the potential of observational research.

Because the standard paradigm has become so dominant, statisticians are now largely unable to see beyond it. Efforts to develop a broader perspective have attracted little interest within the field of statistics. Recent developments in causal modeling are starting to change this situation. However, most causal analyses are still predicated on the principles underlying the standard statistical paradigm. So far, methodologists have not ventured far into the terra incognita outside this familiar terrain. This is not surprising, since they lack a compass to point them in any another direction.

The principles listed in Section 11.2 sit at one intellectual pole, where tightly controlled experiments are the ideal and confirmation of a predefined hypothesis is the goal. At the opposite pole are studies that occur in more natural settings, are usually less definitive in their findings, and require the creative application of critical judgment. Can we identify a set of principles that might apply to such studies? These principles could represent a start toward systematizing and expanding our criteria of judgment, as Susser suggests. The following list is offered as a provisional attempt:

- The purpose is to obtain evidence about a causal effect.
- The study must draw upon all relevant information.
- A causal effect can be suggested, refined, or possibly confirmed.
- Variability of individual effects is assumed to be the norm.
- Bias and chance are equally important explanations to consider.
- Internal validity is critical, but external validity is also important.
- The effect may depend on the circumstances in which exposure occurs.

Would an alternative research paradigm embodying such principles be preferable to the current standard? That depends on the research issues being

Table 11.1   Possible Guiding Principles for Comparative Studies

Standard Paradigm	Alternative Perspective
Primary aim is to reach a decision.	Primary aim is to obtain evidence.
Study must be self-contained.	Study draws on all sources of information.
Prespecified hypothesis is tested.	Study may suggest or refine hypotheses.
Aggregate effects are main product.	Individual effect variability is important.
Statistical significance is critical.	Bias and random error must be analyzed.
Internal validity is absolutely critical.	Internal and external validity both matter.
SUTVA must hold.	Effect could depend on circumstances.

studied. Any actual study being planned might fall somewhere between these extremes. Thus, in principle, we may need to develop a range of methodologies that are tailored to the particular conditions that define the research problem. Let us consider how such alternatives might complement the standard paradigm. Table 11.1 contains a convenient summary of the two alternative perspectives.

A decision is not always the main product of a comparative study. A decision must be based on a number of considerations, some of which can be psychological, social, or economic, in addition to the value of a causal effect. By itself, a scientific study can rarely guide a decision. Rather, the study can provide relevant *evidence* about an effect. This evidence can be "packaged" in various forms and can draw upon various sources of information in addition to the data generated from the study itself. In this way, evidence may be relevant to future decisions that are currently only vaguely apprehended.

If future applications of the evidence obtained are not fully known, a mathematically precise decision-theoretic analysis would not be appropriate. Some might argue that such a situation is not amenable to causal analysis. An alternative view is that the kinds of inferences likely to prove useful would take a different form. These inferences would be more open, in the sense of making connections to insights from theory and from other research. The result would rarely rise to the level of proof, but it might provide valuable evidence that could inform future research as well as professional practice.

In the standard paradigm, the causal effect of interest must be specified in advance. That way, a well-defined probabilistic model can be used to calculate p-values (or posterior probabilities) and confidence intervals (or Bayesian credible intervals) to inform decisions. This requirement makes sense if the purpose of the study is seen as exclusively confirmatory. However, confirmation represents only one possible objective. Rather than supporting a decision, a comparative study might be able to provide evidence pertaining to *indications* or *conclusions* (Tukey, 1962).

In the context of comparative studies, an indication is a pattern in the data that *suggests* a causal relationship. The strength of the evidence within the study may be weak or strong. However, the plausibility of the suspected effect may rest in part on theory or data from beyond the study's confines. Indications are unanticipated findings that may be worthy of later refinement and testing. Conclusions, on the other hand, lie at the other end of the inferential spectrum. A conclusion is a broad and general premise that has gained wide acceptance within a scientific community. Conclusions represent working knowledge that, at least for the time being, is included within the scientific canon:

> Like any other human endeavor, science involves many decisions, but it progresses by the building up of a fairly well established body of knowledge. (One whose relevance is supposed to be broad.) This body grows by the reaching of conclusions— by acts whose essential characteristics differ widely from the making of decisions. Conclusions are established with careful regard to evidence, but without regard to consequences of specific actions in specific circumstances. (They are, of course, based on specific experiments or observations.) *Conclusions are withheld until adequate evidence has accumulated.* (Tukey, 1960, 425, emphasis added)

These scientific building-blocks are usually the result of a line of research comprising multiple studies. Many individual studies may provide only *indications* of varying evidentiary strength; a few may reach *conclusions* based on a synthesis of the available evidence. In stressing the importance of indications and conclusions, Tukey was recognizing that validation of a theory could depend on evidence from multiple sources accumulating over time. Replication of indications, possibly under a wide range of conditions, plays an important role in achieving the confidence to reach a conclusion.

An *apparent* correlation between the strength of an effect and a particular factor is a common discovery in many studies. However, determining whether such a post hoc effect is real or a product of chance variation is difficult for two reasons. First, the relevant data may pertain to a small subgroup of the total sample, so that the precision of estimates and the power of statistical tests can be quite low. Second, the observed interaction may be only one of many such indications that *could* have been identified. Consequently, the validity of an "interaction effect" that may indicate causation tends to be heavily discounted. In part for these reasons, it is widely *assumed* that interactive effects are unusual occurrences relative to main effects.

The reluctance to acknowledge interactive effects is inherently conservative. As Susser (1977, 14) wisely advises: "negative judgments require as much caution as positive judgments." Thus, an apparent interactive effect should not necessarily be ignored, just because it fails to reach statistical significance in a single study. In the aggregate, across many studies, the suppression of potentially valuable leads may far outweigh the errors from provisionally entertaining possible effects that are ultimately refuted.

The standard paradigm treats *chance* as the primary competing theory to explain an apparent effect. Failure to reach statistical significance, usually at the conventional 0.05 level, is interpreted as evidence that chance is a plausible alternative. As a practical matter, absence of significance often consigns a potential causal effect to oblivion. But failure to achieve significance means that chance is *not refuted* in the study, not that chance is necessarily the best explanation.

Suppose that we have performed an RCT and obtained a p-value of 0.06 for the main effect. Assume that the study sample size was adequate to assure a statistical power of 0.90. That is, the probability of finding a significant effect for a prespecified effect-size was 0.90. The actual result would ordinarily be interpreted as a negative finding. Further development of an innovative medical technology or educational program might well be halted based on such a study outcome. However, suppose we also knew that the outcome was measured with some amount of error, and that half the subjects did not fully comply with the assigned intervention. Assume that both these sources of bias were believed to reduce the empirical effect, but by an unknown amount.

Clearly, the nominal significance level of 0.06 does not tell the full story. Given the "handicaps" resulting from the known sources of bias, it might seem likely that the "true" p-value is much smaller. On the other hand, there may be other sources of bias that ought to be factored in as well. Whether by one of the formal techniques described in Section 11.1 or by a more subjective evaluation, both random and systematic sources of variability need to be considered. The common practice of treating the significance level in isolation as a threshold criterion may entirely preclude a more balanced and comprehensive assessment.

In the standard paradigm, internal validity is considered the sine qua non of causal inference. However, we have seen that a measure of effect can vary across different populations, because the individual effects (response patterns) typically vary across individuals. Consequently, the idea of a causal effect that is meaningful in some general sense, albeit with minor variations across different populations, is often unrealistic. For example, suppose that a study population happens to contain a roughly equal proportion of *causal* and *preventive* individuals ($Q_2 = Q_3$). Then the risk difference (RD) in the study population would be zero. On the other hand, the RD in the target population might be substantially positive, because $P_2 > P_3$.

Conventional logic would attempt to establish first that there exists *some* effect, and then to generalize this effect to a broader population. In our example, the null effect would not generalize in this way. In fact, the null effect found in the study would be qualitatively different from the effect in the target population. If the study population has been severely restricted in order to facilitate internal validity at the expense of external validity, then what has really been accomplished? To obtain causal inferences of practical importance, we must be concerned about *both* internal and external validity.

The principle of SUTVA essentially decrees that the individual's causal effect must not vary with the experimental conditions. In particular, SUTVA would preclude a situation in which the response pattern could depend on how exactly the choice of treatment modality was determined. For example, a woman might respond differently to a therapy suggested by her trusted physician than to the same treatment assigned by a roll of the dice. For unbiased causal inference to be possible in a randomized experiment, it is certainly correct that no violations of SUTVA must be allowed to exist. However, in many practical situations, the manner of administration can be of vital importance.

The causal analysis of noncompliance in Section 8.3 provides some additional insight. A subject who fails to comply modifies the treatment modality in some unknown manner. So, the treatment actually received is not well defined. We explained how only the causal effect among "compliers" could be estimated from a randomized experiment. The compliers are individuals who would always "follow orders" with respect to an assigned modality. An RCT can thus provide valid information about a subset of the population that is potentially malleable. Conversely, it cannot necessarily tell us about the effect on a population of individuals who are firmly committed to a particular course of action (always-takers or never-takers). Yet, the potential effects on such currently intractable individuals may perhaps be of great interest.

In the studies of employment training programs, such as the National Supported Work Demonstration (NSW), we might be especially interested in people who are highly motivated to participate in the experimental program. Such individuals could well be those for whom the largest benefits of the intervention would accrue. Ironically, however, if assigned to a control group, such individuals might attempt to subvert the researcher's intentions by finding alternative ways to obtain some of the benefits provided by the experimental program. Consequently, the true effect on the program for these "true believers" would be underestimated.

11.4   REAL-WORLD BIAS REVISITED

In the concluding section to follow, some speculations about the future role of statistics in dealing with bias will be offered. Before that, it will be useful to reconsider the case studies presented in Section 1.3. These vignettes have provided context for many of the concepts developed throughout this book. Now we will review these again as illustrations of the practical and theoretical challenges posed by comparative studies. Before proceeding further, the reader might find it useful to reread Section 1.3.

It is commonly appreciated that an "ideal" study is rarely feasible, for a variety of practical reasons. Most of these constraints relate to the sources of bias discussed in Chapter 10. From the traditional perspective, the resulting biases are errors that distort the true effect. The aim of analysis is to undo this distortion, so that the result that *would have been obtained in a properly executed RCT* is approximated. In the previous section, we have suggested why the standard statistical paradigm may not be ideal in all circumstances. Many sources of bias reflect essential and unavoidable aspects of the natural or social conditions. Therefore, the aim of research might often be to appraise important causal effects as they operate *in fact*, rather than as they would operate in an idealized experimental setting.

In this section, we consider the applicability of the standard paradigm and of randomization in particular, in the context of our illustrative examples. In what ways have the possibilities for causal inference been impaired by both practical and theoretical constraints? To what extent are these limitations critical, and how might they have been overcome?

Evaluating the Efficacy of Antityphoid Vaccine

Evaluation of a vaccine against typhoid fever at the turn of the twentieth century took place before randomized experimentation was developed. From a modern perspective, it is clear that an RCT would have provided by far the most persuasive evidence in this situation. At the time, the debate over adoption of the new vaccine reflected standards of scientific evidence that did not contemplate randomization as a possibility, but in all other respects, the quality of analysis was remarkably sophisticated. So, this historical case study provides a fascinating window on a world without randomization.

Applying the nascent statistical technology of that era, Pearson invoked two main criteria. One would be quite familiar to us: the effect of the vaccine was statistically significant. The other was pragmatic and to some degree judgmental. Pearson derived an empirical criterion for the required strength of the effect, based on the efficacy of two previous vaccines that had been developed (smallpox and diphtheria). His hesitation about the antityphoid vaccine was based on this criterion. This reference to an external standard, derived from actual experience in somewhat similar circumstances, is consistent with Tukey's dictum that data analysis can and should become more of an empirical science:

> We need to face up to the need for *collecting the results of actual experience with specific data-analytic techniques*. Mathematical or empirical-sampling studies of the behavior of techniques in idealized situations have very great value, but they cannot replace experience with the behavior of techniques in real situations. (Tukey, 1962, 62)

Nowhere is this need greater than in causal inference when randomization is not feasible.

Pearson's particular objection was ultimately overruled, in part because of other, more qualitative, modes of evidence. Wright argued persuasively that his vaccine could not be fairly evaluated by statistical analyses of the data in a vacuum. For example, it was likely that exposure ascertainment errors and outcome measurement errors would have attenuated the apparent effect. Perhaps most important, the vaccine's efficacy was validated by replication in various populations and under different circumstances. Although this reasoned judgment based on the totality of evidence would not have been preferable to evidence from an RCT, it was far superior to reliance either on statistics alone or on clinical judgment alone. As discussed in Section 1.3, the RCT evidence obtained much later was found to be consistent with the observational research. In contrast with the apparent "failures" of the observational studies of hormone replacement therapy and supported work programs, here is an apparent success story. Why are observational studies sometimes valid and sometimes not? Methodological research to answer this critical question has received very little attention.

Racial Disparities in Death Sentencing

In the evaluation of a vaccine, the assumptions underlying the standard statistical paradigm appear eminently reasonable. Now we consider a situation in which even the idea of a causal effect is ambiguous at best, and randomization is meaningless. In a landmark study of the administration of death sentences in Georgia, David Baldus and colleagues concluded that the race of the victim played a causal role. Their statistical approach was based mainly on logistic models in which the dependent variable was whether or not a death sentence was issued. The independent variables included the race-of-victim (black vs. white) and many other case characteristics. It would not be very fruitful here to second-guess the study authors or the courts with respect to the merits of their decisions. However, it is instructive to consider some limitations of the statistical approach employed.

At the outset, an objection can be raised that the victim's race is an immutable characteristic. Can we think seriously about a hypothetical experiment in which everything remains the same, except for the race of the victim? Perhaps no valid study of causation is possible. On the other hand, assessing causation is precisely what the legal system must do on a daily basis. Did an employer base hiring decisions in part on the applicant's race? Did a landlord illegally refuse to rent apartments to members of certain ethnic groups? So, while causal analysis of such questions may seem nonsensical in the ivory tower, the exercise can be deadly serious in the halls of justice.

A potential criticism of the Baldus study that has been raised by some critics pertains to the sample selection. The database of cases analyzed consisted of those defendants who were convicted of voluntary murder. Those who were convicted of lesser offenses, or whose cases were settled by a plea bargain or were otherwise disposed, were not included. In Chapter 5, we showed how sampling from a larger universe could lead to substantial bias. Moreover, in Section 10.1 we discussed how the relevant target population may be difficult to define. In the situation considered here, these problems are exacerbated by the complexity of the decision-making process, which entails a series of decisions by prosecutors, judges, and jurors.

These conditions make it very hard to speculate about the possible implications of the sampling process employed. For instance, based on the discussion in Section 5.4, we can say that exclusion of *doomed* individuals (those who would have received a death-sentence regardless of the race of the victim) would have increased the empirical risk ratio. To understand whether such a bias occurred, we would need to first think through the issue of which part of the process is alleged to entail racial discrimination. Is it the entire process or the process beyond a particular point (e.g., being indicted for capital murder)? Then we would need to consider the target population of those cases that satisfied this definition. Of these cases, were those defendants who were *doomed* more likely than others to be excluded? Causal modeling can help to structure the analysis, but the basis of ultimate conclusions must be judgmental.

Setting aside the issues of meaningfulness and sampling bias, we come to the question of exchangeability. Were black-victim and white-victim cases exchangeable, after accounting for measurable characteristics? The logistic regression analyses employed by the researchers estimated an overall risk ratio (actually an odds ratio approximation). The major limitations of this approach, as discussed in Section 7.2, are its reliance on a linear model and assuming a constant effect. Both of these restrictions are especially questionable in light of the many factors that can affect the decision-making process.

When the original study was performed, propensity-score methodology was virtually unknown among social scientists. This approach might have ameliorated the problems in using standard logistic modeling. In this context, exchangeability would need to hold after conditioning on all *legitimate* case characteristics. By a legitimate characteristic, we mean a factor that was permissible for the decision-makers to consider. Conditioning on illegitimate factors in order to achieve exchangeability would be inappropriate. As an extreme example, suppose that the *defendant's* race was a covariate included in the model. This might well assist us to achieve strong ignorability, but the resulting causal effect would be the effect after adjusting for possible racial discrimination based on the race-of-defendant. In effect, one form of discrimination (race-of-defendant) could exonerate the other (race-of-victim).

Assuming strong ignorability (conditional exchangeability) in this context, the propensity score would provide an estimate of the probability that the case involved a white victim, given the full set of legitimate case characteristics. Conditioning on the propensity score would, in theory, provide a means of estimating an unbiased risk ratio or risk difference. Perhaps even more important, the stratification by propensity scores would shed light on the nature of the differences between the black-victim and white-victim cases. In particular, are these differences qualitative (very little overlap, so that most cases fall in strata that contain predominantly one type of case) or quantitative (substantial overlap in most strata)? Examining the characteristics of the cases in the more extreme strata, combined with substantive sociological and legal knowledge, might provide valuable insight. For example, do cases in strata that contain almost exclusively white-victim cases tend to reflect a greater degree of "moral culpability" than other strata? If not, an inference of possible racial bias would be strengthened.

More formal statistical analyses of potential discrimination could be restricted to the range of propensity scores that are not extreme (i.e., the region of support). However, the extent to which a reduction in the sample may reduce statistical power would need to be assessed. Deciding whether the results could be extrapolated beyond the range of support might also be complex—and involve judgment. For example, would a high frequency of death sentences in the type of cases that almost always involve a white victim reflect racial bias? Would we have observed a significant difference in this stratum if there had been enough black-victim cases for a valid comparison?

Finally, we note that potential applications of propensity scoring are all predicated on the assumption of conditional exchangeability. Achieving statistical balance for numerous relevant variables would be inadequate if other important factors were omitted. The propensity score model would have to capture all of the factors that might legitimately influence the judge or jury and would also be predictive of the race of the victim. This would certainly be a tall order. Ultimately, assessing the success of this endeavor would be a matter of informed judgment.

Evaluation of Employment Training Programs

The extensive discussion swirling around the evaluation of supported work interventions is emblematic of a deep divide among quantitative research methodologists. Some argue that only randomized experiments can reliably assess the efficacy of social programs. Others believe that observational studies can often supply valid information. This faith is grounded in confidence that one or another brand of statistical technique can compensate for the lack of random assignment. Despite their differences, most of the protagonists in this imbroglio share a common understanding regarding the *goal* of program evaluation: to obtain an unbiased estimate of the average causal effect that would have resulted from a perfectly executed randomized experiment. Thus, the standard of validity for an observational study is how close its results come to those of an RCT. Put differently, the objective is to eliminate assignment bias.

The different reanalyses of the NSW data, beginning with LaLonde's study (LaLonde, 1986), have almost exclusively been based on this randomization principle. The results of the original RCT represent the bull's-eye that statistical methods applied to observational data are aiming to hit. The wide variability and substantial "errors" of various findings have been widely interpreted as discouraging. But what does all this effort really tell us? Perhaps it is time to step back and ask whether these methodological competitions really address the right issues.

Of the few economists who have adopted this broader methodological perspective, James Heckman and his colleagues are perhaps best known. These iconoclasts have raised several issues that have been discussed in this book. Their starting point is a counterfactual perspective, combined with recognition that individual effects are likely to be variable (Heckman and Smith, 1995). They have explored alternative measures of effect, rejecting the almost universal focus on the ACE (Heckman, Smith, and Clements, 1997). Furthermore, they point out the importance of external validity, and how difficult to attain it may be in randomized experiments. In addition, they highlight a ubiquitous tendency toward noncompliance (substitution bias) on the part of participants assigned to the control group.

Perhaps their most important insight is that randomized experiments may sacrifice valuable contributions to a larger process of knowledge accumulation on the altar of a narrowly conceived methodological rigor:

> In fact, simple black-box evaluations pose a serious threat to the accumulation of knowledge about the behavior of persons and institutions. Because they are not conducted within a behaviorally coherent framework of analysis, the evidence from experiments does not cumulate. The end result of a research program based on experiments is just a list of programs that "work" and "don't work," but no understanding of why they succeed or fail. The long-run value of cumulative knowledge is high, but is neglected by advocates of "short-run" evaluations conducted outside of coherent social science frameworks. The potential of evaluations to add to this store of knowledge, and for this store of knowledge to inform future evaluations, needs to be more widely recognized and should be factored into current discussions regarding evaluation methodology. (Heckman and Smith, 1995, 108)

Black-box program evaluations generally formulate simple hypotheses that are geared primarily toward policy decisions rather than toward the advancement of scientific knowledge. As perceived by most quantitative methodologists, the only alternative is a "softer" brand of qualitative case-study research. The minority who advocate a middle road that is scientifically rigorous but not rigidly procedural see a somewhat different purpose for such evaluations:

> The proper function of evaluations is to speed up the learning process by communicating what might otherwise be overlooked or wrongly perceived. The evaluator, then, is an educator. His success is to be judged by his success in communication; that is, by what he leads others to understand and believe. *Payoff comes from the insight that the evaluator's work generates in others.*

> A study that is technically admirable falls short if what the evaluator learns does not enter the thinking of the relevant political community. That community may include clients, program staffs, taxpayer leagues, environmentalist lobbies, legislators, bureaucrats, and interested citizens. *An evaluation fulfills its function to the extent that it assists participants in the political process to resolve conflicts intelligently.* (Cronbach, 1982, 8, emphasis added)

Phenylpropanolamine and Hemorrhagic Stroke

Case–control studies are especially vulnerable to potential bias. Obviously, there is no randomized assignment of exposure status. Nor is the sample usually selected randomly from some target population, although this can occur in a nested case–control design. Furthermore, lack of coherence between the cases and controls is possible. Finally, because the data on exposure status is collected retrospectively, and often relies on fallible reporting, exposure misclassification can occur. In light of all these serious threats, one might wonder why in the world this approach is ever employed.

One answer is that, in many situations, a case–control study is the only viable option. When the outcome event is something rare and undesirable, both practical and ethical considerations can preclude a prospective study. The Hemorrhagic Stroke Project (HSP) was designed in just such a situation. Phenylpropanolamine (PPA) was suspected of causing hemorrhagic strokes, which in young adults are extremely infrequent but very serious. So, the investigators were virtually forced to adopt the case–control methodology.

A more satisfying answer to the question of why case–control studies are performed is that the results often provide useful *indications* of possible effects. These indications might suggest future confirmatory research and might contribute to the accumulation of evidence pointing toward a *conclusion*. The conclusions about effects of cigarette smoking were based in part on evidence from many relevant case–control studies. Even a single case–control approach can sometimes be dispositive, when the causal mechanism is well understood. For example, investigations of outbreaks caused by bacterial infections may be aimed at identifying a particular contaminated product. Interviewing cases and controls about their recent consumption of food products can narrow the search to a few possibilities. Then physical evidence can be sought in the form of product samples that may actually contain high levels of the bacteria.

In the case of PPA, such favorable conditions did not hold. The etiology of hemorrhagic stroke is complex. There were numerous competing theories raised by critics that could have explained the apparent effect. Perhaps the most defensible posture for the investigators would have been to admit this uncertainty, rather than to feel compelled to reach a conclusion. In Tukey's words, conclusions must be "withheld until adequate evidence has accumulated." This position would have required extraordinary political and academic fortitude; the pressures from regulatory agencies and scientific journals for definitive answers were no doubt intense. To defer a firm conclusion after 6 years of research may have seemed irresponsible.

The main problem in this case was not necessarily with the study's design, but with the expectation that the study would (and should) have been conclusive. But what might be an alternative way of responding to ambiguous findings when a conclusion is needed? One possibility would be to make the study's data available to other researchers for independent analysis. Then the critical issues could be explicated and aired openly in the scientific literature.

Other researchers could bring their own perspectives to bear, and a scientific consensus might emerge. In effect, this is what eventually happened, with a battle of experts in a litigation setting, and with lay jurors rendering the ultimate judgments. It would seem preferable for the issues to be debated by scientists in an academic forum.

Another possibility would be to weigh whatever clinical evidence could be obtained from both internal and external sources. Spontaneous reports of adverse events represent an obvious type of external information. Indeed, the hypothesis that PPA was harmful was based on just such "soft" data. Historically, case reports and postmarketing surveillance systems have, with good reason, been largely ignored, except for suggesting hypotheses. But the rapid evolution of internet-based technologies holds great promise for improvements in both the quantity and quality of these reports. There exists a great opportunity to capitalize on this potential by shaping the reporting process into a form that will make the data more useful.

Postmenopausal Hormone Replacement Therapy and Cardiovascular Risk

Controversy and confusion about hormone replacement therapy (HRT) rages on. Many practitioners and the women they treat are unhappy with the current state of affairs. On the one hand, it is clear that HRT offers effective relief of uncomfortable postmenopausal symptoms, such as hot flashes and night sweats, as well as some benefit in terms of bone preservation. On the other hand, long-term HRT use confers a slight increase in the risk of breast cancer. These facts have been well established. Prior to the Women's Health Initiative (WHI) study results in 2002, HRT was also believed to substantially decrease cardiovascular risks. Since that time, the effect of HRT on coronary heart disease (CHD) has been widely understood to be at best neutral, and possibly harmful. In light of this new risk–reward balance, women are being warned to avoid HRT altogether or to restrict use to the first few years after menopause.

The WHI and Heart and Estrogen/Progestin Replacement Study (HERS) projects were large-scale RCTs that attempted to reach a definitive conclusion regarding the health effects of HRT. The studies were high-quality implementations of the gold-standard approach to research within the standard statistical paradigm. These studies convincingly demonstrated that HRT administered in a standardized form to a particular population did not decrease the overall level of CHD. The general perception of both scientists and the lay public is that the observational research "got it wrong" and that HRT for protection against CHD is ineffective and possibly even harmful. Is this interpretation justified?

In the large-scale RCTs, the type and dosage of HRT were strictly held to a uniform standard. There was no flexibility or discretion allowed to the women or their prescribing physicians. Such tight specification of the intervention is a hallmark of experimental control aimed at insuring internal validity. As a result, the trials were well designed to inform us about the effects of *this particular modality* on a *particular population* of postmenopausal women. An implicit assumption was that a beneficial (or harmful) effect found in the study would be the true effect, although it might vary somewhat in magnitude across different target populations.

Is it really plausible, though, that one size would fit all, when it comes to rectifying a presumed estrogen deficiency? Perhaps, the optimal type and dosage can vary, so that the therapy needs to be tailored to the individual. In that case, we might expect a fixed regimen to be too weak (in some sense) for some women, but too strong for others. Thus, the response patterns for the main endpoints of interest might vary across individuals. As shown in Section 5.4, sampling bias might arise from the process by which women were recruited and screened.

For example, exclusion of those who refused to abstain from HRT for a lengthy "wash-out" period might have eliminated many women who believed they were being helped by HRT in some form. To the extent that these women were *preventive* with respect to CHD, the empirical risk ratio may have been inflated, or even changed in direction. Thus, the apparent elevation of risk for CHD *could* be merely a statistical artifact. Of course, this hypothetical is not meant to prove the existence of sampling bias in randomized studies of HRT, but simply to suggest that analyzing the potential sources of sampling bias could provide valuable insight.

Conversely, we might speculate that the calibration of dosage arrived at by women and their physicians would have a more favorable risk profile. In that case, the findings of the observational research could be correct for the variable treatment regimens these women actually received. In Section 7.2, we suggested that propensity-score methodology might shed light on this issue. Suppose we were to analyze the observational studies by creating strata based on the propensity to use HRT. However, rather than estimating an overall effect, we could focus on the stratum-specific estimates. In particular, is the effect of HRT on CHD beneficial for those women most likely to use HRT? Such a pattern would be an *indication* that HRT might be effective *for those who were choosing to take it*, even though it might be harmful for some other women.

If such an indication were observed, it would be worthwhile to learn more about the women with high propensity scores. What characteristics appear to determine whether a particular woman tended to adopt HRT? For example, the presence of certain menopausal complaints, or severity of these complaints, might predispose some women to use HRT. It has been hypothesized that such signs might also be indicative of "a woman's susceptibility to the benefits of hormone replacement therapy" (van der Schouw and Grobbee, 2005). This hypothesis represents an important competing theory that might be subject to empirical testing. For instance, suppose we could identify the characteristics of those subjects in observational studies with a high propensity to adopt the kind of treatment regimen studied in the major RCTs. Then the competing theory would predict that the stratum consisting of such women in the RCTs ought to fare relatively well. Indeed, if an analysis restricted to this stratum yielded results that were congruent between observational studies and the RCTs, it would have important clinical implications.

Antidepressants and Adolescent Suicide

The flashpoint in the debate over the potential role of antidepressants (principally SSRIs) in adolescent suicidality (suicidal tendencies) was the meta-analysis undertaken by the U.S. Food and Drug Administration (Hammad et al., 2006a). A meta-analysis is a study that attempts to summarize the results of several studies that test the same exposure. The usual aim of a meta-analysis is to produce an overall estimate of the causal effect and its statistical significance, based on combining the estimates of the component studies. Sometimes, characteristics of the effects are also analyzed, to identify possible reasons for variation in the findings of different studies.

The FDA meta-analysis included 24 RCTs and found a risk ratio of 1.95 for the primary endpoint of suicidality. The investigators concluded that "the data suggested a risk of antidepressant drug-induced suicidality in the pediatric population" and recommended that a warning be placed on product labels (Hammad et al., 2006a). To their credit, the investigators were aware that exclusion of high-risk patients from the RCTs may have led to bias. They examined the relationship between the inclusion–exclusion criteria of various studies and the estimated effects: "None was found to consistently explain the observed differences in the risk estimates between trials within or between development programs" (Hammad et al., 2006a).

The potential importance of eligibility criteria stems from criticisms regarding the exclusion of suspected high-risk patients from most of the RCTs. The rationale for this restriction was linked to the ethical concern that the study drugs might precipitate suicidal tendencies. If so, a relatively high-risk study population would be expected to generate more outcomes classified as indicating suicidality. By excluding high-risk individuals and thereby the number of outcome events, this strategy might be deemed conservative, in the sense of making it harder for studies to result in a significantly elevated effect. However, this restriction may have created a bias that artificially inflated the observed causal effect.

The causal model discussed in Section 5.4 can provide a useful prism through which to analyze the potential bias. In particular, individuals who appear to be high-risk will predominantly be either *doomed* or *preventive*. The causal model implies that the RR will be inflated if such individuals tend to be excluded (see Section 5.4). However, it has also been observed that the absolute rate of outcome events in the placebo group will be reduced, and this rate can serve as an empirical index of the degree of high-risk exclusion (Weisberg et al., 2009). One way to exploit this insight is to analyze the relationship between the risk ratio and the placebo event rate. Suppose that all the studies with risk ratios substantially above 1.0 had relatively low placebo event rates. This pattern would be consistent with the competing theory that the apparently elevated suicidality risk was an artifact of the selection criteria. A recent analysis lends some support to the competing theory that sampling bias has played a role (Weisberg et al., 2009).

A major lesson from this example is that some common intuitions may turn out to be misconceptions when examined from a causal modeling perspective. For example, it is commonly believed that a relative risk tells us something about a drug per se, but this parameter actually describes the effect in a particular population:

> Suppose that in an RCT we observe a relative risk of 3.0 for a placebo-control group in a study with a low-risk population. From a conventional perspective, we might expect the RR to be similar, or perhaps even higher, in other studies of the drug in a high-risk population. This prediction would be in sharp contrast to the predicted pattern described previously. For example, if the study with an RR of 3.0 happened to have the lowest event rate and highest RR among all available studies, we would suspect that its results had been distorted by a selection effect. (Weisberg et al., 2009, 114)

The fallacy in the usual thinking results from a confusion between "high-risk" as meaning a high level of absolute risk and "high-risk" as an indication of susceptibility to the risk factor. If the high-risk population consists primarily of *causal* individuals, then it will also be high-risk in the sense of susceptibility. However, suppose the level of risk is primarily related to the proportion of *doomed*. Then, a higher proportion of *doomed* individuals will increase the absolute risk, but decrease the relative risk.

11.5   STATISTICS AND CAUSATION

Throughout this book, we have alluded to the traditional reluctance of statisticians to address the issue of causation explicitly. We have mentioned that classical probability theory, the mathematical basis of statistics, provides no means to distinguish causation from correlation. The adoption of probability theory as the foundation for statistical analysis effectively drove causality underground. This banishment occurred because probability theory implicitly assumed the existence of a stable population of individuals who were indistinguishable in causal terms. It therefore provided no framework for thinking about how an individual might respond to a hypothetical manipulation of some causal factor.

The central problem of statistical inference for over a century has been how to generalize from a relatively small available data set to reach a broadly applicable inference about the world. Like all scientific theories, the solution rested on a simplified model of reality. The set of units on which data were available was conceptualized as a *sample* selected at *random* from a very large (effectively infinite) *population* of units. This hypothetical population was then taken to be the real object of scientific interest. Probability theory served as the link between the limited available data and the desired broader generalization, embodied in parameters of the population, such as a proportion, mean, or correlation.

By the late nineteenth century, statistical associations between variables had become of great intellectual interest, stimulated in large measure by the wide-ranging empirical research of Francis Galton, inventor of the correlation coefficient (Galton, 1889). Because a set of data was imagined to be a random sample from some population, the observed correlation coefficient was an estimate of the "true" correlation in that population. But this true correlation reflected whatever was happening in the population, causal or otherwise. To interpret an association as a *causal* effect, the data analyst would have needed to bring in logical or evidentiary considerations from outside the study itself. In other words, opinions about causation could be inferred only in part from statistical analysis.

One area in which statistics could be particularly relevant to causation was controlled experimentation. In an experiment, it was feasible to manipulate a single factor of interest, while attempting to control for extraneous causes by holding them constant. For example, a medical researcher might try a new remedy on a group of patients and the traditional treatment on a control group *believed* to be very similar. This might be accomplished by finding pairs of subjects who were closely matched on characteristics thought to affect the outcome. Then one member of each pair would be given the new treatment, and one given the traditional. If the pairs were similar enough prior to the allocation of treatments, any observed average outcome difference between the groups could be attributed to the medication.

The basic problem was that there were many unknown ways in which individuals could vary. As a result, the distributions of outcomes in the two study groups could differ in terms of both average value and variability. In analyzing these different observed distributions, there was no way to disentangle the causal effect of interest from the inherent variability attributable to individual differences. Consequently, there was no way to calculate a causal effect reliably. An observed mean difference between the groups might be affected by the causal effect, by systematic preexisting differences between groups (bias), or by random variability.

The difficulties just explained greatly limited the applicability of statistical analysis for causal inference until the 1920s. In 1919, a young geneticist and statistician named Ronald A. Fisher assumed statistical leadership of a venerable agricultural research facility known as the Rothamstead Experimental Station. Within a few years, Fisher invented the field now known as the statistical design of experiments, with random assignment of treatment modalities as the centerpiece. Instead of focusing on equating the individual units (plots of land) in the treatment groups, randomization effectively nullified any natural association between the treatment modality and preexisting individual characteristics. Fisher turned the traditional experimental logic on its head; he argued for *ignoring* all individual characteristics except perhaps for a few that were worth the effort to control directly. Ironically, by giving up the attempt to control for specific confounding factors, randomization was able to exert a form of indirect statistical control over *all* such factors.

Random assignment of treatments ensured that the total study sample in an experiment could, prior to the implementation of the treatments, be regarded as a random sample from a single population, rather than from two separate populations. Thus, randomization precluded any association between group membership and the study outcome in this population, *unless* there was a causal effect. A nonzero difference of means (or other measure of association) in the study population *had* to be a causal effect. Moreover, under the *null hypothesis* of no treatment effect, all of the observations, regardless of which group they were in, could be viewed as being randomly drawn from a single probability distribution. This crucial fact permitted the statistician to apply the laws of probability to take account of random variability through tests of statistical significance. Fisher had thus devised a way to marry standard probability theory and causal inference, without having to invoke any special principles. In effect, randomization could both remove bias and account for random variability, allowing the causal effect to be disentangled from these extraneous influences.

When it could be applied, randomization elegantly solved the problem of causal inference at the level of the population. However, this statistical wizardry was achieved at a cost: the experimental units (e.g., plots of land) had to be regarded as essentially indistinguishable. If causal effects could vary substantially across the units, then the resulting causal effect would not necessarily be applicable to each individual. In the context of agricultural experimentation, this was not a critical handicap, because the individual qualities of different plots would not vary substantially. Moreover, even if the plots varied somewhat, the average yield over a fairly large geographical area was of primary interest.

Randomization quickly became the core design principle of statistics in agricultural and industrial experimentation, but it was not adopted for research on human health and behavior until much later. In medical research, physicians were understandably reluctant to allocate treatments by the toss of a coin, and there were many practical problems of implementation to be overcome (Marks, 1997). It was not until the 1950s that randomized clinical trials began to be widely accepted. In the social sciences, randomized experiments were limited to psychological research in artificially controlled situations until around 1970, and then applied to program evaluation, but only on a very limited basis. Indeed, there is no evidence that Fisher himself was a strong proponent of randomization in these contexts. Perhaps he was influenced by the knowledge that randomization entails a trade-off between what can be learned about a particular population and what must be ignored about the individuals who comprise that population:

This fundamental requirement for the applicability to individual cases of the concept of classical probability shows clearly the role of subjective ignorance, as well as that of objective knowledge in a typical probability statement. It has often been recognized that any probability statement, being a rigorous statement involving uncertainty, has less factual content than an assertion of certain fact would have, and at the same time has more factual content than a statement of complete ignorance. The *knowledge* required for such a statement refers to a well-defined aggregate, or population, of possibilities within which the limiting frequency ratio must be exactly known. The necessary *ignorance* is specified by our inability to discriminate any of the different sub-aggregates having different limiting ratios, such as must always exist. (Fisher, 1959, 33)

Fisher certainly realized that there might be potentially identifiable subpopulations to which the causal effect from a randomized experiment would not apply. However, to obtain the benefits of randomization, the researcher must ignore individual differences and concentrate on an aggregate effect for the entire study population.

In our day, randomization has become firmly established as the cornerstone of causal inference. However, we have largely forgotten "the role of subjective ignorance" that underlies this technique. In the context of the research problems we face, it is important to consider the proper degree of "ignorance" to be maintained. When research relates to human health and behavior, we must expect individual variability to play a much larger role than in agricultural or industrial experimentation. People are not plots of land, and medical or social interventions are not fertilizers. So, is it perhaps time to reevaluate our methodological options?

The standard statistical paradigm is ideal for problems of a certain type, especially those for which confirmation of a well-elaborated specific hypothesis is the objective. On the other hand, there are many causal questions for which alternative modes of data analysis may be needed. These methods could be based on some other set of principles, perhaps along the lines suggested earlier in this chapter (see Table 11.1). It seems safe to suggest that something more than mathematical–statistical modeling alone is needed.

Studying causation will often require a blend of mathematical analysis, substantive knowledge, and critical reasoning. Competing theories will need to be proposed and tested. Case studies and qualitative description can serve as a necessary complement to quantitative data and mathematical modeling. Some of the most compelling and influential research manages to "paint a picture" by combining descriptive data analysis, statistical modeling, in-depth narratives, and background knowledge.

Finally, we need to give up the vain hope that data analysis can be founded upon a logico-deductive system like Euclidean plane geometry (or some form of the propositional calculus) and to face up to the fact that *data analysis is intrinsically an empirical science*. Some may feel let down by this, may feel that if data analysis cannot be a logico-deductive system, it inevitably falls to the state of a crass technology. With them I cannot agree. It will still be true that there will be aspects of data analysis well called technology, but there will also be the hallmarks of stimulating science: intellectual adventure, demanding calls upon insight, and a need to find out "how things really are" by investigation and confrontation of insights with experience. (Tukey, 1962, 63)

Consider the widely acclaimed study by Anthony Bryk and colleagues on Catholic School education (Bryk et al., 1993). In this exemplary research, various modes of quantitative and qualitative analysis are complementary and mutually reinforcing. Competing theories are considered, and the principle of informative complexity is creatively applied in testing them. Through this multimodal triangulation, convincing evidence of causal effects is developed, and the general thesis that Catholic schools have important lessons to offer educational researchers and policy-makers is supported. Importantly, the foundations of the researchers' arguments are transparent. Thus, readers can bring to bear their own perspectives in judging the plausibility of the conclusions reached. In this way, a productive conversation among educational experts and policy-makers around important issues has been facilitated.

In the biomedical sciences, advances in biological research, especially genomics, are leading to personalized medical treatments that increasingly target more specific subclasses of patients and finer subcategories of disease. An eminent geneticist (and former student of R. A. Fisher), Sir Walter Bodmer, has suggested that research in this dynamic environment will require creative approaches that incorporate elements of both randomized and observational study designs (Bodmer, 2003). For example, he calls for "smaller initial studies which may involve only limited randomization, but which will depend on an appropriate choice of perhaps several sets of historical controls." He also advocates greater reliance on the more refined animal models now available, and on an "intrinsic biological understanding of the likely mechanism of action of any agent or procedure."

Observational studies can complement randomized experiments by contributing to patterns of evidence in several ways. An observational study may serve as a prelude to subsequent confirmatory studies, help to refine a tentative causal hypothesis, clarify a possible causal mechanism, or sometimes even provide convincing evidence bordering on proof. Of course, there are pitfalls. We can easily be gulled by promising associations that later prove to be spurious. Statisticians must continue to provide safeguards against false positives that can arise from the over-mining of complex databases. But it would be wrong to err excessively on the side of caution. Sometimes professional judgment and instincts can guide us accurately enough to discern correlations that *do* in fact indicate causation, even though they *could* be merely chance associations or products of bias.

A major challenge of tomorrow will be to extract valid causal information from complex databases in a way that takes much fuller account of personal characteristics. Increasingly, there will be a demand for solutions that are tailored to smaller and smaller subgroups, and even to individuals. To develop methods of data analysis that can meet this challenge, statisticians will need to question many of the assumptions on which traditional approaches have been constructed. This will not be easy or comfortable, but it is essential if real progress is to be made.

> The future of data analysis can involve great progress, the overcoming of real difficulties, and the provision of a great service to all fields of science and technology. Will it? That remains to us, to our willingness to take up the rocky road of real problems in preference to the smooth road of unreal assumptions, arbitrary criteria, and abstract results without real attachments. Who is for the challenge? (Tukey, 1962, 64)

Glossary

ACE: The ACE is the average causal effect, which is the average of the individual causal effects (ICEs) across a particular population.

Adherence: Adherence refers to the extent to which a subject continues to obey the study's requirements. Adherence and compliance are sometimes used interchangeably, but failure to adhere does not have the connotation of being willful, as noncompliance often does. Some forms of nonadherence may be entirely outside the subject's control.

Adjusted effect: An adjusted effect is an estimated effect that has been adjusted by controlling for covariates using a statistical technique such as matching, stratification, or regression.

Assignment bias: Assignment bias is the bias that can result from lack of exchangeability between the study groups. Assignment bias in various contexts has been called confounding, lack of internal validity, or selection bias.

Assignment homogeneity: Assignment homogeneity across strata means that the assignment probabilities are identical for all of the strata. In this book, the term has been used primarily when we assume the existence of exchangeability of the study groups within strata. In this case, there is a single assignment probability that applies to all response types and has the same value for each stratum.

Assignment mechanism: The process by which each study subject is assigned to a particular exposure state. For each possible response pattern, the assignment mechanism induces a corresponding assignment probability.

Assignment probability: For each possible response pattern, the assignment probability is the proportion of subjects in the study population assigned to the exposure (or active treatment). For example, with a dichotomous outcome, the proportion of individuals with response-type 1 (*causal*) who are assigned to exposure is the assignment probability A_1.

As-treated: An as-treated analysis of data from a randomized controlled trial (RCT) restricts the analysis to subjects who qualify as adherent (or compliant) according

---

Bias and Causation: Models and Judgment for Valid Comparisons, By Herbert I. Weisberg
Copyright © 2010 John Wiley & Sons, Inc.

to some prespecified standard of adequate adherence. An as-treated analysis is also called a *per-protocol* analysis.

Attrition: Attrition from a study refers to the process by which some subjects drop out of the study and are lost to follow-up. Attrition can be regarded as an extreme form of lack of adherence.

Bayesian network: A Bayesian network is a set of causally interrelated variables represented graphically in which the input information is generally subjective and can be updated in light of empirical data, by using Bayes' theorem.

Berkson's bias: Berkson's bias (also Berkson's fallacy, Berksonian bias) was originally posed as a problem arising from the selection of a hospital-based study population. In Berkson's original hypothetical example, there was no effect in the general population, but a positive effect was induced by the particular way that hospital patients were referred to the hospital. Others have interpreted Berkson's bias more broadly to include various situations in which selection of a subpopulation can distort an effect, especially if the apparent effect is nonzero but there is no "real" effect in the population.

Bias: Bias is a systematic error in the estimation of a causal effect. Formally, bias is the difference between the empirical effect and the causal effect. Social scientists tend to include bias within a broader class of *threats to validity*.

Blinding: The subjects and/or investigators in a study are said to be *blinded* if the nature of the treatment modality received by any subject had been masked. Blinding of both the subjects and investigators is termed double-blinding.

CACE: The CACE is the complier average causal effect. This is the ACE for the principal stratum consisting of individuals who always comply with the assigned treatment modality, whether or not exposed. The CACE is the "direct" causal effect for this subset of individuals—and the only causal effect that can be estimated in an unbiased manner.

Case: A case is an individual who experiences the outcome event under study. The term is used primarily in the context of case–control studies. In the notation of this book, a case is often indicated as having an outcome value of D (Disease) as opposed to H (Healthy).

Case–control study: A case–control (or case–referent) study is a type of retrospective study design in which a case series is obtained along with a sample of controls that are often derived from a different source. This method of obtaining data would lead to bias if analyzed normally, so special methods based on the odds-ratio as the measure of effect are typically employed.

Causal (response type): A *causal* individual is one who experiences the outcome event if and only if exposed.

Causal configuration: For any individual (and set of circumstances), the causal configuration is the entire set of values for all causal factors.

Causal context: The causal context is the background pattern of values for the set of causal factors that, in conjunction with the causal factor of interest (exposure, risk factor, intervention, treatment), determine the outcome value.

Causal effect: For an individual, a causal effect is the response pattern (or a specified function of it) that measures the difference between the outcome value when exposed vs. when unexposed. The arithmetic difference (ICE) is the most common measure of an individual effect. For a population, the causal effect is a summary measure based on the distribution of response patterns. The ACE is the most common measure of causal effect when the outcome is numerical. The risk ratio is most common when the outcome is dichotomous. A causal effect for the study population might differ from the corresponding effect in the target population (sampling bias).

Causal factor: A causal factor (or simply a factor) is one of a set of relevant conditions that together determine the value of the outcome. A *relevant condition* is one such that changing its value would alter the outcome value under at least one set of values for all the other causal factors. A causal factor is equivalent to a component cause in the sufficient component cause (SCC) modeling framework.

Causal homogeneity: Causal homogeneity characterizes a stratification in which the strata are all exchangeable (i.e., they all have identical distributions of response patterns) and are exchangeable with the total study population. Causal homogeneity implies causal effect homogeneity, but the converse does not necessarily hold.

Causality: Causality is the aspect or quality of a theory that involves a cause–effect relationship.

Causal mediator: A causal mediator (or intermediate causal variable) of the effect of an exposure on a particular outcome is a causal factor that is both a causal modifier of the exposure and is an outcome that can be caused by the exposure.

Causal model: Generally, a causal model is a mathematical representation of causal relationships that describes the generation of data in counterfactual terms. In this book, a causal model focuses explicitly on distributions of individual response patterns for a study population.

Causal modifier: A causal modifier is a causal factor whose value can affect the response pattern for a given exposure. A causal modifier for a factor F has the property that changing the value of the modifier, while holding all other causal factors fixed, would cause the response pattern of F to change for at least some individuals.

Causal synergy: Causal synergy is the relationship that exists between two causal factors that are causal modifiers of each other.

Causation: Causation is the action of exerting a causal effect. That primary aim of a comparative study is usually to determine whether it is reasonable to attribute an observed effect of a specified factor to causation or to some other reason.

Coherence: In a case–control study, a group of controls is coherent with the cases if it is from essentially the same population that gave rise to the cases (study base).

Cohort study: A cohort study is a study based on observing the progression of a large group of subjects (the cohort) over an extended period of time. The terms *longitudinal study* and *follow-up study* are sometimes used as synonyms.

Collapsibility: In its strong form, collapsibility occurs when the empirical effect for the study is identical to the empirical effects for each of the strata formed by conditioning on the covariates used for statistical adjustment. In a weaker form, collapsibility refers to a situation in which the value of the estimated effect after statistical adjustment remains the same. Collapsibility is a commonly used criterion for deciding whether a proposed additional covariate is necessary.

Comparative study: A comparative study involves the estimation of a causal effect based on the outcome data for two or more study groups that vary in terms of exposure status.

Competing theory: A competing theory, or plausible rival hypothesis, is an alternative explanation for a putative causal effect.

Compliance: Compliance is sometimes used as a synonym for adherence. Compliance generally has a narrower connotation of being under the control of the subject, who may choose whether or not to comply with the protocol.

Conditioning: Conditioning refers to restricting a statistical analysis to that subset of the total population which has a certain specified value (or set of values) of a particular variable or set of variables.

Confounding: At one level, confounding refers to the bias in estimating a causal effect that results from lack of exchangeability between study groups. At a deeper level, this lack of exchangeability reflects the different distributions of causal factors that exist in the two groups.

Construct validity: Construct validity is a term developed by psychometricians to describe the ability of a variable to represent accurately an underlying characteristic of interest.

Control: In a cohort study, a control subject is a member of the unexposed study group, sometimes called the *control group*. In a case–control study, a control subject is one who does not experience the outcome event (i.e., is not a case). *Statistical control* (for covariates) is used as a synonym for conditioning on covariates, or statistical adjustment.

Correlation: Correlation is sometimes used informally to mean a statistical association between two variables, or perhaps the strength of such an association. Technically, the correlation can be interpreted as the degree to which a linear relationship between the variables exists (i.e., each variable is a linear function of the other) as measured by the correlation coefficient.

Counterfactual: A counterfactual condition is a circumstance that could have occurred but did not. The counterfactual framework, or model, pertains to a way of expressing causal situations in terms of the *potential outcomes* that would occur under different conditions (e.g., exposed vs. unexposed).

Covariance: The covariance is a measure of the strength of a statistical association between two variables. Mathematically, it is equal to the expected value of the product of the variables minus the product of the expected values, which can also be expressed as the correlation coefficient times the product of the standard deviations of the variables.

Covariate: A covariate is an ancillary variable that is measured along with the exposure and outcome. Typically, covariates are meant to capture individual characteristics that are causally related to the outcome, and on which the study groups differ. These covariates are traditionally known as *confounding factors*, or confounders. From a counterfactual perspective, a set of covariates is potentially useful because conditioning on this set may result in strata within which the groups are exchangeable.

Depletion of susceptibles: The depletion of susceptibles refers to a source of bias related to nonrandom sampling from the target population. It occurs when the composition of the target population is changing in such a way that the proportion of individuals who would be affected by the intervention or risk factor is declining.

Dichotomous variable: A dichotomous variable, or dichotomy, is a variable that can assume one of two possible outcomes.

Direct effect: The direct effect is the causal effect that is net of the effects of a causal mediator. The direct effect can be defined unambiguously only for members of a principal stratum consisting of individuals with a fixed value of the mediator.

Distinguishable: Two individuals are causally distinguishable if they have different response patterns and can be differentiated on the basis of some observable characteristic(s). Two or more subpopulations are distinguishable if they are not exchangeable and their members can be identified on the basis of some observable characteristic(s).

Distribution: For any variable, a statistical distribution is the breakdown of a total population according to the proportions who have each possible value of the variable. A distribution can be regarded as a probability distribution, because the proportion with each value is also the probability that a randomly selected individual will have that value. For a hypothetical infinite population, methods of calculus may be necessary to express the distribution.

Doomed (response type): A *doomed* individual is one who always experiences the outcome event, whether or not exposed.

Effect measure: An effect measure is a mathematical function (e.g., risk ratio, risk difference, odds ratio, ACE) that summarizes the distribution of response patterns in a specified population.

Effect modification: Effect modification occurs when the causal effect can vary across different strata. This can occur either because the stratification is based on a causal modifier or because of the way that the strata have been selected from the total population. In a statistical model, effect modification is expressed as an *interaction* between the exposure variable and some other variable.

Effectiveness: The effectiveness of an intervention refers to the causal effect that would occur when the intervention is applied in actual situations.

Efficacy: The efficacy of an intervention means the causal effect under ideal circumstances, as when all subjects adhere fully to a study's protocol.

Eligibility criteria: The eligibility criteria for a study are the conditions that must be met by a prospective subject in order to qualify for entry into the study. Eligibility criteria are often categorized as either inclusion criteria or exclusion criteria.

Empirical effect: The empirical effect is the observed effect that would be approached in a given comparative study as the sample sizes of the study groups increase without limit. In this book, we often assume that the empirical effect is in fact the observed effect, because we do not deal with issues that pertain to finite-sampling (i.e., random) error.

Encouragement trial: An encouragement trial is an RCT in which the intervention is an *inducement* to accept the real treatment modality of interest.

Endpoint: An endpoint is a prespecified outcome to be analyzed in a comparative study. It is often employed as a synonym for outcome in an RCT.

Enrichment strategy: An enrichment strategy is a deliberate attempt to overrepresent certain types of individuals in the study in order to obtain certain methodological advantages. Typically, enrichment strategies enhance internal validity at the possible expense of external validity.

Equipoise: Equipoise is the condition of a priori equivalence between the treatment modalities being compared in an RCT or quasi-experiment. Equipoise is usually considered a requirement for conducting an RCT, because if the modalities are believed to differ in efficacy or safety, it may be unethical to assign some study participants to the inferior modality.

***Estimation*:** Estimation refers broadly to the calculation of a statistical parameter (e.g., mean, risk ratio) to estimate the causal effect based on data collected from a study. An estimate is subject to random error as well as bias. In this book, we are assuming an effectively infinite sample size, so estimation is simply the calculation of an empirical effect.

***Exchangeable*:** Different populations are exchangeable if their response-pattern distributions are identical. If two study groups are exchangeable, the assignment mechanism is said to be strongly ignorable. More generally, exchangeability can be defined as the counterfactual condition sufficient to assure unbiased estimation of a causal effect. Weaker versions of exchangeability have been defined to express necessary and sufficient conditions for unbiased estimation of particular effect measures.

***Explicit and implicit selection*:** Explicit selection occurs when the individual's outcome value determines whether or not his or her outcome value is observed. Implicit selection occurs when a variable that is associated with the outcome value determines whether or not the individual's outcome value is observed.

***Exposure*:** The term "exposure" is used in two different ways. First, the exposure is the particular causal factor (risk factor, intervention, etc.) that is being studied. For example, we could say: "Tobacco smoke is a highly toxic exposure." Second, the term might refer to the condition of being exposed. For example, we might say: "Exposure to tobacco smoke is harmful."

***Exposure misclassification*:** Exposure misclassification occurs when a subject's exposure state (treatment modality) is incorrectly recorded. Misclassification is *differential* if the probability of being misclassified depends on the outcome value, and *nondifferential* if not.

***Exposure state*:** The exposure state is the condition to which an individual is assigned in a comparative study. With two study groups, the exposure state is simply "exposed" or "unexposed." In an RCT or quasi-experiment, an exposure state may be called a treatment modality, or simply a treatment.

***External validity*:** A study has external validity when its results are generalizable to the target population of interest. Formally, external validity means that the causal effect based on the study population equals the causal effect in the target population. In counterfactual terms, external validity requires that the study population be exchangeable with the target population.

***False negative*:** A subject who is identified as failing to have experienced the event of interest (e.g., exposure, disease) but has truly experienced the event is termed a false negative.

***False positive*:** A subject who is identified as having experienced the event of interest (e.g., exposure, disease) but has not truly experienced the event is termed a false positive.

***ICE*:** The individual causal effect (ICE) is the arithmetic difference between the potential outcomes under exposure and nonexposure.

***Identifiability*:** A parameter of a statistical model is identifiable if its value would be uniquely determined if the sample size could be increased without limit.

Ignorability: The assignment mechanism is weakly ignorable, or simply ignorable, if the marginal distributions of the potential outcomes in different study groups are identical. Ignorability allows unbiased estimation of the usual measures of causal effect. Strong ignorability means that the response-pattern distributions in the groups are identical, which is equivalent to our definition of exchangeability In statistical parlance, strong ignorability means that the joint distribution of the potential outcomes is independent of study group membership.

Immune (response type): An *immune* individual is one who never experiences the outcome event, whether or not exposed.

Immutable characteristic: An immutable characteristic (attribute) is a causal factor that cannot actually be manipulated, even hypothetically. Typical examples of immutable characteristics are gender and ethnicity.

Indirect effect: The indirect effect refers to the influence of a causal mediator in producing the effect of the risk factor. This influence is not a causal effect in the usual sense, which we have defined as the reponse pattern, assuming a fixed causal context. Rather, it is the *change* in the response pattern that results from effective modification of the causal context by the risk factor.

Information bias: Information bias is the traditional category of bias that includes all types of bias related to errors of measurement, including measurement of the exposure state, the outcome value, and sometimes covariate values.

Informative complexity: Informative complexity is the ability to obtain data regarding a pattern of results from multiple analyses, within and across studies, that sheds light on causation. In particular, comparisons of studies based on competing theories, each with different implications, can help to determine whether an empirical effect is unbiased.

Instrumental variable: In the context of comparative studies, an instrumental variable (IV) is a causal factor for which exposure status has been determined effectively at random and which is indirectly related to the outcome through a causal mediator. This causal mediator determines whether or not the individual actually does adopt the treatment modality of real interest. Under certain assumptions, an IV can be used to obtain an unbiased estimate of the causal effect of interest.

Intermediate response pattern: For a causal factor, an intermediate response pattern is the response pattern with respect to the causal mediator considered as the outcome. For example, an *I-doomed* individual will always experience the intermediate event, regardless of exposure status.

Internal validity: A study has internal validity when it provides an unbiased estimate of the causal effect of interest. Formally, internal validity means that the empirical effect from the study is equal to the causal effect in the study population.

INUS: INUS stands for an Insufficient but Necessary part of an Unnecessary but Sufficient condition. This characterization of a cause as one of a set of factors that together bring about an event was introduced by the philosopher J. L. Mackie.

ITT: ITT stands for intention-to-treat or intent-to-treat. The ITT approach to analysis is commonly used in an RCT when some subjects fail to comply fully with the assigned modality. In the ITT approach, all subjects who are randomized are included in the analysis.

LATE: LATE stands for the local average treatment effect, which is the particular causal effect that can be estimated using an instrumental variable.

Matching: Matching is a technique for statistical control of confounding. In the simplest form, individuals from the two study groups are paired on the basis of similar values of one or more covariates. Matching can be viewed as a special case of stratification in which each stratum consists of only two individuals.

Misclassification: When the outcome is dichotomous, measurement error takes the form of misclassification. In a comparative study, misclassification can apply to either or both of the exposure and outcome. If the occurrence of exposure misclassification is independent of the occurrence of outcome misclassification, then misclassification is *independent*; otherwise the misclassification is *dependent*.

Modality: A treatment modality is one of the alternative conditions to which subjects are assigned in a comparative study. In most studies, there are two modalities, which can be designated generally as "exposed" and "unexposed."

Monotonic effect: A monotonic effect is one that can only operate in the same direction for all individuals. For a numerical outcome, monotonicity implies either that $V_{iE} \geq V_{iU}$ for all individuals or that $V_{iE} \leq V_{iU}$ for all. For a dichotomous outcome, either $P_2 > 0$ and $P_3 = 0$ or $P_3 > 0$ and $P_2 = 0$.

Numerical variable: A numerical variable is one whose possible values are numbers. More formally, such a variable is measured on either an "interval scale" or a "ratio scale."

Odds ratio: The odds ratio is a measure of effect that is the ratio of the odds for the exposed group and the unexposed group. The odds for each group is the risk (proportion of subjects with outcome event) divided by 1 minus the risk.

Outcome: In a comparative study, the outcome (endpoint) is the response of interest.

Outcome misclassification: Outcome misclassification occurs when the outcome value is incorrectly observed. Misclassification is considered differential if the probability of being misclassified depends on the exposure status, and nondifferential if not.

Placebo: A placebo is an inactive treatment modality that is meant to be perceived as identical to the modality whose effect is being studied.

Placebo effect: A placebo effect is the individual's response to the placebo. It can be viewed as the causal effect of simply being treated.

Potential outcomes: The potential outcomes are the possible values of the outcome variable that would occur under each exposure state.

Pragmatic trial: A pragmatic trial is an RCT that attempts to estimate the causal effect that would occur in practice (effectiveness), rather than the effect under perfect compliance (efficacy). The ITT approach is generally employed.

Preventive (response type): A *preventive* individual is one who experiences the outcome event if unexposed, but not if exposed.

Principal stratification: When there is an intermediate causal variable, a principal stratification is formed by conditioning on the intermediate response patterns.

Propensity score: The propensity score is a mathematical expression that represents the conditional probability of assignment to exposure as a function of observable covariates. If study groups are exchangeable conditional on the covariates, it is possible to obtain an unbiased estimate of the causal effect by conditioning on any value of the propensity score.

Quasi-experiment: A quasi-experiment is a type of observational study in which alternative treatment modalities are explicitly assigned to the subjects, and certain aspects of experimental control are maintained, but randomization is not considered feasible. Quasi-experiments are common in applied social science for evaluation of proposed educational programs or other interventions.

Randomization: Randomization is the process of assigning subjects to treatment modalities by a mechanical or computerized mechanism that is effectively random.

Rate ratio: The rate ratio is a measure of effect that divides the rate of outcome events in the exposed (treatment) group by the corresponding rate in the unexposed (control) group. The rate for each group is the number of events divided by the time at risk.

RCT: RCT stands for a randomized controlled trial, sometimes called a randomized clinical trial. An RCT is characterized by random assignment (randomization) of exposure status, assuring that study groups are exchangeable.

Regression analysis: Regression analysis is a statistical technique for estimating the relationship between a set of predictors (independent variables) and an outcome variable (dependent variable). Linear least-squares regression, in which the relationship is expressed in a linear form, is the most common type of regression analysis. The mathematical model used in least-squares linear regression is often called the general linear model (GLM).

Regression effect: The regression effect refers to a phenomenon that occurs when the same variable is observed on two different occasions or when two correlated variables are observed. If we consider a set of individuals who have a certain value on one of the occasions (variables), then the average value for the set on the second occasion (variable) will (under usual probability distributions) tend to be closer to the mean for the population from which the individuals were sampled. In a comparative study, these *regressions to the mean* can cause a bias that is called a *regression artifact*.

Relative risk: Relative risk is a general term to describe a measure of how frequently an event occurs in a population that is exposed vs. an unexposed population. It is often used synonymously with either the risk ratio or rate ratio, and it sometimes refers to an odds ratio or hazard ratio.

Residual confounding: Residual confounding refers to the assignment bias that remains after adjustment for covariates.

Response pattern: The response pattern for a particular individual is the set of potential outcomes. In most studies, there are two treatment modalities (or exposure states), and the response pattern consists of a pair of outcome values.

Risk difference: The risk difference is a measure of effect that subtracts the risk of experiencing an outcome event in the exposed (treatment) group from the corresponding risk in the unexposed (control) group. The risk for each group is the number of events divided by the total number of subjects. The risk is sometimes called the rate, frequency, or proportion of events.

Risk factor: The risk factor is the exposure of interest in an epidemiological study and often has the connotation that the outcome event is harmful or in some way undesirable.

***Risk ratio*:** The risk ratio is a measure of effect that divides the risk of outcome events in the exposed (treatment) group by the corresponding risk in the unexposed (control) group. The risk for each group is the number of events divided by the total number of subjects. The risk is sometimes called the rate, frequency, or proportion of events.

***Sampling bias*:** Sampling bias is the bias that results from lack of exchangeability between the distributions of response patterns in the target population and the study population.

***Sampling mechanism*:** The sampling mechanism is the process by which individuals are selected (sampled) from the target population.

***Sampling probability*:** For each response pattern, the sampling probability is the proportion of individuals in the target who are selected for the study population.

***SCC*:** SCC stands for the sufficient component cause model. A sufficient component cause is one necessary condition in a set of conditions that together are sufficient (but not necessary) to bring about the outcome event. This model was introduced by Rothman in the context of epidemiology, but it is essentially identical to Mackie's INUS concept.

***Selection bias*:** Selection bias is a broad term that has traditionally been used variously to describe bias that results from nonrandom sampling, nonrandom assignment, and explicit selection.

***SEM*:** SEM stands for structural equation modeling. SEM is an approach to the statistical modeling of causal relationships among a network of variables. To identify the model's parameters, SEM relies on strong mathematical assumptions. Some practitioners have recently begun to incorporate explicit counterfactual assumptions into SEM model specification.

***Sensitivity*:** For a particular event of interest (e.g., exposure, disease), sensitivity is the probability that the event will be detected, given that it has actually occurred.

***Sensitivity analysis*:** Sensitivity analysis is a methodology for assessing whether an empirical effect is a valid causal effect. The basic idea is to simulate the change in the empirical effect that would result under plausible assumptions about the possible impact of the most likely sources of bias.

***Sharp null hypothesis*:** The sharp null hypothesis refers to the absence of any causal effect (i.e., change in response pattern) for any individual.

***Simpson's paradox*:** Simpson's paradox refers to situations in which the direction of an association can be the same in two subgroups (strata), but the association in the overall population is different and can even be reversed.

***Source of bias*:** A source of bias is some aspect of the circumstances, design, or implementation of a comparative study that can possibly result in a discrepancy between the empirical effect and the causal effect.

***Specificity*:** For a particular event of interest (e.g., exposure, disease), specificity is the probability that the event will not be detected, given that it has not actually occurred.

***Statistical adjustment*:** Statistical adjustment, or statistical control, is the application of any statistical method in order to compensate for confounding, and thus to obtain an unbiased estimate of causal effect.

Statistical association: A statistical association, or statistical relationship, refers to any lack of statistical independence between two variables. Existence of an association implies that the value of one variable provides some information about the distribution of the other. Commonly employed measures of statistical association for a numerical variable are the covariance and correlation. Informally, the terms "statistically independent" and "uncorrelated" are often used interchangeably, although correlation represents a special type of dependence.

Stratification: Stratification is the subdivision of a study population into subgroups, or strata, of subjects that are homogeneous in some respect. Each stratum is usually defined by conditioning on the value of some covariate.

Study base: In a case–control study, the study base is the (hypothetical) population from which the case series arose. It is the population of individuals who *would* have become members of the case series *if* they had experienced the outcome event.

Study groups: The study groups are the different groups of subjects who receive the various treatment modalities. In an actual study, these groups comprise finite samples of individuals. Throughout this book, we have ignored finite-sampling considerations and have treated these groups as if they were actually populations.

Study population: The study population is the entire group of individuals who are sampled from the target population for inclusion in the study. In an actual study, this group is a finite sample. Throughout this book, we have ignored finite-sampling considerations and have treated the group as if it were a population.

Successful adjustment: Successful adjustment occurs when the statistical adjustment results in a conditionally unbiased empirical effect. Adjustment is successful when conditioning on covariates results in exchangeability of the study groups within each stratum.

Support region: For a covariate (or set of covariates) on which matching is based, the region of support is the set of values for which there is sufficient overlap in the covariate distributions for matching to be a viable analytic approach.

Surrogate endpoint: A surrogate endpoint is an outcome variable that serves as a proxy for the true endpoint of interest. The surrogate is typically utilized when the actual outcome is difficult to obtain or would require a very long time to develop.

SUTVA: SUTVA stands for the stable unit–treatment value assumption, which states that, for each combination of experimental unit (subject) and treatment modality, the outcome value is uniquely defined.

Target population: The target population is the actual or hypothetical total population of individuals for whom the causal effect of interest is intended to apply.

Total effect: In a study, the causal effect of a given factor may be partly attributable to the "effect" on one or more intermediate variables. The total effect of the factor includes any "indirect" effects that happen to occur in the study, regardless of whether similar indirect effects are likely to occur in practical applications.

Treatment: The term "treatment" is used sometimes to refer to the particular treatment modality (exposure) being studied and sometimes to mean any of the different treatment modalities being compared.

***True score*:** Classical measurement theory defines the true score as the hypothetical value around which the scores observed on various occasions (under identical conditions) are assumed to vary randomly.

***Validity*:** Validity is the ability of the study to produce correct results. There are various specific types of validity (see internal validity, external validity, construct validity). Threats to validity include primarily what we have termed bias, but encompass a wider range of methodological problems, including random error and lack of construct validity.

***Variable*:** A variable is an individual characteristic that can be expressed as a value (categorical or numerical) for each individual in a population. The variable has a distribution for the population. A causal factor for a particular outcome is a type of variable, but many variables are not causal factors.

Bibliography

Abbott, E. A. (1884). *Flatland: A Romance of Many Dimensions*. London: Seeley & Company.

Allard, R., and Boivin, J.-F. (2007). Using causal models to show the effect of untestable assumptions on effect estimates in randomized controlled trials. *Clinical Trials, 4*, 611–620.

Al-Shahi, R., Vousden, C., and Warlow, C. (2005). Bias from requiring explicit consent from all participants in observational research: prospective population based study. *British Medical Journal, 331*, 942–947.

Amery, W., and Dony, J. (1975). A clinical trial design avoiding undue placebo treatment. *Journal of Clinical Pharmacology, 10*, 674–679.

Anderson, S., Auquier, A., Hauck, W., Oakes, D., Vandaele, W., and Weisberg, H. (1980). *Statistical Methods for Comparative Studies*. New York: John Wiley & Sons.

Angrist, J. D., Imbens, G. W., and Rubin, D. B. (1996). Identification of causal effects using instrumental variables (with discussion). *Journal of the American Statistical Association, 91*, 444–472.

Antithrombotic Trialists Collaboration (ATT). (2009). Aspirin in the primary prevention of vascular disease: collaborative meta-analysis of individual participant data from randomised trials. *Lancet, 373*, 1849–1860.

Austin, P. C. (2008). A critical appraisal of propensity-score matching in the medical literature (with discussion). *Statistics in Medicine, 27*, 2037–2049.

Baldus, D. C., Woodworth, G. G., and Pulaski, C. A. (1990). *Equal Justice and the Death Penalty: A Legal and Empirical Analysis*. Boston: Northeastern University Press.

Barnard, J., Frangakis, C. E., Hill, J. L., and Rubin, D. B. (2003). Principal stratification approach to broken randomized experiments: a case study of school choice vouchers in New York City. *Journal of the American Statistical Association, 98*, 299–323.

Barnow, B. (1987). The impact of CETA programs on earnings. *Journal of Human Resources, 22*, 157–193.

Baron, R. M., and Kenny, D. A. (1986). The moderator–mediator variable distinction in social psychological research: conceptual, strategic and statistical considerations. *Journal of Personality and Social Psychology, 51,* 1173–1182.

Barrett-Connor, E. (1998). Hormone replacement therapy. *British Medical Journal, 317,* 457–461.

Barrett-Connor, E., and Grady, D. (1998). Hormone replacement therapy, heart disease and other considerations. *Annual Review of Public Health, 19,* 55–72.

Barron, B. A. (1977). The effects of misclassification on the estimation of relative risk. *Biometrics, 33,* 414–418.

Becher, H. (1992). The concept of residual confounding in regression models and some applications. *Statistics in Medicine, 11,* 1747–1758.

Bell, J. (1998). The new genetics in clinical practice. *British Medical Journal, 316,* 618–620.

Berger, V. W. (2005). *Selection Bias and Covariate Imbalances in Randomized Clinical Trials.* Chichester: John Wiley & Sons.

Berk, R. A. (1983). Introduction to sample selection bias in sociological data. *American Sociological Review, 48,* 386–398.

Berk, R. A., and Ray, S. C. (1982). Selection biases in sociological data. *Social Sciences Review, 11,* 352–398

Berkson, J. (1946). Limitations of the application of fourfold table analysis to hospital data. *Biometrics Bulletin, 2,* 47–53.

Bitler, M. P., Gelbach, J. B., and Hoynes, H. W. (2006). What mean impacts miss: distributional effects of welfare reform experiments. *American Economic Review, 96,* 988–1012.

Bodmer, W. (2003). RA Fisher, statistician and geneticist extraordinary: a personal view. *International Journal of Epidemiology, 32,* 938–942.

Boissel, J. (1998). Individualizing aspirin therapy for prevention of cardiovascular events. *Journal of the American Medical Association, 280,* 1949–1950.

Boissel, J., Collet, J., Lievre, M., and Girard, P. (1993). An effect model for the assessment of drug benefit: example of antiarrhythmic drugs in postmyocardial infarction patients. *Journal of Cardiovascular Pharmacology, 22,* 356–363.

Box, G. E. P., Hunter, J. S., and Hunter, W. G. (1978). *Statistics for Experimenters: An Introduction to Design, Data Analysis, and Model Building.* New York: John Wiley & Sons.

Brenner, H. (1996). How independent are multiple "independent" diagnostic classifications? *Statistics in Medicine, 15,* 1377–1386.

Bridge, J. A., Barbe, R. P., Birmaher, B., Kolko, D. J., and Brent, D. A. (2005). Emergent suicidality in a clinical psychotherapy trial for adolescent depression. *American Journal of Psychiatry, 162,* 2173–2175.

Bridge, J. A., Iyengar, S., Salary, C. B., Barbe, S. P., Birmaher, B., Pincus, H. A., Ren, L., and Brent, D. A. (2007). Clinical response and risk for reported suicide ideation and suicide attempts in pediatric antidepressant treatment. *Journal of the American Medical Association, 297,* 1683–1696.

Brittain, E., and Wittes, J. (1990). The run-in period in clinical trials: the effect of misclassification on efficacy. *Controlled Clinical Trials, 11,* 327–338.

Britton, A., McKee, M., Black, N., McPherson, K., Sanderson, C., and Bain, C. (1999). Threats to applicability of randomised trials: exclusions and selective participation. *Journal of Health Services Research Policy*, *4*, 112–121.

Bross, I. D. (1954). Misclassification in 2×2 tables. *Biometrics*, *10*, 478–486.

Bryk, A. S., and Raudenbush, S. W. (1992). *Hierarchical Linear Models: Applications and Data Analysis Methods*. Newbury Park, CA: Sage Publications.

Bryk, A. S., and Weisberg, H. I. (1977). Use of the nonequivalent control group design when subjects are growing. *Psychological Bulletin*, *84*, 950–962.

Bryk, A. S., Lee, V. E., and Holland, P. B. (1993). *Catholic Schools and the Common Good*. Cambridge, MA: Harvard University Press.

Burtless, G., and Orr, L. (1986). Are classical experiments needed for manpower policy? *Journal of Human Resources*, *21*, 606–639.

Campbell, D. T. (1957). Factors relevant to the validity of experiments in social settings. *Psychological Bulletin*, *54*, 297–312.

Campbell, D. T. (1969). Reforms as experiments. *American Psychologist*, *24*, 409–429.

Campbell, D. T., and Kenny, D. A. (1999). *A Primer on Regression Artifacts*. New York: The Guilford Press.

Campbell, D. T., and Stanley, J. C. (1963). *Experimental and Quasi-experimental Designs for Research*. Chicago: Rand McNally.

Chalmers, T. C. (1975). Effects of ascorbic acid on the common cold: an evaluation of the evidence. *American Journal of Medicine*, *58*, 532–536.

Charlton, B. (1996). The scope and nature of epidemiology. *Journal of Clinical Epidemiology*, *49*, 623–626.

Chin, R., and Lee, B. Y. (2008). *Principles and Practice of Clinical Trial Medicine*. London: Elsevier.

Choi, B. (2000). Bias, overview. In M. Gail and J. Benichou (Eds.), *Encyclopedia of Epidemiologic Methods* (pp. 74–82). Chichester: John Wiley & Sons.

Choi, B., and Noseworthy, A. (1992). Classification, detection and prevention of bias in epidemiologic research. *Journal of Occupational Medicine*, *34*, 265–271.

Christenfeld, N. J., Sloan, R. P., Carrol, D., and Greenland, S. (2004). Risk factors, confounding and the illusion of control. *Psychosomatic Medicine*, *66*, 868–875.

Cvjetanovic, B. B. (1957). Field trials of typhoid vaccines. *American Journal of Public Health*, *47*, 578–581.

Cochran, W. G. (1965a). *Sampling Techniques* (3rd ed.). New York: John Wiley & Sons.

Cochran, W. G. (1965b). The planning of observational studies of human populations. *Journal of the Royal Statistical Society (Series A)*, *128*, 234–266.

Cochran, W. G. (1968a). Errors of measurement in statistics. *Technometrics*, *10*, 637–666.

Cochran, W. G. (1968b). The effectiveness of adjustment by subclassification in removing bias in observational studies. *Biometrics*, *24*, 295–313.

Cochran, W. G., and Cox, G. M. (1957). *Experimental Designs* (2nd ed.). New York: John Wiley & Sons.

Cockburn, W. C. (1955). The early history of typhoid vaccination. *Journal of the Royal Army Medical Corps*, *101*, 171–185.

Cole, P. (1979). The evolving case–control study (with comment by E. D. Acheson and discussion). *Journal of Chronic Diseases, 32*, 15–34.

Cook, R. J., and Wei, W. (2002). Selection effects in randomized trials with count data. *Statistics in Medicine, 21*, 515–531.

Cook, T. D., and Campbell, D. T. (1979). *Quasi-experimentation: Design and Analysis Issues for Field Settings*. Boston: Houghton Mifflin.

Copeland, K. T., Checkoway, H., Holbrook, H. R., and McMichael, A. J. (1977). Bias due to misclassification in the estimation of relative risk. *American Journal of Epidemiology, 105*, 488–495.

Cornfield, J. (1951). A method of estimating comparative rates from clinical data: applications to cancer of the lung, breast, and cervix. *Journal of the National Cancer Institute, 11*, 1269–1275.

Cox, D. R. (1958). *Planning of Experiments*. New York: John Wiley & Sons.

Cox, D. R. (1972). Regression models and life-tables (with discussion). *Journal of the Royal Statistical Society (Series B), 34*, 187–220.

Cox, D. R., and Oakes, D. (1984). *Analysis of Survival Data*. London: Chapman & Hall.

Cox, D. R., and Wermuth, N. (2004). Causality: a statistical view. *International Statistical Review, 72*, 285–305.

Cronbach, L. J. (1970). *Essentials of Psychological Testing*. New York: Harper & Row.

Cronbach, L. J. (1982). *Designing Evaluations of Educational and Social Programs*. San Francisco: Jossey-Bass.

Cronbach, L. J., and Meehl, P. E. (1955). Construct validity in psychological tests. *Psychological Bulletin, 52*, 281–302.

Cuddleback, G., Wilson, E., Orme, J. G., and Combs-Orme, T. (2004). Detecting and statistically correcting sample selection bias. *Journal of Social Services Research, 30*, 19–33.

D'Agostino, R. B. (1998). Propensity score methods for bias reduction in the comparison of a treatment to a non-randomized control group. *Statistics in Medicine, 17*, 2265–2281.

Dawid, A. P. (2000). Causal inferences without counterfactuals (with discussion). *Journal of the American Statistical Association, 95*, 407–448.

Deeks, J. J., Dinnes, J., D'Amico, R., Sowden, A. J., Sakarovitch, C., Song, F., Petticrew, M., and Altman, D. G. (2003). Evaluating non-randomised intervention studies. *Health Technology Assessment, 7*, 1–173.

Dehejia, R. H. (2005). Practical propensity score matching: a reply to Smith and Todd. *Journal of Econometrics, 125*, 355–364.

Dehejia, R. H., and Wahba, S. (1999). Causal effects in nonexperimental studies: reevaluating the evaluation of training programs. *Journal of the American Statistical Association, 94*, 1053–1062.

Delgado-Rodriguez, M., and Llorca, J. (2004). Bias. *Journal of Epidemiology and Community Health, 58*, 635–641.

Dempster, A. P. (2002). John W. Tukey as "Philosopher." *Annals of Statistics, 30*, 1619–1628.

Dickinson, K., and Maynard, R. (1981). *The Impact of Supported Work on Ex-Addicts*. New York: Manpower Demonstration Research Corporation.

Dubicka, B., and Goodyer, I. (2005). Should we prescribe antidepressants to children? *Psychiatric Bulletin*, *29*, 164–167.

Dubicka, B., Hadley, S., and Roberts, C. (2006). Suicidal behavior in youths with depression treated with new generation antidepressants. *British Journal of Psychiatry*, *189*, 393–398.

Elashoff, J. D. (1969). Analysis of covariance: a delicate instrument. *American Educational Research Journal*, *6*, 383–402.

Ellenberg, J. H. (1994). Selection bias in observational and experimental studies. *Statistics in Medicine*, *13*, 557–567.

Feinstein, A. R. (1973). Clinical biostatistics, XX: the epidemiologic trohoc, the ablative risk ratio, and "retrospective" research. *Clinical Pharmacology and Therapeutics*, *14*, 291–307.

Feinstein, A. R. (1985). *Clinical Epidemiology: The Architecture of Clinical Research*. Philadelphia: WB Saunders Company.

Feinstein, A. R., Walter, S. D., and Horwitz, R. I. (1986). An analysis of Berkson's bias in case–control studies. *Journal of Chronic Diseases*, *39*, 495–504.

Fisher, R. A. (1925). *Statistical Methods for Research Workers*. Edinburgh: Oliver & Boyd.

Fisher, R. A. (1959). *Statistical Methods and Scientific Inference* (2nd ed.) Edinburgh: Oliver & Boyd.

Flanders, W. D. (2006). On the relationship of sufficient component cause models with potential outcome (counterfactual) models (with discussion). *European Journal of Epidemiology*, *21*, 855–878.

Flanders, W. D., Boyle, C. A., and Boring, J. R. (1989). Bias associated with differential hospitalization rates in incident case–control studies. *Journal of Clinical Epidemiology*, *42*, 395–401.

Flegal, K. M., Browne, C., and Haas, J. D. (1986). The effects of exposure misclassification on estimates of relative risk. *American Journal of Epidemiology*, *123*, 736–751.

Fleming, T. R., and DeMets, D. (1996). Surrogate end points in clinical trials: are we being misled? *Annals of Internal Medicine*, *125*, 605–613.

Fossa, S. D., and Skovlund, E. (2002). Selection of patients may limit the generalizability of results from cancer trials. *Acta Oncologia*, *41*, 131–137.

Fraker, T., and Maynard, R. (1987). The adequacy of comparison group design for evaluations of employment-related programs. *Journal of Human Resources*, *22*, 194–227.

Frangakis, C. E. (2009). The calibration of treatment effects from clinical trials to target populations. *Clinical Trials*, *6*, 136–140.

Frangakis, C. E., and Rubin, D. B. (2002). Principal stratification in causal inference. *Biometrics*, *58*, 21–29.

Frankel, A. (2006). The mass tort bonanza that wasn't. *The American Lawyer*, January 6.

Freedman, D. A. (2006). Statistical methods for causation: a critical review. In *Encyclopedia of Statistics in Behavioral Science*. New York: John Wiley & Sons.

Freedman, D. A., and Berk, R. A. (2008). Weighting regressions by propensity scores. *Evaluation Review*, *32*, 392–409.

Fugh-Berman, A., and Bythrow, J. (2007). Bioidentical hormones for menopausal hormone therapy: variation on a theme. *Journal of General Internal Medicine, 22,* 1030–1034.

Galton, F. (1889). Co-relations and their measurement. *Proceedings of the Royal Society, 45,* 135–145.

Garbe, E., and Suissa, S. (2004). Hormone replacement therapy and acute coronary outcomes: methodological issues between randomized and observational studies. *Human Reproduction, 19,* 8–13.

Gelman, A., and Hill, J. (2007). *Data Analysis Using Regression and Multivariate/ Hierarchical Models.* New York: Cambridge University Press.

Geng, Z., Guo, J., and Fung, W. (2002). Criteria for confounders in epidemiological studies. *Journal of the Royal Statistical Society (Series B), 64,* 3–15.

Geng, Z., Guo, J., Lau, T. S., and Fung, W. (2001). Confounding, homogeneity and collapsibility for causal effects in epidemiologic studies. *Statistica Sinica, 11,* 63–75.

Gerhard, T. (2008). Bias: considerations for research practice. *American Journal of Health System Pharmacists, 65,* 2159–2168.

Gibbons, R. D., Brown, C. H., Hur, K., Marcus, S. M., Bhaumik, D. K., Erkens, J. A., Herings, R., and Mann, J. J. (2007). Early evidence on the effects of regulators' suicide warnings on SSRI prescriptions and suicide in adolescents. *American Journal of Psychiatry, 164,* 1356–1363.

Gladen, B., and Rogan, W. J. (1979). Misclassification and the design of environmental studies. *American Journal of Epidemiology, 109,* 607–616.

Glasziou, P. P., and Irwig, L. M. (1995). An evidence based approach to individualising treatment. *British Medical Journal, 311,* 1356–1359.

Gluud, L. L. (2006). Bias in clinical intervention research. *American Journal of Epidemiology, 163*(6), 493–501.

Glymour, C. (1998). Review of "The Art of Causal Conjecture," by G Shafer. *Journal of the American Statistical Association, 93,* 1513–1515.

Goldberger, A. S. (1981). Linear regression after selection. *Journal of Econometrics, 15,* 357–366.

Gordis, L. (2004). *Epidemiology* (3rd ed.). Philadelphia: Elsevier Saunders.

Grady, D., Herrington, D., Bittner, V., Blumenthal, R., Davidson, M., Hlatky, M., Hsia, J., Hulley, S., Herd, A., Khan, S., Newby, L. K., Waters, D., Vittinghoff, E., and Wenger, N. (2002). Cardiovascular disease outcomes during 6.8 years of hormone therapy. *Journal of the American Medical Association, 288,* 49–57.

Grayson, D. A. (1987). Confounding confounding. *American Journal of Epidemiology, 126,* 546–553.

Greenhouse, J. B., Kaizar, E. E., Kelleher, K., Seltman, H., and Gardner, W. (2008). Generalizing from clinical trial data: a case study. The risk of suicidality among pediatric antidepressant users. *Statistics in Medicine, 27,* 1801–1813.

Greenland, S. (1980). The effect of misclassification in the presence of covariates. *American Journal of Epidemiology, 112,* 564–569.

Greenland, S. (Ed.). (1987a). *Evolution of Epidemiologic Ideas: Annotated Readings on Concepts and Methods.* Chestnut Hill, MA: Epidemiology Resources.

Greenland, S. (1987b). Interpretation and choice of effect measures in epidemiologic analysis. *American Journal of Epidemiology, 125,* 761–768.

Greenland, S. (1988). Statistical uncertainty due to misclassification: implications for validation substudies. *Journal of Clinical Epidemiology, 41*, 1167–1174.

Greenland, S. (1989). Modeling and variable selection in epidemiologic analysis. *American Journal of Public Health, 79*, 340–349.

Greenland, S. (1990). Randomization, statistics and causal inference. *Epidemiology, 1*, 421–429.

Greenland, S. (1993). Basic problems in interaction assessment. *Environmental Health Perspectives, 101* (Suppl. 4), 59–66.

Greenland, S. (1996a). Absence of confounding does not correspond to collapsibility of the rate ratio or rate difference. *Epidemiology, 7*, 498–501.

Greenland, S. (1996b). Basic methods for sensitivity analysis of biases. *International Journal of Epidemiology, 25*, 1107–1116.

Greenland, S. (2005a). Epidemiologic measures and policy formulation: lessons from potential outcomes. *Emerging Themes in Epidemiology, 2*, 1–7.

Greenland, S. (2005b). Multiple-bias modelling for analysis of observational data. *Journal of the Royal Statistical Society (Series A), 168*, 267–306.

Greenland, S., and Brumback, B. (2002). An overview of relations among causal modelling methods. *International Journal of Epidemiology, 31*, 1030–1037.

Greenland, S., and Poole, C. (1988). Invariants and noninvariants in the concept of interdependent effects. *Scandinavian Journal of Work and Environmental Health, 14*, 125–129.

Greenland, S., and Robins, J. M. (1986). Identifiability, exchangeability, and epidemiological confounding. *International Journal of Epidemiology, 15*, 413–419.

Greenland, S., and Robins, J. M. (1988). Conceptual problems in the definition and interpretation of attributable fractions. *American Journal of Epidemiology, 128*, 1185–1197.

Greenland, S., Pearl, J., and Robins, J. M. (1999). Causal diagrams for epidemiological research. *Epidemiology, 10*, 37–48.

Greenland, S., Robins, J. M., and Pearl, J. (1999). Confounding and collapsibility in causal inference. *Statistical Science, 14*, 29–46.

Greenwood, M., and Yule, G. U. (1915). The statistics of anti-typhoid and anti-cholera inoculation, and the interpretation of such statistics in general. *Proceedings of the Royal Society of Medicine, 8*, 113–194.

Griffin, B. A., McCaffery, D. F., and Morral, A. R. (2008). An application of principal stratification to control for institutionalization at follow-up in studies of substance abuse programs. *Annals of Applied Statistics, 2*, 1034–1055.

Grodstein, F., Clarkson, T. B., and Manson, J. E. (2003). Understanding the divergent data on postmenopausal hormone therapy. *New England Journal of Medicine, 348*, 645–650.

Grodstein, F., Stampfer, M. J., Colditz, G. A., Willett, W. C., Manson, J. E., Joffe, M., Rosner, B., Fuchs, C., Hankinson, S. E., Hunter, D. J., Hennekens, C. H., and Speizer, F. E. (1997). Postmenopausal hormone therapy and mortality. *New England Journal of Medicine, 336*, 1769–1775.

Gullen, W. H., Bearman, J. E., and Johnson, E. A. (1968). Effects of misclassification in epidemiologic studies. *Public Health Reports, 53*, 1956–1965.

Gulliksen, H. (1950). *Theory of Mental Tests.* New York: John Wiley & Sons.

Guyatt, G. H., Sackett, D. L., and Cook, D. J. (1993). Evidence-Based Medicine Working Group. Users' guides to the medical literature, II: how to use an article about therapy or prevention; A: are the results of the study valid? *Journal of the American Medical Association, 270,* 2598–2601.

Guyatt, G. H., Sackett, D. L., and Cook, D. J. (1994). Evidence-Based Medicine Working Group. Users' guides to the medical literature, II: how to use an article about therapy or prevention; B: what were the results and will they help me in caring for my patients? *Journal of the American Medical Association, 271,* 59–63.

Haavelmo, T. (1943). The statistical implications of a system of simultaneous equations. *Econometrica, 11,* 1–12.

Hafeman, D. M. (2008). A sufficient cause based approach to the assessment of mediation. *European Journal of Epidemiology, 23,* 711–721.

Halloran, M. E., and Struchiner, C. J. (1995). Causal inference for infectious diseases. *Epidemiology, 6,* 142–151.

Hamilton, M. A. (1979). Choosing a parameter for a 2 × 2 table or a 2 × 2 × 2 table analysis. *American Journal of Epidemiology, 109,* 362–375.

Hammad, T. A., Laughren, T., and Racoosin, J. (2006a). Suicidality in pediatric patients treated with antidepressant drugs. *Archives of General Psychiatry, 63,* 332–339.

Hammad, T. A., Laughren, T. P., and Racoosin, J. A. (2006b). Suicide rates in short-term randomized controlled trials of newer antidepressants. *Journal of Clinical Psychopharmacology, 26*(2), 203–207.

Hastie, T., Tibshirani, R., and Friedman, J. (2009). *The Elements of Statistical Learning: Data Mining, Inference and Prediction.* (2nd ed.). New York: Springer.

Hauck, W. W., and Anderson, S. A. (1986). A proposal for interpreting and reporting negative studies. *Statistics in Medicine, 5,* 203–209.

Hauck, W. W., Neuhaus, J. M., Kalbfleisch, J. D., and Anderson, S. (1991). A consequence of omitted covariates when estimating odds ratios. *Journal of Clinical Epidemiology, 44,* 77–81.

Haynes, R. B., Sackett, D. L., Guyatt, G. H., and Tugwell, P. (2006). *Clinical Epidemiology: How to Do Clinical Practice Research* (3rd ed.). Philadelphia: Lippincott Williams & Wilkins.

Heckman, J. J., and Hotz, V. J. (1989). Choosing among alternative nonexperimental methods for estimating the impact of social programs: the case of manpower training (with discussion). *Journal of the American Statistical Association, 84,* 862–880.

Heckman, J. J., and Robb, R. (1985). Alternative methods for evaluating the effect of interventions. In J. Heckman and B. Singer (Eds.), *Longitudinal Analysis of Labor Market Data* (pp. 156–246). New York: Cambridge University Press.

Heckman, J. J., and Smith, J. A. (1995). Assessing the case for social experiments. *Journal of Economic Perspectives, 9,* 85–110.

Heckman, J. J., Ichimura, H., and Todd, P. E. (1997). Matching as an econometric evaluation estimator: evidence from evaluating a job training programme. *Review of Economic Studies, 64,* 605–654.

Heckman, J. J., Smith, J., and Clements, N. (1997). Making the most out of social experiments: accounting for heterogeneity in programme impacts. *Review of Economic Studies, 64,* 487–536.

Hernan, M. A. (2004). A definition of causal effect for epidemiological research. *Journal of Epidemiology and Community Health*, *58*, 265–271.

Hill, A. B. (1965). The environment and disease: association or causation? *Proceedings of the Royal Society of Medicine*, *58*, 295–300.

Hill, J. (2008). Comment: can nonrandomized experiments yield accurate answers? A randomized experiment comparing random and nonrandom assignments. *Journal of the American Statistical Association*, *103*, 1346–1350.

Hoffman, G. E., and Zup, S. L. (2003). Good versus evil: changing the approach to hormone therapy. *Endocrinology*, *144*, 4698–4699.

Höfler, M. (2005a). Causal inference based on counterfactuals. *BMC Medical Research Methodology*, *5*, 1–12.

Höfler, M. (2005b). The effect of misclassification on the estimation of association: a review. *International Journal of Methods in Psychiatric Research*, *14*, 92–101.

Höfler, M., Lieb, R., and Wittchen, H. (2007). Estimating causal effects from observational data with a model for multiple bias. *International Journal of Methods in Psychiatric Research*, *16*, 77–87.

Holford, T. R. (1995). Study design for epidemiologic studies with measurement error. *Statistical Methods in Medical Research*, *4*, 339–358.

Holland, P. W. (1986). Statistics and causal inference (with discussion). *Journal of the American Statistical Association*, *81*, 945–970.

Holland, P. W. (1988). Causal inference, path analysis, and recursive structural equation models. *Sociological Methodology*, *18*, 449–484.

Holland, P. W., and Rubin, D. B. (1988). Causal inference in retrospective studies. *Evaluation Review*, *12*, 203–231.

Horwitz, R. I., and Feinstein, A. R. (1978). Alternate analytic methods for case–control studies of endometrial cancer. *New England Journal of Medicine*, *299*, 1089–1094.

Hosmer, D., and Lemeshow, S. (2000). *Applied Logistic Regression* (2nd ed.). New York: John Wiley & Sons.

Hulley, S., Furberg, C., and Barrett-Conner, C. (2002). Noncardiovascular disease outcomes during 6.8 years of hormone therapy, Heart and Estrogen/Progestin Replacement Study Follow-up (HERS II). *Journal of the American Medical Association*, *288*, 58–66.

Hulley, S., Grady, D., and Bush, T. E. (1998). Randomized trial of estrogen plus progestin for secondary prevention of coronary heart disease in postmenopausal women. *Journal of the American Medical Association*, *280*, 605–613.

Hume, D. (1748). *An Enquiry Concerning Human Understanding*. Repr., Lasalle, Ill.: Open Court Press, 1988.

Ibrahim, M. A., and Spitzer, W. O. (1979). The case–control study: the problem and the prospect. *Journal of Chronic Diseases*, *32*, 139–144.

Imai, K., King, G., and Stuart, E. A. (2008). Misunderstandings between experimentalists and observationalists about causal inference. *Journal of the Royal Statistical Society (Series A)*, *171*, 481–502.

Imbens, G. W., and Rubin, D. B. (1997). Bayesian inference for causal effects in randomized experiments with noncompliance. *The Annals of Statistics*, *6*, 34–58.

Jick, H., and Vessey, M. P. (1978). Case–control studies in the evaluation of drug-induced illnesses. *American Journal of Epidemiology, 107,* 1–7.

Jin, H., and Rubin, D. B. (2008). Principal stratification for causal inference with extended partial compliance. *Journal of the American Statistical Association, 103,* 101–111.

Joffe, M. M., and Colditz, G. A. (1998). Restriction as a method for reducing bias in the estimation of direct effects. *Statistics in Medicine, 17,* 2233–2249.

Johnston, J. (1963). *Econometric Methods.* New York: McGraw Hill.

Kamstrup, P. R., Tybjaerg-Hansen, A., and Steffensen, R. (2009). Genetically-elevated Lipoprotein(a) and increased risk of myocardial infarction. *Journal of the American Medical Association, 301,* 2331–2339.

Kaptchuk, T. J. (2001). The double-blind, randomized, placebo-controlled trial: gold standard or golden calf? *Journal of Clinical Epidemiology, 54,* 541–549.

Kaufman, J. S., and Cooper, R. S. (1999). Seeking causal explanations in social epidemiology. *American Journal of Epidemiology, 150,* 113–120.

Kaufman, J. S., MacLehose, R. F., and Kaufman, S. (2004). A further critique of the analytic strategy of adjusting for covariates to identify biologic mediation. *Epidemiologic Perspectives & Innovations, 1,* 1–13.

Kernan, W. N., Viscoli, C. M., Brass, L. M., Broderick, J. P., Brott, T., Feldmann, E., Morgenstern, L. B., Wilterdink, L. D., and Horwitz, R. I. (2000). Phenylpropanolamine and the risk of hemorrhagic stroke. *New England Journal of Medicine, 343,* 1826–1832.

Keys, A., and Kihlberg, J. K. (1963). The effect of misclassification on estimated relative prevalence of a characteristic. *American Journal of Public Health, 53,* 1656–1665.

Kish, L. (1965). *Survey Sampling.* New York: John Wiley & Sons.

Kleinbaum, D. G., Kupper, L. L., and Morgenstern, H. (1982). *Epidemiologic Research.* Belmont, CA: Lifetime Learning Publications.

Kleinbaum, D. G., Morgenstern, H., and Kupper, L. L. (1981). Selection bias in epidemiologic studies. *American Journal of Epidemiology, 113,* 452–463.

Klungel, O. H., Martens, E. P., Psaty, B. M., Grobbee, D. E., Sullivan, S. D., Sticker, B. H., Leufkens, H. G. M., and de Boer, A. (2004). Methods to assess intended effects of drug treatment in observational studies are reviewed. *Journal of Clinical Epidemiology, 57,* 1223–1231.

Knoke, J. D., Burns, D. M., and Thun, M. J. (2008). The change in excess risk of lung cancer attributable to smoking following smoking cessation: an examination of different analytic approaches using CPS-I data. *Cancer Causes Control, 19,* 207–219.

Koh, K. K., Han, S. H., Shin, M., Ahn, J. Y., Lee, Y., and Shin, E. K. (2005). Significant differential aspects of lower doses of hormone therapy or tibolone on markers of cardiovascular disease in postmenopausal women: a randomized, double-blind, crossover study. *European Heart Journal, 26,* 1362–1368.

Koopmans, T. C., and Reiersol, O. (1950). The identification of structural characteristics. *Annals of Mathematical Statistics, 21,* 165–181.

Krieger, N. (1994). Epidemiology and the web of causation: has anyone seen the spider? *Social Science Medicine, 39,* 887–903.

Kristensen, P. (1992). Bias from nondifferential but dependent misclassification of exposure and outcome. *Epidemiology, 3,* 210–215.

Kuhn, T. S. (1962). *The Structure of Scientific Revolutions.* Chicago: University of Chicago Press.

Lachin, J. M. (2000). Statistical considerations in the intent-to-treat principle. *Controlled Clinical Trials, 21,* 167–189.

LaLonde, R. J. (1986). Evaluating the econometric evaluations of training programs with experimental data. *American Economic Review, 76,* 604–620.

Lane-Claypon, J. E. (1926). *A Further Report on Cancer of the Breast.* Reports of Public Health and Medical Subjects, Ministry of Health. London: H.M.S.O.

Last, J. M. (*Ed.*). (2001). *A Dictionary of Epidemiology* (4th ed.). New York: Oxford University Press.

Laupacis, A., Sackett, D. L., and Roberts, R. S. (1988). An assessment of clinically useful measures of the consequences of treatment. *New England Journal of Medicine, 318,* 1728–1733.

Leishman, W. B. (1910). Anti-typhoid inoculation. *Journal of the Royal Institute of Public Health and Hygiene, 10,* 385.

Leslie, L. K., Newman, T. B., Chesney, P. J., and Perrin, J. M. (2005). The Food and Drug Administration's deliberations on antidepressant use in pediatric patients. *Pediatrics, 116,* 195–202.

Lewbel, A. (2007). Estimation of average treatment effects with misclassification. *Econometrica, 75,* 537–551.

Little, R. J., and Rubin, D. B. (2000). Causal effects in clinical and epidemiological studies via potential outcomes: concepts and analytical approaches. *Annual Review of Public Health, 21,* 121–145.

Lonn, E. (2001). The use of surrogate endpoints in clinical trials: focus on clinical trials in cardiovascular disease. *Pharmacoepidemiological Drug Safety, 10,* 497–508.

Lord, F. M. (1967). A paradox in the interpretation of group comparisons. *Psychological Bulletin, 68,* 304–305.

Lord, F. M., and Novick, M. R. (1968). *Statistical Theories of Mental Test Scores.* Reading, MA: Addison Wesley.

Lyles, R. H. (2002). A note on estimating crude odds ratios in case–control studies with differentially misclassified responses in binary regression. *Biometrics, 58,* 1034–1037.

Machens, K., and Schmidt-Gollwitzer, K. (2003). Issues to debate on the women's health initiative (WHI) study. Hormone replacement therapy: an epidemiological dilemma? *Human Reproduction, 18* (10), 1992–1999.

Mackie, J. L. (1965). Causes and conditions. *American Philosophical Quarterly, 2,* 245–264.

Mackie, J. L. (1974). *The Cement of the Universe: A Study of Causation.* Oxford, U.K.: Clarendon Press.

MacKinnon, D. P., Lockwood, C. M., Hoffman, J. M., West, S. G., and Sheets, V. (2002). A comparison of methods to test mediation and other intervening variable effects. *Psychological Methods, 7,* 83–104.

Maclure, M., and Schneeweiss, S. (2001). Causation of bias: the episcope. *Epidemiology, 12,* 114–122.

Maldonado, G., and Greenland, S. (2002). Estimating causal effects. *International Journal of Epidemiology*, *31*, 422–429.

Manpower Demonstration Research Corporation. (1983). *Summary and Findings of the National Supported Work Demonstration*. Cambridge, MA: Ballinger.

Manski, C. F. (1989). Anatomy of the selection problem. *Journal of Human Resources*, *24*, 343–360.

Manski, C. F. (1990). Nonparametric bounds on treatment effects. *American Economic Review, Papers and Proceedings*, *80*, 319–323.

Manski, C. F. (1999). Comment: Choice as an alternative to control in observational studies, by P Rosenbaum. *Statistical Science*, *14*, 279–281.

Manski, C. F. (2007). *Identification for Prediction and Decision*. Cambridge, MA: Harvard University Press.

Manski, C. F., Sandefur, G. D., and McLanahan, S. (1992). Alternative estimates of the effect of family structure during adolescence on high school graduation. *Journal of the American Statistical Association*, *87*, 25–37.

Manson, J. E., Hsia, J., Johnson, K. C., Roussouw, J. E., Assaf, A. R., Lasser, N. L., Trevisan, N., Black, H. R., Heckbert, S. R., Detrano, R., Strickland, O. L., Wong, N. D., Crouse, J. R., Stein, E., Cushman, M., and the Women's Health Initiative Investigators (2003). Estrogen plus progestin and risk of coronary heart disease. *New England Journal of Medicine*, *349*, 523–534.

Mant, D. (1999). Can randomised trials inform clinical decisions about individual patients? *Lancet*, *353*, 743–748.

Mark, D. H. (1997). Interpreting the term selection bias in medical research. *Family Medicine*, *29*, 132–136.

Marks, H. M. (1997). *The Progress of Experiment*. Cambridge, U.K.: Cambridge University Press.

Masters, S., and Maynard, R. (1981). *The Impact of Supported Work on Long-Term Recipients of AFDC Benefits*. New York: Manpower Demonstration Research Corporation.

McFarlane, M. J., Brennan, T. A., Horwitz, R. I., and Feinstein, A. R. (1985). The role of susceptibility bias in epidemiological research. *Archives of Internal Medicine*, *145*, 909–912.

McNamee, R. (2003). Confounding and confounders. *Occupational and Environmental Medicine*, *60*, 227–234.

Meehl, P. E. (1977). Specific etiology and other forms of strong influence: some quantitative meanings. *The Journal of Medicine and Philosophy*, *2*, 33–53.

Menzies, P. (2004). Difference-making in context. In J. Collins, N. Hall, and L. A. Paul (Eds.), *Causation and Counterfactuals* (pp. 139–180). Cambridge, MA: The MIT Press.

Meyer, B. D. (1995). Natural and quasi-experiments in economics. *Journal of Business and Economic Statistics*, *13*, 151–161.

Michels, K. B. (2001). A renaissance for measurement error. *International Journal of Epidemiology*, *30*, 421–422.

Michels, K. B. (2003). Hormone replacement therapy in epidemiologic studies and randomized clinical trials: are we checkmate? *Epidemiology*, *14*, 3–5.

Miettinen, O. S. (1985). The "case control" study: valid selection of subjects. *Journal of Chronic Diseases, 38*, 543–548.

Miettinen, O. S., and Cook, E. F. (1981). Confounding: essence and detection. *American Journal of Epidemiology, 114*, 593–603.

Mill, J. S. (1874). *A System of Logic* (8th ed.). New York: Harper and Brothers.

Million Women Study Collaborators. (2003). Breast cancer and hormone replacement therapy in the Million Women Study. *Lancet, 362*, 419–427.

Moffitt, R. (1991). Program evaluation with nonexperimental data. *Evaluation Review, 15*, 291–314.

Morgan, S. L., and Winship, C. (2007). *Counterfactuals and Causal Inference: Methods and Principles for Social Research.* New York: Cambridge University Press.

Moride, Y., and Abenhaim, L. (1994). Evidence of the depletion of susceptibles effect in non-experimental pharmacological research. *Journal of Clinical Epidemiology, 47*, 731–737.

Morrison, D. (2008). Clinical inference: critically weighing the evidence from trials and registries to make clinical decisions. *Catheterization and Cardiovascular Interventions, 72*, 381–385.

Mosca, L., Collins, P., Herrington, D. M., Mendelsohn, M. E., Pasternack, R. C., Robertson, R. M., Schenck-Gustaffson, K., Smith, S. C., Taubert, K. A., and Wenger, N. K. (2001). Hormone replacement therapy and cardiovascular disease: a statement for healthcare professionals from the American Heart Association. *Circulation, 104*, 499–503.

Moses, L. E. (1995). Measuring effects without randomized trials? Options, problems and challenges. *Medical Care, 33*, AS8-AS14.

Moskowitz, D. (2006). A comprehensive review of the safety and efficacy of bioidentical hormones for the management of menopause and related health risks. *Alternative Medicine Review, 11*, 208–223.

Naftolin, F., Taylor, H. S., Karas, R., Brinton, E., Newman, I., Clarkson, T. B., Mendelsohn, M., Lobo, R. A., Judelson, D. R., Nachtigall, L. E., Heward, C. B., Hecht, H., Jaff, M. R., and Harman, S. M. (2004). The Women's Health Initiative could not have detected cardioprotective effects of starting hormone therapy during the menopausal transition. *Fertility and Sterility, 81*, 1498–1450.

Neuhaus, J. M. (1999). Bias and efficiency loss due to misclassified responses in binary regression. *Biometrika, 86*, 843–855.

Newell, D. J. (1962). Errors in the interpretation of errors in epidemiology. *American Journal of Public Health, 52*, 1925–1928.

New York Times. (1910). Wholesale vaccination to prevent typhoid fever. June 5.

New York Times. (1991). Editorial. September 29.

Neyman, J. (1923). On the application of probability theory to agricultural experiments: essay on principles. *Annals of Agricultural Science.* Translated *(1990)* in *Statistical Science, 5*, 465–480.

Olfson, M., Shaffer, D., Marcus, S. C., and Greenberg, T. (2003). Relationship between antidepressant medication treatment and suicide in adolescents. *Archives of General Psychiatry, 60*, 978–982.

Paradis, C. (2008). Bias in surgical research. *Annals of Surgery*, *248*, 180–188.

Pearce, N., Checkoway, H., and Kriebel, D. (2007). Bias in occupational epidemiological studies. *Occupational and Envrionmental Medicine*, *64*, 562–568.

Pearl, J. (2000). *Causality: Models, Reasoning and Inference*. New York: Cambridge University Press.

Pearson, K. (1900). *Grammar of Science*. London: Adam and Charles Black.

Pearson, K. (1904). Report on certain enteric fever inoculation statistics. *British Medical Journal*, *2*, 1243–1246.

Penston, J. (2005). Large-scale randomised trials—a misguided approach to clinical research. *Medical Hypotheses*, *64*, 651–657.

Pentti, K., Honkanon, R., Tuppurainen, M. T., Sandini, L., Kroger, H., and Saarikoski, S. (2006). Hormone replacement therapy and mortality in 52- to 70-year-old women: the Kuopo Osteoporosis Risk Factor and Prevention Study. *European Journal of Endocrinology*, *154*, 101–107.

Petitti, D. B. (1998). Hormone replacement therapy and heart disease prevention: experimentation trumps observation. *Journal of the American Medical Association*, *280*, 650–652.

Phillips, C. V., and Goodman, K. J. (2006). Causal criteria and counterfactuals; nothing more (or less) than scientific common sense. *Emerging Themes in Epidemiology*, *3*, 1–7.

Popper, K. R. (1959). *The Logic of Scientific Discovery*. New York: Basic Books.

Prentice, R. L., Langer, R. D., Stefanick, M. L., Howard, B. V., Pettinger, M., Anderson, G. L., Barad, D., Curb, J. D., Kotchen, J., Kuller, L., Limacher, M., Wactawski-Wende, J., and the Womens Health Initiative Investigators (2006). Combined analysis of Women's Health Initiative observational and clinical trial data on postmenopausal hormone therapy and cardiovascular disease. *American Journal of Epidemiology*, *163*, 589–599.

President's Council of Advisors on Science and Technology. (2008). *Priorities for Personalized Medicine*. Washington, D.C.: U.S. Government Printing Office.

Reynolds, K. D., and West, S. G. (1987). A multiplist strategy for strengthening nonequivalent control group designs. *Evaluation Review*, *11*, 691–714.

Rice, K. (2003). Full-likelihood approaches to misclassification of a binary exposure in matched case–control studies. *Statistics in Medicine*, *22*, 3177–3194.

Roberts, R. S., Spitzer, W. O., Delmore, T., and Sackett, D. L. (1978). An empirical demonstration of Berkson's bias. *Journal of Chronic Diseases*, *31*, 119–128.

Robins, J. M. (1989). The control of confounding by intermediate variables. *Statistics in Medicine*, *8*, 679–701.

Robins, J. M., and Greenland, S. (1986). The role of model selection in causal inference from nonexperimental data. *American Journal of Epidemiology*, *123*, 392–402.

Robins, J. M., and Greenland, S. (1992). Identifiability and exchangeability for direct and indirect effects. *Epidemiology*, *3*, 143–155.

Robins, J. M., and Morgenstern, H. (1987). The mathematical foundations of confounding in epidemiology. *Computers and Mathematical Applications*, *14*, 869–916.

Rogosa, D., Brandt, D., and Zimowski, M. (1982). A growth curve approach to the measurement of change. *Psychological Bulletin*, *92*, 726–748.

Rosenbaum, P. R. (1984). From association to causation in observational studies: the role of tests of strongly ignorable treatment assignment. *Journal of the American Statistical Association, 79*, 41–48.

Rosenbaum, P. R. (1999). Choice as an alternative to control in observational studies (with discussion). *Statistical Science, 14*, 259–304.

Rosenbaum, P. R. (2001). Replicating effects and biases. *American Statistician, 55*, 223–227.

Rosenbaum, P. R. (2002). *Observational Studies*. New York: Spinger-Verlag.

Rosenbaum, P. R., and Rubin, D. B. (1983a). The central role of the propensity score in observational studies for causal effects. *Biometrika, 70*, 41–55.

Rosenbaum, P. R., and Rubin, D. B. (1983b). Assessing sensitivity to an unobserved covariate in an observational study with binary outcome. *Journal of the Royal Statistical Society (Series B), 45*, 212–218.

Rothman, K. J. (1976). Causes. *American Journal of Epidemiology, 104*, 587–592.

Rothman, K. J. (Ed.). (1988). *Causal Inference*. Chestnut Hill, MA: Epidemiology Resources Inc.

Rothman, K. J., and Greenland, S. (1998). *Modern Epidemiology* (2nd ed.). Boston: Lippincott, Williams & Wilkins.

Rothman, K. J., and Greenland, S. (2005). Causation and causal inference. *American Journal of Public Health, 95* (Suppl. 1), S144-S150.

Rothman, K. J., Greenland, S., and Lash, T. L. (2008). *Modern Epidemiology* (3rd ed.). Boston: Lippincott, Williams & Wilkins.

Rothwell, P. M. (2005). External validity of randomised controlled trials: to whom do the results of this trial apply? *Lancet, 365*, 82–93.

Roussouw, J. E., Prentice, R. L., Manson, J. E., Wu, L., Barad, D., Barnabei, V. M., Ko, M., LaCroix, A. Z., Margolis, K. L., and Stefanick, M. L. (2007). Postmenopausal hormone therapy and risk of cardiovascular disease by age and years since menopause. *Journal of the American Medical Association, 297*, 1465–1477.

Roy, J., Hogan, J. W., and Marcus, B. H. (2008). Principal stratification with predictors of compliance for randomized trials with 2 active treatments. *Biostatistics, 2*, 277–289.

Rubin, D. B. (1974). Estimating causal effects of treatments in randomized and non-randomized studies. *Journal of Educational Psychology, 66*, 688–701.

Rubin, D. B. (1976). Inference and missing data (with Discussion and Reply). *Biometrika, 63* (3), 581–592.

Rubin, D. B. (1977). Assignment to treatment group on the basis of a covariate. *Journal of Educational Statistics, 2*, 1–26.

Rubin, D. B. (1978). Bayesian inference for causal effects: the role of randomization. *Annals of Statistics, 6*, 34–58.

Rubin, D. B. (1984). William G. Cochran's contributions to the design, analysis and evaluation of observational studies. In P. S. R. S. Rao and J. Sedransk (Eds.), *W. G. Cochran's Impact on Statistics* (pp. 37–69). New York: John Wiley & Sons.

Rubin, D. B. (1986). Which ifs have causal answers? *Journal of the American Statistical Association, 81*, 961–962.

Rubin, D. B. (1990a). Formal models of statistical inference for causal effects. *Journal of Statistical Planning and Inference, 25*, 279–292.

Rubin, D. B. (1990b). Neyman (1923) and causal inference in experiments and observational studies. *Statistical Science, 5*, 472–489.

Rubin, D. B. (1997). Estimating causal effects from large data sets using propensity scores. *Annals of Internal Medicine, 127*, 757–763.

Rubin, D. B. (2004). Direct and indirect causal effects via potential outcomes. *Scandinavian Journal of Statistics, 34*, 161–170.

Rubin, D. B. (2006). Causal inference through potential outcomes and principal stratification: application to studies with "censoring" due to death. *Statistical Science, 21*, 299–309.

Rubin, D. B. (2007). The design versus the analysis of observational studies for causal effects: parallels with the design of randomized trials. *Statistics in Medicine, 26*, 20–36.

Rubin, D. B., and Thomas, N. (1996). Matching using estimated propensity scores: relating theory to practice. *Biometrics, 52*, 249–264.

Rubin, D. B., Stuart, E. A., and Zanutto, E. L. (2004). A potential outcomes view of value-added assessment in education. *Journal of Educational and Behavioral Statistics, 29*, 103–116.

Sackett, D. L. (1979). Bias in analytic research. *Journal of Chronic Diseases, 32*, 51–63.

Sackett, D. L., Rosenberg, W. M., Gray, J. A., Haynes, R. B., and Richardson, W. S. (1996). Evidence based medicine: what it is and what it isn't. *British Medical Journal, 312*, 71–72.

Salas, M., Hofman, A., and Stricker, B. H. (1999). Confounding by indication: an example of variation in the use of epidemiologic terminology. *American Journal of Epidemiology, 149*, 981–983.

Savitz, D. A. (2003). *Interpreting Epidemiologic Evidence: Strategies for Study Design and Analysis*. New York: Oxford University Press.

Schlesselman, J. J. (1982). *Case–Control Studies: Design, Conduct, Analysis*. New York: Oxford University Press.

Schulz, K. F. (1995). Subverting randomization in controlled trials. *Journal of the American Medical Association, 274*, 1456–1458.

Schulz, K. F., and Grimes, D. A. (2002). Allocation concealment in randomised trials: defending against deciphering. *Lancet, 359*, 614–618.

Schulz, K. F., Chalmers, I., Hayes, R. J., and Altman, D. G. (1995). Empirical evidence of bias: dimensions of methodological quality associated with estimates of treatment effects in controlled trials. *Journal of the American Medical Association, 273*, 408–412.

Shadish, W. R., and Cook, T. D. (1999). Comment: Design rules: more steps toward a complete theory of quasi-experimentation. *Statistical Science, 14*, 294–300.

Shadish, W. R., Clark, M. H., and Steiner, P. M. (2008). Can nonrandomized experiments yield accurate answers? A randomized experiment comparing random and nonrandom assignments. *Journal of the American Statistical Association, 103*, 1334–1343.

Shadish, W. R., Cook, T. D., and Campbell, D. T. (2002). *Experimental and Quasi-Experimental Designs for Generalized Causal Inference*. Boston: Houghton Mifflin.

Shafer, G. (1996). *The Art of Causal Conjecture*. Cambridge, MA: MIT Press.

Shah, B. R., Laupacis, A., Hux, J. E., and Austin, P. C. (2005). Propensity score methods gave similar results to traditional regression modeling in observational studies: a systematic review. *Journal of Clinical Epidemiology, 58*, 550–559.

Shonkoff, J. P., and Phillips, D. A. (Eds.). (2000). *From Neurons to Neighborhoods: The Science of Early Childhood Development.* Washington, D.C.: National Academy Press.

Siddiqui, A. E., Sikorskii, A., Given, C. W., and Given, B. (2008). Early participant attrition from clinical trials: role of trial design and logistics. *Clinical Trials, 5*, 328–335.

Siegmund, K. D., and Langholz, B. (2002). Ascertainment bias in family-based case-control studies. *American Journal of Epidemiology, 155*, 875–880.

Simpson, E. H. (1951). The interpretation of interaction in contingency tables. *Journal of the Royal Statistical Society (Series B), 13*, 238–241.

Sloman, S. (2005). *Causal Models: How People Think about the World and Its Alternatives.* New York: Oxford University Press.

Smith, G. D., and Ebrahim, S. (2003). "Mendelian Randomization": can genetic epidemiology contribute to understanding environmental determinants of disease? *International Journal of Epidemiology, 32*, 1–22.

Smith, J., and Todd, P. (2005a). Does matching overcome LaLonde's critique of non-experimental estimators? (with reply by R Dehejia and rejoinder). *Journal of Econometrics, 125*, 305–375.

Smith, J., and Todd, P. (2005b). Rejoinder (to Dehejia reply to Smith and Todd). *Journal of Econometrics, 125*, 365–375.

Sobel, M. E. (1990). Effect analysis and causation in linear structural equation models. *Psychometrika, 55*, 495–515.

Sommer, A., and Zeger, S. L. (1991). On estimating efficacy from clinical trials. *Statistics in Medicine, 10*, 45–52.

Steineck, G., and Albohm, A. (1992). A definition of bias founded on the concept of the study base. *Epidemiology*, 477–482.

Steyer, R. (2005). Analyzing individual and average causal effects via structural equation models. *Methodology, 1*, 39–54.

Stone, R. (1993). The assumptions on which causal inference rest. *Journal of the Royal Statistical Society (Series B), 55*, 455–486.

Strenio, J., Weisberg, H. I., and Bryk, A. S. (1983). Empirical Bayes estimation of individual growth parameters and their relationship to covariates. *Biometrics, 39*, 71–88.

Strom, B. (2007). Methodologic challenges to studying patient safety and comparative effectiveness. *Medical Care, 45* (10 Suppl. 2), S13–S15.

Susser, M. (1977). Judgment and causal inference: criteria in epidemiologic studies. *American Journal of Epidemiology, 105*, 1–15.

Susser, M. (1988). Falsification, verification, and causal inference in epidemiology. In K. J. Rothman (Ed.), *Causal Inference.* Chestnut Hill, MA: Epidemiology Resources.

Szklo, M., and Nieto, F. H. (2007). *Epidemiology: Beyond the Basics.* Sudbury, MA: Jones and Bartlett.

Thanassoulis, G., and O'Donnell, C. J. (2009). Mendelian randomization: nature's randomized trial in the post-genome era. *Journal of the American Medical Association*, *301*, 2386–2388.

Tukey, J. W. (1960). Conclusions vs decisions. *Technometrics*, *2*, 423–433.

Tukey, J. W. (1962). The future of data analysis. *Annals of Mathematical Statistics*, *33*, 1–67.

Tukey, J. W. (1977). *Exploratory Data Analysis*. Reading, MA: Addison-Wesley.

U.S. Food and Drug Administration. (2006). Relationship between psychotropic drugs and pediatric suicidality. November 16, 2006. Retrieved January 29, 2008, from http://www.fda.gov/ohrms/dockets/ac/briefing/2004-4065b1-10-TAB08-Hammads-Review.pdf

Valabhji, J., and Elkeles, R. S. (2000). Debate: are surrogate end-point studies worth the effort? *Cardiovascular Medicine*, *1*, 72–75.

van der Schouw, Y. T., and Grobbee, D. E. (2005). Menopausal complaints, oestrogens, and heart disease risk: an explanation for discrepant findings on the benefits of post-menopausal therapy. *European Heart Journal*, *26*, 1358–1361.

Van Spall, H. G., Toren, A., Kiss, A., and Fowler, R. A. (2007). Eligibility criteria of randomized controlled trials published in high-impact general medical journals: a systematic sampling review. *Journal of the American Medical Association*, *297*, 1233–1240.

Vandenbroucke, J. P. (1987). Should we abandon statistical modeling altogether? *American Journal of Epidemiology*, *126*, 10–13.

VanderWeele, T. J., and Robins, J. M. (2007). The identification of synergism in the SCC framework. *Epidemiology*, *18*, 329–339.

VanderWeele, T. J., and Robins, J. M. (2008). Empirical and counterfactual conditions for sufficient cause interactions. *Biometrika*, *95*, 49–61.

Vickers, M. R., MacLennan, A. H., Lawton, B., Ford, D., Martin, J., Meredith, S. K., DeStavola, B. L, Rose, S., Dowell, A., Wilkes, H. C., Darbyshire, J. H., and Meade, T. W. (2007). Main morbidities recorded in the women's international study of long duration oestrogen after menopause (WISDOM): a randomized controlled trial of hormone replacement therapy in postmenopausal women. *British Medical Journal*, *338*, 239–250.

Vineis, P. (2003). Causality in epidemiology. *Praventivmed*, *48*, 80–87.

Vineis, P., and McMichael, A. J. (1998). Bias and confounding in molecular epidemiological studies: special considerations. *Carcinogenesis*, *19*, 2063–2067.

Wacholder, S., McLaughlin, J. K., Silverman, D. T., and Mandel, J. S. (1992). Selection of controls in case–control studies, I: principles. *American Journal of Epidemiology*, *135*, 1019–1028.

Wacholder, S., Silverman, D. T., McLaughlin, J. K., and Mandel, J. S. (1992a). Selection of controls in case–control studies, II: types of controls. *American Journal of Epidemiology*, *135*, 1029–1041.

Wacholder, S., Silverman, D. T., McLaughlin, J. K., and Mandel, J. S. (1992b). Selection of controls in case–control studies, III: design options. *American Journal of Epidemiology*, *135*, 1042–1050.

Wallace, A. E., Neily, J., Weeks, W. B., and Friedman, M. J. (2006). A cumulative meta-analysis of selective serotonin reuptake inhibitors in pediatric depression: did unpublished studies influence the efficacy/safety debate? *Journal of Child and Adolescent Psychopharmacology, 16,* 37–58.

Weinberg, C. R. (1993). Toward a clearer definition of confounding. *American Journal of Epidemiology, 137,* 1–8.

Weinberg, C. R., Umbach, D. M., and Greenland, S. (1994). When will misclassification of an exposure preserve the direction of a trend? *American Journal of Epidemiology, 140,* 565–571.

Weisberg, H. I. (1979). Statistical adjustments and uncontrolled studies. *Psychological Bulletin, 86,* 1149–1164.

Weisberg, H. I. (2004). Expert report on phenylpropanolamine and hemorrhagic stroke. Unpublished.

Weisberg, H. I., Hayden, V. C., and Pontes, V. P. (2009). Selection criteria and generalizability within the counterfactual framework: explaining the paradox of antidepressant-induced suicidality? *Clinical Trials, 6,* 109–118.

Wheeler, B. W., Metcalfe, C., Martin, R. M., and Gunnell, D. (2009). International aspects of regulatory action to limit antidepressant prescribing on rates of suicide in young people. *Pharmacoepidemiology and Drug Safety, 18,* 579–588.

Whitehead, A. N. (1933). *Adventures of Ideas.* New York: The Free Press.

Whittington, C. J., Kendall, T., Fonagy, P., Cottrell, D., Cotgrove, A., and Boddington, E. (2004). Selective serotonin reuptake inhibitors in childhood depression: systematic review of published versus unpublished data. *Lancet, 363,* 1342–1345.

Wickramaratne, P. J., and Holford, T. R. (1987). Confounding in epidemiologic studies. *Biometrics, 43,* 751–765.

Winship, C., and Mare, R. D. (1992). Models for sample selection bias. *Annual Review of Sociology, 18,* 327–350.

Winship, C., and Morgan, S. L. (1999). The estimation of causal effects from observational data. *Annual Reviews of Sociology, 25,* 659–706.

Witteman, J. C., D'Agostino, R. B., Stijnen, C., Kannel, W. B., Cobb, J. C., de Ridder, M. A., Hofman, A., and Robins, J. M. (1998). G-estimation of causal effects: isolated systolic hypertension and cardiovascular death in the Framingham heart study. *American Journal of Epidemiology, 148,* 390–401.

Wright, S. (1921). Correlation and causation. *Journal of Agricultural Research, 20,* 557–585.

Writing Group for the Women's Health Initiative Investigators. (2002). Risks and benefits of estrogen plus progestin in healthy postmenopausal women, principal results from the Women's Health Initiative randomized controlled trial. *Journal of the American Medical Association, 288,* 321–333.

Yule, G. U. (1903). Notes on the theory of association of attributes in statistics. *Biometrika, 2,* 121–134.

Yusuf, S., Held, P., and Teo, K. K. (1990). Selection of patients for randomized controlled trials: implications of wide or narrow eligibility criteria. *Statistics in Medicine, 9,* 73–86.

Zhang, J. L., Rubin, D. B., and Mealli, F. (2009). Likelihood-based analysis of causal effects of job-training programs using principal stratification. *Journal of the American Statistical Association, 104,* 166–176.

Index

WILEY SERIES IN PROBABILITY AND STATISTICS
ESTABLISHED BY WALTER A. SHEWHART AND SAMUEL S. WILKS

The *Wiley Series in Probability and Statistics* is well established and authoritative. It covers many topics of current research interest in both pure and applied statistics and probability theory. Written by leading statisticians and institutions, the titles span both state-of-the-art developments in the field and classical methods.

Reflecting the wide range of current research in statistics, the series encompasses applied, methodological and theoretical statistics, ranging from applications and new techniques made possible by advances in computerized practice to rigorous treatment of theoretical approaches.

This series provides essential and invaluable reading for all statisticians, whether in academia, industry, government, or research.

---

*Now available in a lower priced paperback edition in the Wiley Classics Library.

†Now available in a lower priced paperback edition in the Wiley–Interscience Paperback Series.

*Now available in a lower priced paperback edition in the Wiley Classics Library.
†Now available in a lower priced paperback edition in the Wiley–Interscience Paperback Series.

*Now available in a lower priced paperback edition in the Wiley Classics Library.
†Now available in a lower priced paperback edition in the Wiley–Interscience Paperback Series.

HOEL · Introduction to Mathematical Statistics, *Fifth Edition*
HOGG and KLUGMAN · Loss Distributions
HOLLANDER and WOLFE · Nonparametric Statistical Methods, *Second Edition*
HOSMER and LEMESHOW · Applied Logistic Regression, *Second Edition*
HOSMER, LEMESHOW, and MAY · Applied Survival Analysis: Regression Modeling
   of Time-to-Event Data, *Second Edition*
† HUBER and RONCHETTI · Robust Statistics, *Second Edition*
HUBERTY · Applied Discriminant Analysis
HUBERTY and OLEJNIK · Applied MANOVA and Discriminant Analysis,
   Second Edition
HUNT and KENNEDY · Financial Derivatives in Theory and Practice, *Revised Edition*
HURD and MIAMEE · Periodically Correlated Random Sequences: Spectral Theory
   and Practice
HUSKOVA, BERAN, and DUPAC · Collected Works of Jaroslav Hajek—
   with Commentary
HUZURBAZAR · Flowgraph Models for Multistate Time-to-Event Data
IMAN and CONOVER · A Modern Approach to Statistics
† JACKSON · A User's Guide to Principle Components
JOHN · Statistical Methods in Engineering and Quality Assurance
JOHNSON · Multivariate Statistical Simulation
JOHNSON and BALAKRISHNAN · Advances in the Theory and Practice of Statistics: A
   Volume in Honor of Samuel Kotz
JOHNSON and BHATTACHARYYA · Statistics: Principles and Methods, *Fifth Edition*
JOHNSON and KOTZ · Distributions in Statistics
JOHNSON and KOTZ (editors) · Leading Personalities in Statistical Sciences: From the
   Seventeenth Century to the Present
JOHNSON, KOTZ, and BALAKRISHNAN · Continuous Univariate Distributions,
   Volume 1, *Second Edition*
JOHNSON, KOTZ, and BALAKRISHNAN · Continuous Univariate Distributions,
   Volume 2, *Second Edition*
JOHNSON, KOTZ, and BALAKRISHNAN · Discrete Multivariate Distributions
JOHNSON, KEMP, and KOTZ · Univariate Discrete Distributions, *Third Edition*
JUDGE, GRIFFITHS, HILL, LÜTKEPOHL, and LEE · The Theory and Practice of
   Econometrics, *Second Edition*
JUREČKOVÁ and SEN · Robust Statistical Procedures: Aymptotics and Interrelations
JUREK and MASON · Operator-Limit Distributions in Probability Theory
KADANE · Bayesian Methods and Ethics in a Clinical Trial Design
KADANE AND SCHUM · A Probabilistic Analysis of the Sacco and Vanzetti Evidence
KALBFLEISCH and PRENTICE · The Statistical Analysis of Failure Time Data, *Second
   Edition*
KARIYA and KURATA · Generalized Least Squares
KASS and VOS · Geometrical Foundations of Asymptotic Inference
† KAUFMAN and ROUSSEEUW · Finding Groups in Data: An Introduction to Cluster
   Analysis
KEDEM and FOKIANOS · Regression Models for Time Series Analysis
KENDALL, BARDEN, CARNE, and LE · Shape and Shape Theory
KHURI · Advanced Calculus with Applications in Statistics, *Second Edition*
KHURI, MATHEW, and SINHA · Statistical Tests for Mixed Linear Models
KLEIBER and KOTZ · Statistical Size Distributions in Economics and Actuarial Sciences
KLEMELÄ · Smoothing of Multivariate Data: Density Estimation and Visualization
KLUGMAN, PANJER, and WILLMOT · Loss Models: From Data to Decisions,
   Third Edition
KLUGMAN, PANJER, and WILLMOT · Solutions Manual to Accompany Loss Models:
   From Data to Decisions, *Third Edition*

---

*Now available in a lower priced paperback edition in the Wiley Classics Library.
†Now available in a lower priced paperback edition in the Wiley–Interscience Paperback Series.

*Now available in a lower priced paperback edition in the Wiley Classics Library.
†Now available in a lower priced paperback edition in the Wiley–Interscience Paperback Series.

MICKEY, DUNN, and CLARK · Applied Statistics: Analysis of Variance and Regression, *Third Edition*
* MILLER · Survival Analysis, *Second Edition*
MONTGOMERY, JENNINGS, and KULAHCI · Introduction to Time Series Analysis and Forecasting
MONTGOMERY, PECK, and VINING · Introduction to Linear Regression Analysis, *Fourth Edition*
MORGENTHALER and TUKEY · Configural Polysampling: A Route to Practical Robustness
MUIRHEAD · Aspects of Multivariate Statistical Theory
MULLER and STOYAN · Comparison Methods for Stochastic Models and Risks
MURRAY · X-STAT 2.0 Statistical Experimentation, Design Data Analysis, and Nonlinear Optimization
MURTHY, XIE, and JIANG · Weibull Models
MYERS, MONTGOMERY, and ANDERSON-COOK · Response Surface Methodology: Process and Product Optimization Using Designed Experiments, *Third Edition*
MYERS, MONTGOMERY, VINING, and ROBINSON · Generalized Linear Models. With Applications in Engineering and the Sciences, *Second Edition*
† NELSON · Accelerated Testing, Statistical Models, Test Plans, and Data Analyses
† NELSON · Applied Life Data Analysis
NEWMAN · Biostatistical Methods in Epidemiology
OCHI · Applied Probability and Stochastic Processes in Engineering and Physical Sciences
OKABE, BOOTS, SUGIHARA, and CHIU · Spatial Tesselations: Concepts and Applications of Voronoi Diagrams, *Second Edition*
OLIVER and SMITH · Influence Diagrams, Belief Nets and Decision Analysis
PALTA · Quantitative Methods in Population Health: Extensions of Ordinary Regressions
PANJER · Operational Risk: Modeling and Analytics
PANKRATZ · Forecasting with Dynamic Regression Models
PANKRATZ · Forecasting with Univariate Box-Jenkins Models: Concepts and Cases
* PARZEN · Modern Probability Theory and Its Applications
PEÑA, TIAO, and TSAY · A Course in Time Series Analysis
PIANTADOSI · Clinical Trials: A Methodologic Perspective
PORT · Theoretical Probability for Applications
POURAHMADI · Foundations of Time Series Analysis and Prediction Theory
POWELL · Approximate Dynamic Programming: Solving the Curses of Dimensionality
PRESS · Bayesian Statistics: Principles, Models, and Applications
PRESS · Subjective and Objective Bayesian Statistics, *Second Edition*
PRESS and TANUR · The Subjectivity of Scientists and the Bayesian Approach
PUKELSHEIM · Optimal Experimental Design
PURI, VILAPLANA, and WERTZ · New Perspectives in Theoretical and Applied Statistics
† PUTERMAN · Markov Decision Processes: Discrete Stochastic Dynamic Programming
QIU · Image Processing and Jump Regression Analysis
* RAO · Linear Statistical Inference and Its Applications, *Second Edition*
RAUSAND and HØYLAND · System Reliability Theory: Models, Statistical Methods, and Applications, *Second Edition*
RENCHER · Linear Models in Statistics
RENCHER · Methods of Multivariate Analysis, *Second Edition*
RENCHER · Multivariate Statistical Inference with Applications
* RIPLEY · Spatial Statistics
* RIPLEY · Stochastic Simulation
ROBINSON · Practical Strategies for Experimenting

*Now available in a lower priced paperback edition in the Wiley Classics Library.
†Now available in a lower priced paperback edition in the Wiley–Interscience Paperback Series.

*Now available in a lower priced paperback edition in the Wiley Classics Library.
†Now available in a lower priced paperback edition in the Wiley–Interscience Paperback Series.

THOMPSON and SEBER · Adaptive Sampling

THOMPSON, WILLIAMS, and FINDLAY · Models for Investors in Real World Markets

TIAO, BISGAARD, HILL, PEÑA, and STIGLER (editors) · Box on Quality and Discovery: with Design, Control, and Robustness

TIERNEY · LISP-STAT: An Object-Oriented Environment for Statistical Computing and Dynamic Graphics

TSAY · Analysis of Financial Time Series, *Third Edition*

UPTON and FINGLETON · Spatial Data Analysis by Example, Volume II: Categorical and Directional Data

† VAN BELLE · Statistical Rules of Thumb, *Second Edition*

VAN BELLE, FISHER, HEAGERTY, and LUMLEY · Biostatistics: A Methodology for the Health Sciences, *Second Edition*

VESTRUP · The Theory of Measures and Integration

VIDAKOVIC · Statistical Modeling by Wavelets

VINOD and REAGLE · Preparing for the Worst: Incorporating Downside Risk in Stock Market Investments

WALLER and GOTWAY · Applied Spatial Statistics for Public Health Data

WEERAHANDI · Generalized Inference in Repeated Measures: Exact Methods in MANOVA and Mixed Models

WEISBERG · Applied Linear Regression, *Third Edition*

WEISBERG · Bias and Causation: Models and Judgment for Valid Comparisons

WELSH · Aspects of Statistical Inference

WESTFALL and YOUNG · Resampling-Based Multiple Testing: Examples and Methods for *p*-Value Adjustment

WHITTAKER · Graphical Models in Applied Multivariate Statistics

WINKER · Optimization Heuristics in Economics: Applications of Threshold Accepting

WONNACOTT and WONNACOTT · Econometrics, *Second Edition*

WOODING · Planning Pharmaceutical Clinical Trials: Basic Statistical Principles

WOODWORTH · Biostatistics: A Bayesian Introduction

WOOLSON and CLARKE · Statistical Methods for the Analysis of Biomedical Data, *Second Edition*

WU and HAMADA · Experiments: Planning, Analysis, and Parameter Design Optimization, *Second Edition*

WU and ZHANG · Nonparametric Regression Methods for Longitudinal Data Analysis

YANG · The Construction Theory of Denumerable Markov Processes

YOUNG, VALERO-MORA, and FRIENDLY · Visual Statistics: Seeing Data with Dynamic Interactive Graphics

ZACKS · Stage-Wise Adaptive Designs

ZELTERMAN · Discrete Distributions—Applications in the Health Sciences

* ZELLNER · An Introduction to Bayesian Inference in Econometrics

ZHOU, OBUCHOWSKI, and McCLISH · Statistical Methods in Diagnostic Medicine

Printed in the United States
By Bookmasters